I0061073

Vasileios Fotopoulos, Gholamreza Gohari (Eds.)
Engineered Nanoparticles in Agriculture

Also of interest

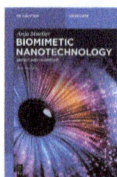

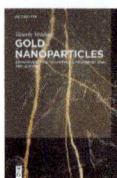

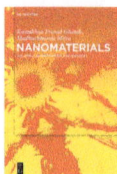

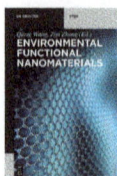

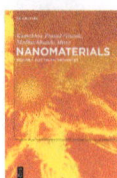

Engineered Nanoparticles in Agriculture

From Laboratory to Field

Edited by
Vasileios Fotopoulos and Gholamreza Gohari

DE GRUYTER

Editors
Dr. Vasileios Fotopoulos
Cyprus University of Technology
Kitiou Kyprianou Street 38
3041 Lemesos
Cyprus
vassilis.fotopoulos@cut.ac.cy

Dr. Gholamreza Gohari
Faculty of Agriculture
University of Maragheh
5518183111 Maragheh
Iran
gohari.gh@maragheh.ac.ir

ISBN 978-1-5015-2315-1
e-ISBN (PDF) 978-1-5015-2322-9
e-ISBN (EPUB) 978-1-5015-1552-1

Library of Congress Control Number: 2023936460

Bibliographic information published by the Deutsche Nationalbibliothek
The Deutsche Nationalbibliothek lists this publication in the Deutsche Nationalbibliografie;
detailed bibliographic data are available on the internet at http://dnb.dnb.de.

© 2023 Walter de Gruyter GmbH, Berlin/Boston
Cover image: wasja/iStock/Getty Images
Typesetting: Integra Software Services Pvt. Ltd.
Printing and binding: CPI books GmbH, Leck

www.degruyter.com

Preface

Nanotechnology has revolutionized many fields, including agriculture, by providing unique tools for addressing challenges such as climate change, environmental degradation, and food security. Engineered nanoparticles, in particular, have emerged as promising candidates for improving crop production and mitigating the negative impacts of abiotic and biotic stresses on plant growth and yield.

The book *Engineered Nanoparticles in Agriculture: From Laboratory to Field* aims to provide a comprehensive overview of the current state of the art in the use of nanoparticles in agriculture, with a focus on their practical applications from laboratory to field. The book covers a range of topics, including the synthesis and characterization of nanoparticles, their mechanisms of action in plant growth and stress responses, and their potential applications in crop production and soil management.

The book is divided into 10 chapters. Chapter 1 covers the synthesis and characterization of nanoparticles, including various methods for their synthesis, properties, and characterization techniques. Chapters 2 to 6 cover the practical applications of nanoparticles in agriculture, including their use in plant protection, seed priming, soil amendments, and nanofertilizers. The chapters in this section also discuss the potential risks associated with the use of nanoparticles in agriculture and the need for responsible nanotechnology. Chapters 7 to 10 focus on the mechanisms of action of nanoparticles in plant growth and horticultural plant production and application of new engineered nanoparticles to improve the postharvest quality of horticultural crops. The chapters in this section cover topics such as nanoparticle uptake and translocation, their effects on plant physiology, and their potential to mitigate the negative impacts of abiotic and biotic stresses.

This book is intended for researchers, students, and professionals in the fields of agriculture, nanotechnology, and environmental science, as well as policymakers and stakeholders interested in the future of sustainable agriculture. We hope that this book will serve as a valuable resource for those interested in exploring the potential of engineered nanoparticles in agriculture and their practical applications in the field.

Professor Vasileios Fotopoulos
Dr. Gholamreza Gohari

https://doi.org/10.1515/9781501523229-202

Contents

L. A. Martínez-Chávez, A. Rosales-Pérez, R. Hernández-Rangel,
Karen Esquivel*

Chapter 1
Nanomaterials: classification, synthesis methods, and physicochemical characterization

Abstract: The nanomaterials (NMs) possess unique physicochemical properties which allow them to be used in the fields of medicine, pharmaceuticals, biotechnology, energy, cosmetics, electronics, and environmental remediation processes. Due to their size and surface reactivity, NMs can interact and even move along cells, intracellular structures, and metabolic pathways, and enter all plant structures producing some beneficial or toxic effects by different mechanisms promoting changes in plant growth. The changes observed in different vegetal species are produced by NMs on their nature, concentration, crystallinity, surface area, and morphology. To produce NM with specific properties, synthesis methods based on physical, chemical, and biosynthetic principles have been developed. This chapter recapitulates results and progress of NM synthesis methods and their characterization showing the importance of its understanding to suggest safe agroindustry NM applications to achieve sustainable progress.

Keywords: Nanomaterials, Characterization, synthesis methods

1.1 Introduction

The agriculture sector has been the most important and stable field because it produces and provides materials to be used in different industrial processes, and it is considered the base field of social and technological development [1]. However, the world food production and distribution is facing massive stress due to increasing population, climate change, environmental contamination, and higher demands of water and energy [2]. The incoming challenges of sustainable production and food security have resulted in

---

*Corresponding author: Karen Esquivel, Graduate and Research Division, Engineering Faculty, Universidad Autónoma de Querétaro, Cerro de las campanas, C.P., Santiago de Querétaro 76010, México, e-mail: karen_esq_2001@yahoo.com; karen.esquivel@uaq.mx, URL: http://www.uaq.mx
L. A. Martínez-Chávez, A. Rosales-Pérez, Graduate and Research Division, Engineering Faculty, Universidad Autónoma de Querétaro, Cerro de las campanas, C.P., Santiago de Querétaro 76010, México
R. Hernández-Rangel, Chemistry Department, Universidad Autónoma de Aguascalientes, Avenida Universidad No. 94, Ciudad Universitaria, Aguascalientes 20131, México

https://doi.org/10.1515/9781501523229-001

technological advancements and innovations in agriculture [3, 4]. Recently, massive research has been carried out focusing on nanotechnology, emphasizing different applications in the agriculture sector. Nanotechnology can provide practical solutions to various agriculture-related issues [1, 5].

Nanotechnology is a field of science related to investigating and developing structures and materials on a nanoscale with different applications in areas like health care, agriculture, and technology [6]. For example, nanotechnology has shown promising potential to promote sustainable agriculture, and the development of new techniques applied in agriculture is based on nanomaterials (NMs) [1]. Nanotechnology provides scientific interest in understanding the properties of materials in their bulk and atomic-scale form [7]. In addition, nanotechnology focuses on developing and applying materials with a size ranging from 1 to 100 nm; materials in this size range are well known as nanomaterials. If the size of at least one of the dimensions of the material is in the range from 1 to 100 nm, it is considered an NM; otherwise, it is classified as a non-NM. Due to their chemical and physical properties like particle size, quantum effect, surface area, mechanical, thermal, and electrical conductivities, NMs have versatile uses and different applications [8]. Researchers have worked and modified the properties of NMs to add new characteristics, and as a consequence, these materials have been applied in different fields of science, engineering, and technology [9].

The study and application of these materials have been in all possible areas through the past decades from medicine, pharmaceutics, cosmetology, environment remediation, electronics, aerospace sciences, automotive industry, materials engineering, textiles, food sciences, biotechnology, and agriculture. For the agri-food sector, many studies apply NMs as fertilizers, pesticides, additives, and elicitors, all of which improve yield production and, as a consequence, generate better yields of functional food efficiently and rapidly [1, 2, 10].

1.2 Nanomaterials

As a definition, a nanometer (nm) is one-billionth of a meter; thus, NMs are objects with at least one of their dimensions (length, width, and depth) in the nanometric scale, which for the area of nanoscience is defined as 1–100 nm. These structures can be classified into multiple structures depending on the dimensions (Figure 1.1) present in the nanoscale [8]. The 0D NMs are those in which none of their dimensions is outside the scale already mentioned. The 1D classification encompasses materials whose one dimension exceeds the nanoscale, such as nanotubes, nanorods, and nanowires. Materials with two larger dimensions (2D) contain structures such as sheets, thin films, and coatings, and, in the 3D classification, the tree dimensions exceed 100 nm. Structures found in this classification are nanocomposites, powders, sets of nanotubes or multilayers, and MEMS [11, 12].

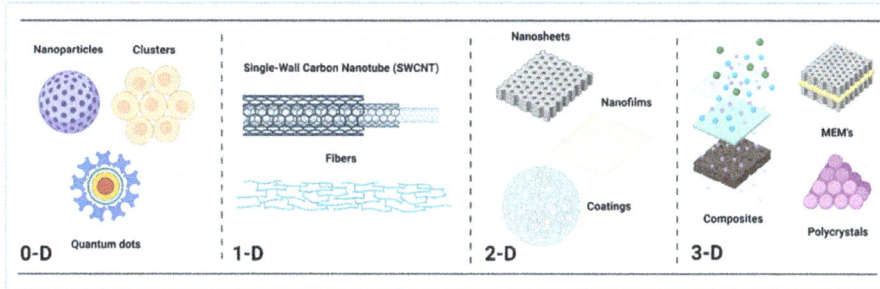

Figure 1.1: Nanomaterial classification by dimension.

Why do NMs change their properties concerning their bulk form? The reason is an alteration in the atoms of the material, and subsequently, it develops a magnetic power between them. It is well known that a smaller size of NM, a larger surface area, and more activity are exhibited. Thus, nanotechnology plays a vital role in the physical and chemical perception and understanding of materials. Therefore, the modified properties of NMs are more reactive in different and specific processes [13, 14]. For example, NMs possess unique and distinct electronic associations; the plasmonic and optical properties are related to the quantum confinement effects; also, the alteration of the electronic energy levels may appear due to the surface area concerning volume ratio [15, 16]. Recent research has shown a significant request for fast, steady, and low-cost systems for sensing, monitoring, and diagnosing different molecules in agricultural sectors where NMs can be applied [17]. NMs can also be classified by their chemical composition; the most used categories are organic materials, metallic materials, metallic oxides, semiconductors, and carbon-based materials, as shown in Figure 1.2 [8].

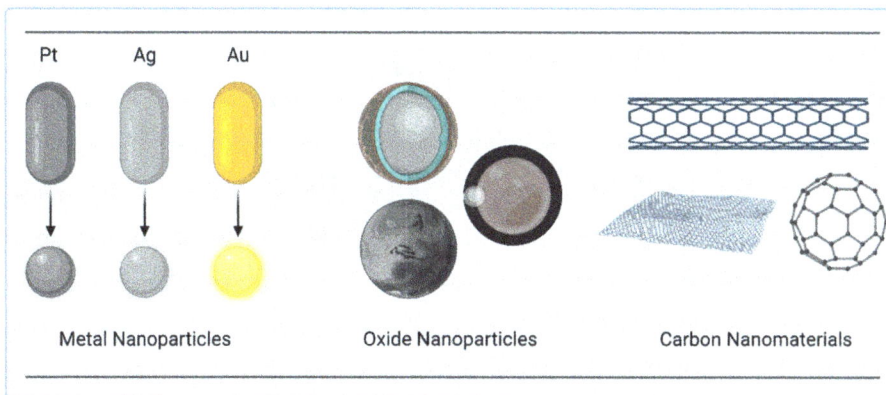

Figure 1.2: Nanoparticle classification by chemical composition.

Each of these materials finds interesting applications in an infinity of fields; hence, nanoscience has given the possibility of finding new properties in materials.

1.2.1 Metallic nanoparticles

The interest in the synthesis and application of metallic nanoparticles has increased significantly due to the unique properties that metals present in a nanometric range, making them viable for their application in different fields of science and technology [18]. In nanotechnology, metallic nanoparticles have shown several properties and have unlocked many new pathways in different fields [19]. Almost all the metals can be synthesized into their nanoparticles [20]. The most common materials for creating metallic nanoparticles are gold (Au), silver (Ag), and platinum (Pt), which find various applications in different fields [21–23]. Also, metals such as cobalt (Co), copper (Cu), iron (Fe), titanium (Ti), and zinc (Zn) can be used to prepare nanoparticles [24]. Some of these metallic NMs are applied in catalysis, disinfection, energy storage and harvest, medicine, and other different applications [25]. Metallic NPs could be used as an antimicrobial agent to inhibit or prevent bacterial or fungal diseases in crops [26, 27], fertilizers [2], or even sensors for identifying disease presence or different organic molecules in plants [28].

1.2.2 Oxides nanoparticles

Metal oxide nanoparticles have shown different physical and chemical properties from their bulk forms, such as morphology, surface area, optical absorption, and thermal and electrical conductivities [29]. These nanoparticles can be prepared by adding a reducing or oxidizing agent during the synthesis process [30]. Metal oxides are among the most produced nanoparticles due to their properties and applications. They are formed by the chemical union of a metal and oxygen, such as TiO_2, ZnO, CeO_2, Fe_3O_4, FeO, SiO_2, and Al_2O_3; these ones are most commonly used in industrial processes and are therefore mainly studied for their impacts on agriculture and its effect on plants [29, 31]. In addition, these materials have semiconductor properties, making them suitable materials for photocatalytic reactions such as TiO_2 and ZnO that are used for photocatalytic water treatment [32–34]. These materials can also eliminate pathogens, which promise to be an area of application in medicine and in protection of crops against bacteria [35].

1.2.3 Carbon-based nanomaterials

Carbon-based NMs (CNMs) have become an exciting research field because of their properties, such as being lightweight, high strength, and high conductivity [36]. CNMs are composed of carbon elements (Figure 1.3), and they have at least one dimension within the nanometric scale. Carbon nanotubes (CNTs), graphene (GRA), fullerenes (C_{60}), and carbon dots are the carbon-based materials of most significant interest in the research field due to their properties and applications [37].

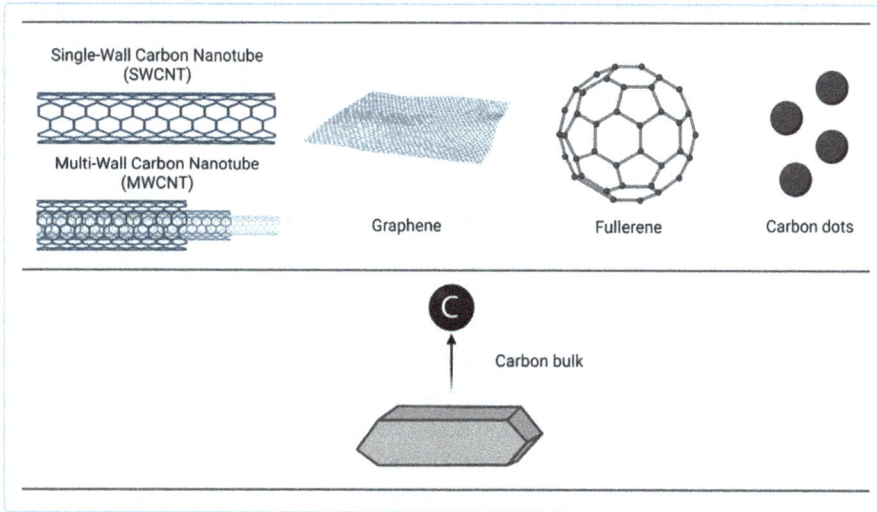

Figure 1.3: Types of carbon-based nanomaterials.

The structure of CNTs consists of GRA sheets rolled into a cylindrical shape; these can be classified according to the number of walls/layers that make up their structure. The simplest CNTs with a single wall are known as single-walled CNTs (SWCNTs), whereas if their structure has two or more walls, then they are known as multiwalled CNTs (MWCNTs) [38]. This type of carbon-based material has become an important research topic due to its properties and is applied in different processes where nano-technology is involved. Due to their one-dimensional structure, these materials ex-hibit interesting and different electronic and optical properties concerning other types of carbon-based materials and other types of nanoparticles [39]. Also, CNTs have small sizes, biocompatibility, high surface area, and consequently high reactiv-ity; thus, this type of materials can be used in different fields of research such as en-ergy [40], biomedicine [41], agriculture [42], electronics [43], photoelectricity [44], and analysis and catalysis [45].

GRA was the first carbon-based synthesized material with a two-dimensional structure; in particular, this material has been of great interest for the properties such as high strength and rigidity, excellent elasticity, and mechanical properties, which make it a material with good stability [46]. Concerning CNTs, both materials are considered isomers of carbon; therefore, they share similarities between their structure and properties. The structure of the GRA consists of a hexagonal plane composed of carbon atoms bonded with an sp^2 electronic configuration [47]. GRA exhibits much better mechanical properties than some NMs; this allows it to manufacture compounds with high mechanical resistance based on GRA [47]. Its structure is complex, with a layer of carbon atoms packed into a honeycomb network [48]. It is considered a one-dimensional material due to its single-layer structure [49] and among its properties that have made it an exciting material are its electrical and thermal conductivities [50], optical transparency, mechanical flexibility, and a low coefficient of thermal expansion [51]. Due to its properties and structure, this type of carbon-based material has found applications in electronic devices [52], batteries [37], agriculture [42], wastewater treatment [47], and sensors [36].

Fullerene (C_{60}) is considered the third type of carbon isomer, and its structure is very similar to that of a soccer ball. Different research works have shown that this carbon-based material has aromatic properties and is adsorbed on organic molecules [53]. The structure of C_{60} molecules determine their physical and chemical properties. This material is essential in development and research for its application in modern technologies in different research fields [54]. At room temperature and atmospheric pressure conditions, C_{60} can be found in its solid form; it is considered a suitable electron acceptor in the ground and the excited states [55]. Research results on C_{60} have shown that they are good prospects for application in electronic devices, light sensors, superconductors, catalysis, luminescent materials, molecular magnets, and biomedicine [56]. This carbon-based material can be classified as crystals with low melting points, high hardness, and insulating properties. When altered by energy stimuli such as photons or electrons of high intensity, C_{60} goes from a ground state to an excited one and can produce polymeric materials [57].

Carbon dots (C dots) are a carbon-based material; they are considered nanoparticles with fluorescent properties, similar to quantum dots, and can be easily synthesized by low-cost synthesis mechanisms [58]. C-dots exhibit optical properties like good absorption and emission that position them as competitors to semiconductor-based quantum dots. The properties of carbon structures on the nanoscale are considerably different from those of bulk carbon. These nanoparticles possess substantial oxygen and hydrogen fractions [59]. Its optical properties, its capability to emit fluorescence, and its different applications have made it a material of great scientific interest [60]. As a consequence, these fluorescent materials have been positioned as promising tools for cell labeling and imaging [61], bioimaging [62], and also have found applications in photocatalysis [60], agriculture [42], optoelectronics [63], drug delivery [64], and probes [65].

1.3 Nanomaterial's synthesis methods

NM synthesis has two main approaches: top-down, in which larger molecules get smaller down to nanometric size (1–100 nm), and the other is bottom-up approach, in which smaller molecules or particles, even atoms, combine to produce NMs [66]. These approaches classify different synthesis methods and focus on different categories, such as chemical, physical, mechanical, and biological. For example, the top-down approach to reducing particle size involves synthesis techniques such as high-energy ball milling, plastic deformation, etching, mechanical alloying, lithography, and micromachining methods. At the same time, the bottom-up approach concerns synthesis techniques to obtain NMs such as molecular self-assembly, sol-gel process, laser ablation (LA), vacuum arc deposition, inert gas condensation, molecular beam epitaxy, electrodeposition, physical vapor deposition (PVD)/chemical vapor deposition (CVD), and ultrasonic dispersion [67].

1.3.1 Physical synthesis methods

PVD consists of a process in which the material used as a target evaporates into atoms, molecules, or ions (plasmas) under high vacuum conditions. This vapor, composed of atoms or ions, condenses on the surface of the substrate to form a thin film coating of the target material. According to the principle they work with, vapor deposition techniques can be classified into CVD and PVD [68]; these two categories of deposition concentrate different synthesis techniques; for example, CVD techniques includes evaporation, condensation, chemical reaction, coalescence, sediment, and coagulation. At the same time, PVD comprises techniques of sublimation, condensation, coalescence, postprocessing, coagulation, deposition, plasma production, and so on [69]. In addition, PVD is subcategorized into sputtering, LA, and vacuum arc deposition [70].

An essential factor in the synthesis of NMs through PVD methods is to explore the optimal conditions to improve the quality of the material obtained and the reproducibility of the technique. Table 1.1 shows an overview, advantages, and disadvantages of some synthesis methods for obtaining materials by physical processes.

The sputtering technique is a kinetically balanced process in which the target (cathode) is sputtered by energetic ions generated in a discharge plasma, resulting in numerous particles deposited on a substrate (anode) [71, 72]. The process consists of an electric discharge between two electrodes (anode and cathode) in the presence of noble gas and a magnetic field; the magnetic field (B) forces the electrons to follow a spiral path, and as a consequence, prolongs the time of residence of electrons in the plasma, while ions are directly affected by the magnetic field due to their weight compared to electrons. During this process, the probability of collision between the electrons and the noble gas atoms increases dramatically, causing many gas atoms to ionize to form a noble gas plasma. The presence of an electric field (E) forces the gas

Table 1.1: Various approaches for the physical synthesis methods of nanocomposites.

Synthesis method	Description	Advantages	Disadvantages	References
PVD	– It is defined as various vacuum deposition techniques that can be used to make films and coatings of different thicknesses – The material used as a target becomes vapor composed of atoms and ions; this vapor is transferred at low pressure from the target to the substrate and undergoes condensation on the substrate to form a thin film/coating – Physical evaporation and sputtering are the two main techniques for forming thin films and coatings	– Economic and ecological approach – The materials obtained show durability, strength, and longtime life. – In contrast to CVD methods, PVD methods show safer handling of equipment and reagents	– A system is necessary to produce ultrahigh vacuum in the equipment – PVD equipment needs constant cooling, translating into higher energy costs in long operating times	[68, 70]
Sputtering	– The method is well known as sputtering plasma production – In this technique, the target is sputtered with energized ions of a noble gas; therefore, the target atoms detach and pass into the gaseous medium deposited on a substrate in the presence of electric and magnetic fields	– A sputtering process can be applied to form films/coatings with almost all types of materials – The product's properties can be modified according to the conditions in which the sputtering process is carried out	– It is an expensive technique due to the prices of the sputtering targets, the ultrahigh vacuum, and the amount of energy required in the process – The equipment needs constant maintenance	[71, 72]

Table 1.1 (continued)

Synthesis method	Description	Advantages	Disadvantages	References
	– Different sputtering processes are used for the development of nanomaterials and among which the direct current sputtering, radiofrequency sputtering, and the magnetron stand out	– Materials can be modified through decoration or doping using the sputtering technique	– To avoid contamination of the sputtering targets, the gas used must be strictly a noble gas, and the lowest sputtering rates must be considered	[71, 72]
Laser ablation	– It is a widely used technique for obtaining thin-film nanomaterials through the production of plasmas – The method is also known as laser pulse deposition – Obtaining different nanomaterials through this technique can be done both in liquid and gas states	– It is a very efficient technique due to the formation and distribution of the plasma plume on the substrate – Highly recommended for obtaining crystalline materials – No harmful reagents are required – The products obtained are of high purity	– Poor performance over large areas, limiting technique for larger scale or industrial applications – A high-energy laser is required to ablate the targets – It is an expensive technique because of the targets, the laser, and the ultrahigh vacuum – Efficiency of ablation decreases with long ablation time	[72–74]

Table 1.1 (continued)

Synthesis method	Description	Advantages	Disadvantages	References
Electric arc deposition (cathodic arc evaporation)	– The method was initially discovered during the production of carbon nanotubes in 1991 – This technique produces an electric arc when a very high electric field force is applied to nonconductive materials. In this process, the gas causes an electrical breakdown of the material, which tilts the conductivity in the environment	– Inexpensive equipment that does not require constant maintenance concerning others – Low contaminants and high performance – They have high performance and the potential to produce various materials in the form of nanoparticles	– It is necessary to use noble gases and low-pressure systems – Complex homogeneous particle size distribution – It takes a large consumption of energy to produce the electric arc	[75, 76]

PVD, physical vapor deposition; CVD, chemical vapor deposition.

ions to accelerate, which results in a bombardment of the target (cathode), and the particles that are finally released condense on the surface of the substrate (cathode) to form a thin film of the desired material [77, 78]. The diagram that exemplifies the magnetron sputtering technique is shown in Figure 1.4.

LA is a physical technique for the formation of thin films by plasmas that use a high-power pulsed laser to ablate a substrate to form structures ranging from 5 nm to 1 μm. It has several applications in the manufacture of metallic nanoparticles, ceramics, coatings for glasses, and polymers. In addition, it has generated scientific interests to obtain different structures with applications in engineering, chemistry, biology, and medicine [79]. The LA technique combines vaporization and expulsion of the molten atoms, as represented in Figure 1.5. When a beam of laser radiation strikes a surface (target), the electrons present in the substrate surface are excited by laser photons [80]. This excitation of the target electrons results from the generation of heat by absorbing the energy of the photons, which is consistent with Beer-Lambert's law [81, 82]. This law states that the amount of light absorbed depends on the thickness of the materials and the intensities of the light source. As a result of the heating of the target surface, the material's fusion or vaporization is caused, resulting in the melting and separation of atoms from the substrate. The transition from solid to gas results in forming a plasma. High pressure is created during the vaporization process,

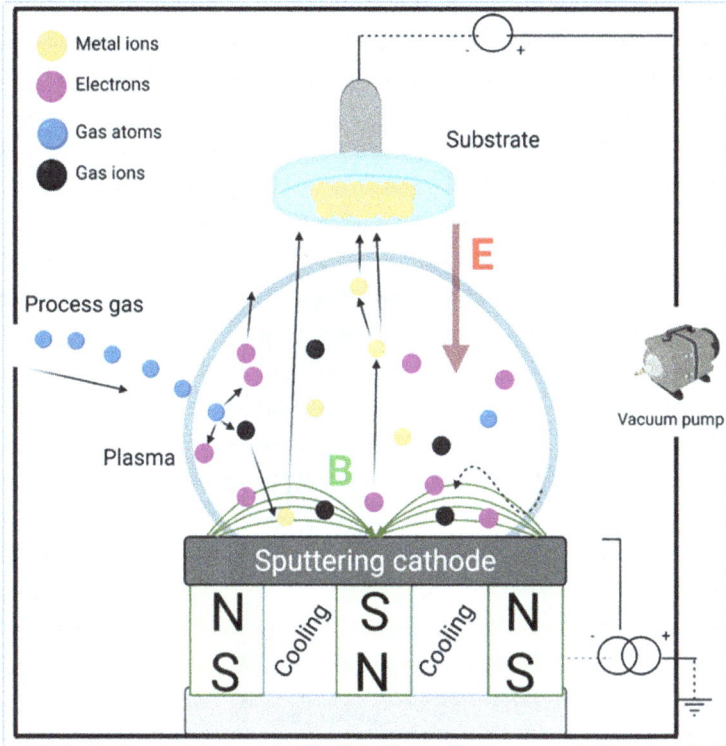

Figure 1.4: Principle of magnetron sputtering technology.

also called recoil pressure, which pushes the target atoms toward the substrate [83]. Ablated particles that are not deposited on the substrate are concerned due to their redeposition [84, 85].

The cathode arc evaporation (CAE) technique consists of a high-current and low-voltage plasma discharge (Figure 1.6). This discharge begins between the target of the cathode made of material to be evaporated and the anode (usually the substrate); high vacuum is required to evaporate the target particles and condense them on the surface of the substrate. This technique does not require the presence of magnetic fields applied to the cathode, so the electric arc points move randomly on the cathode surface. The residence time of the electric arc on the spot is in brief periods close to microseconds [86], and the current density is generally between 10^6 and 1,012 A/m^2 [87]. The high energy density during the process leads to the transition from the solid target (cathode) in the arc spot to the metallic vapor plasma, producing target droplets. The relationship between the ionization of the target and the number of drops is correlated with its melting temperature [87].

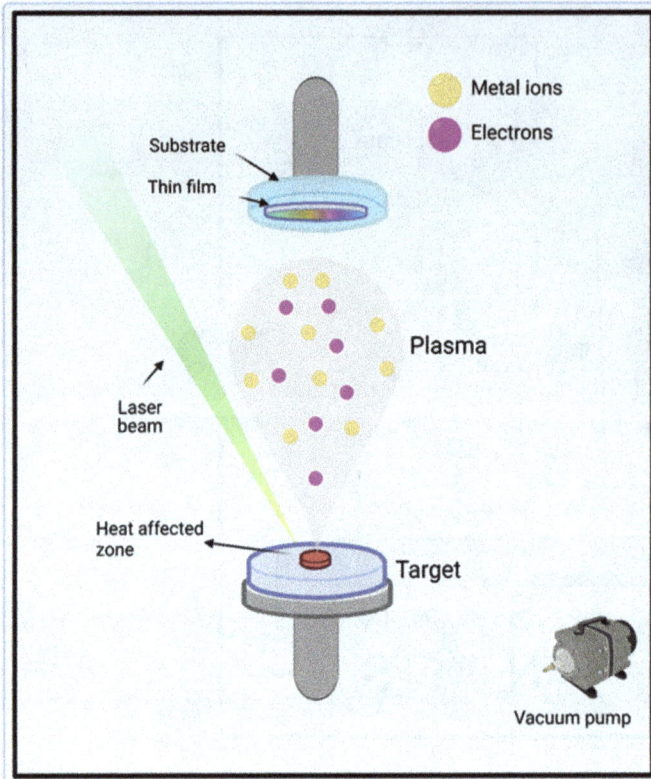

Figure 1.5: Principle of magnetron laser ablation technology.

As a result, the material expelled from the surface of the target is condensed into particles of different sizes, and finally, they are deposited on the surface of the substrate. Due to the nature of the technique and the conditions in which it is carried out, only materials (targets) with good electrical conductivity (primarily metals) can be used as sources of evaporation; otherwise, the process for the formation of the electric arc could not be carried out, and the use of materials with a high or low melting point and with low mechanical resistance is ruled out [88].

Through the physical methods of formation of nanoparticles and NMs mentioned above, it is possible to obtain different materials such as NMs based on metallic particles (Table 1.2), NMs based on metallic oxides (Table 1.3), and carbon-based materials (Table 1.4), each type of material with different applications.

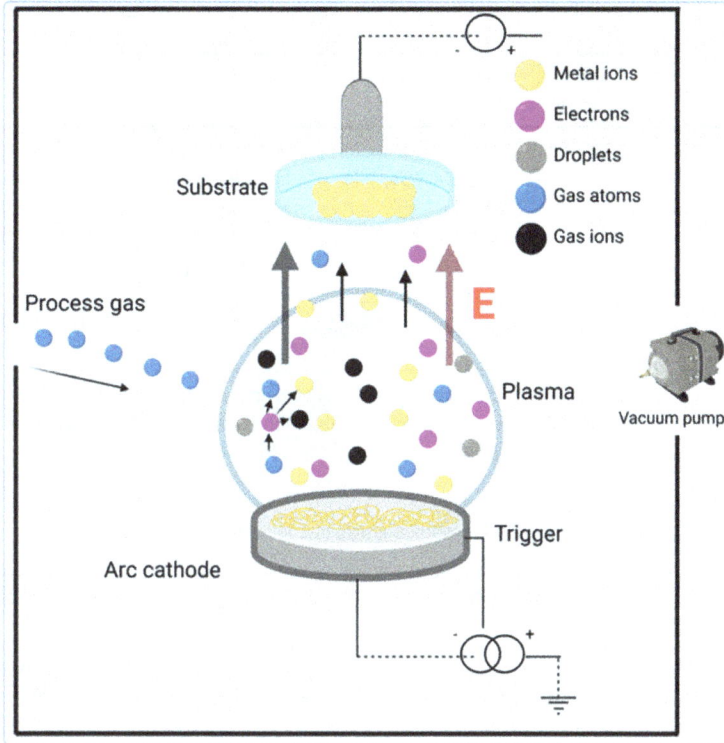

Figure 1.6: Principle of cathodic arc evaporation.

Table 1.2: Metallic nanomaterials obtained through physical methods and their applications.

Materials	Synthesis methods	Size and shape	Characterization	Applications	References
Ag Au Pt Cu Fe Ti	Sputtering/ laser ablation/ cathode arc evaporation	2–90 nm	Raman spectroscopy, XRD, EDX, TGA, SEM, HR-TEM, SAED, microfluorescence, TEM, UV-vis, XPS, cyclic voltammetry, linear sweep voltammetry	Water treatment, photocatalysis, energy storage, optoelectronics, agriculture, bioimaging	[89–91] [34, 92, 93] [93–95] [96–98] [99–101] [102–105]

XRD, X-ray diffraction; EDX, energy-dispersive X-ray spectroscopy; TGA, thermogravimetric analysis; SEM, scanning electron microscopy, HR-TEM, high-resolution transmission electron microscopy; SAED, selected area electron diffraction; XPS, X-ray photoelectron spectroscopy.

Table 1.3: Metallic oxide nanomaterials obtained through physical methods and their applications.

Materials	Synthesis methods	Size and shape	Characterization	Applications	References
TiO_2	Sputtering/laser ablation/cathode arc evaporation	5–200 nm	Raman spectroscopy, XRD, EDX, TGA, SEM, HR-TEM, SAED, microfluorescence, photoluminescence spectroscopy, TEM, UV-vis, XPS	Water treatment, photocatalysis, energy storage, optoelectronics, agriculture, markers	[72, 106, 107]
ZnO					[108–110]
CeO_2					[111–113]
Fe_3O_4					[114–116]
FeO					
SiO_2					[117–119]
Al_2O_3					[120–122]

XRD, X-ray diffraction; EDX, energy-dispersive X-ray spectroscopy; TGA, thermogravimetric analysis; SEM, scanning electron microscopy, HR-TEM, high-resolution transmission electron microscopy; SAED, selected area electron diffraction; XPS, X-ray photoelectron spectroscopy.

Table 1.4: Carbon-based nanomaterials obtained through physical methods and their applications.

Materials	Synthesis methods	Characterization	Applications	References
CNTs	Sputtering/laser ablation/cathode arc evaporation	Raman spectroscopy, XRD, EDX, TGA, SEM, HR-TEM, SAED, microfluorescence, photoluminescence spectroscopy, TEM, UV-vis, XPS	Energy storage devices (batteries, supercapacitors, fuel cells), electrochemical energy, drug/protein/gene delivery, and antibacterial and antifungal	[123–126]
C60				[127, 128]
Carbon dots				[44, 129, 130]
Graphene				[131–133]

XRD, X-ray diffraction; EDX, energy-dispersive X-ray spectroscopy; TGA, thermogravimetric analysis; SEM, scanning electron microscopy, HR-TEM, high-resolution transmission electron microscopy; SAED, selected area electron diffraction; XPS, X-ray photoelectron spectroscopy.

1.3.2 Chemical synthesis methods

Chemical synthesis is a process where chemical and physical processes were carried out using different chemical reactions, which is possible to control and reproduce the principal product. It is commonly used as liquid-phase precursors to synthesize NMs.

Chemical synthesis involves chemical reactions in a liquid phase, mainly the chemical reaction of the precursor species to generate a nucleation process, and posteriorly the formation of the nanoparticles. To enhance the properties, morphology, and reproducibility of the material, it is important to realize a careful selection of precursors and concentration, reaction dynamics, and synthesis parameters. It is essential to control the basic parameters such as pH, temperature, heating, agitation, and reaction time to avoid the modification over selectivity of the reaction, performance, and properties of the material.

Compared to the previous physical synthesis methods, these methods are a more popular choice, allowing a good control of stoichiometry due to the facility of the techniques utilizing ambient conditions, low temperatures, and simple equipment, reducing the cost, and making it more attractive than the physical methods. This synthesis category includes chemical precipitation method, hydrothermal method, sol-gel method, microwave-coupled sol-gel method, and sonochemistry (SQ)-coupled sol-gel method.

1.3.2.1 Chemical precipitation methods

The chemical precipitation technique is one of the most popular and straightforward techniques with high performance in metal-chalcogenides controlling composition. The main reaction is coprecipitation, which consists of the simultaneous process of nucleation, growth, coarsening, and agglomeration. The nucleation process is key to carrying out the small particles. However, a secondary process as aggregation or Ostwald ripening strongly affects the material's size, morphology, and properties. The technique uses a variety of organic and inorganic reducing agents [134]. When conducting the coprecipitation experiment, the inorganic salts are dissolved in an appropriate medium to form a uniform solution. It is necessary to prepare a basic solution of the precursors through dropwise mixing of solvents for better pH control since most materials only precipitate in basic solvents. Chemical composition is an essential aspect of the structure and properties of NPs. For that reason, it is necessary to establish a proper stoichiometric ratio of precursors [135].

The chemical precipitation technique allows the synthesis of different NPs with diverse morphology, shape, and properties, depending on the temperature, salt concentration, and pH modifications. Furthermore, the technique is a simple method with rapid preparation control on particle size, composition, and homogeneity, with low temperature, energy, and cost. However, the technique is inappropriate for reactants with different solubility and precipitation rates. Furthermore, it is a batch process, where the reproducibility and impurities over precipitated material could be a problem [136].

The nanoparticles commonly synthesized for precipitation chemicals are metallic NPs (Ag, Au, and Ni), metallic oxide NPs (ZnO, CuO, CeO_2, AgO, PtO, TiO_2, SiO_2, and Fe_3O_2), and carbon quantum dots [137–140], as presented in Table 1.5.

Zhang et al. synthesized Ce_2O NPs employing the hydrothermal technique by preparing two solutions. The first is a solution of $Ce(NO_3)_3 \cdot 6H_2O$ as the precursor and the second solution was hexamethylenetetramine with different concentrations. The final solutions were stirred for 3 h at 75 °C. Two different particle sizes of 7 and 25 nm were obtained with a truncated octahedral shape. The CeO_2 NPs were investigated using a radiotracer method to evaluate the absorption of the nanoparticles over cucumber plants [141]. In this way, Cui et al. employed the same methodology and obtained CeO_2 NPs with an average size of 7.1 nm and a truncated octahedral shape, demonstrating the method's reproducibility. The phytotoxicity of NPs was evaluated over lettuce seeds [136].

On the other hand, Shilova et al. employed three synthesis routes for obtaining magnetic iron oxide utilizing aqueous solutions of iron (II, III) salts as precursors and modifying the ultrasonic action, argon bubbling, and heating at 60 °C. The crystalline analysis showed a polycrystalline nanopowder and a mix of phases of magnetite and maghemite in the three nanoparticles. The morphological analysis showed that there were different nanoparticle sizes. Furthermore, the effects of NPs as aqueous suspensions with different concentrations were assessed, showing positive effects and development indices of plant seedlings [134]. Meanwhile, Mahanty et al. synthesized magnetite through the chemical coprecipitation method, using ferric chloride as a precursor, obtaining a spherical shape. The nanotoxicity was evaluated over a soil matrix employing a suspension of nanoparticles and *E. coli* at different conditions [142].

1.3.2.2 Hydrothermal method

The hydrothermal technique involves high temperatures and pressures above the ambient conditions to carry out the chemical reactions. The method is based on the ability of water and aqueous solution to dilute at high temperatures (500 °C) and pressure (up to 300 MPa), and substances are practically insoluble under normal conditions. Water is employed as a solvent in the hydrothermal technique; thus, the technique's name. This process is like solvothermal; however, the difference is in using any other solvent rather than water. Water is a strong solvent in which it is possible to dissolve even nonionic compounds at high temperatures and pressures [143].

There are two classes of hydrothermal method: namely, the batch hydrothermal system, in which the process is performed with needful ratio phases; and the continuous hydrothermal system with a high rate of reaction at a shorter time. It is important to note that it is possible to control the particle morphology and structure utilizing various precursors and hydrothermal conditions [144].

The hydrothermal technique works at low energy and avoids wasting materials, making it a preferred choice. The clear advantages of the process are its low cost, and controllable and reproducible conditions to control the nucleation process. However, it is necessary to employ autoclaves, high temperatures, and, is unfeasible to observe the process during growth [135].

Some of the most representative nanoparticles synthesized by the hydrothermal technique are listed in Table 1.5. ZnO NPs were synthesized through the hydrothermal technique by Younes et al., employing a solution of zinc acetate and polyvinyl pyrrolidone, with a pH solution of 9. The mixture was processed in an autoclave at 70 °C and 50 kpsi for 6 h. The results show the morphology of nanorods with a ratio of 7 nm and a specific surface area of 19.2 m^2/g. The effects of ZnO NPs on plant seedlings were evaluated by applying gel coating over the surface. The results show a beneficial effect over germination time and mean germination rate [145].

Konate et al. conducted a hydrothermal process to prepare nano-Fe_3O_4 employing an aqueous solution of $FeCl_3 \cdot 6H_2O$ and $NaHCO_3$. The final mix was processed in an autoclave at 150 °C for 4 h, obtaining a particle size of 6.7 nm. Furthermore, a comparative study of the effects of nano- and bulk Fe_3O_4 on the growth of cucumber was conducted. The materials exhibit different effects in bulk and nanoforms. However, nano-Fe_3O_4 increases biomass, and antioxidant enzymes superoxide and peroxidase, and has less severe phytotoxicity than the bulk Fe_3O_4 [146].

The technique allows the synthesis of composite materials like cerium-doped carbon quantum dots (CDs:Ce). Gong et al. employed a solution of citric acid, urea, and cerium acetate hydrate, which was stirred and then transferred into an autoclave with Teflon lining for 12.5 h at 180 °C. The hydrothermal method demonstrated stable fluorescence efficiency, water solubility, and biocompatibility. The evaluation of CDs: Ce over wheat seed promotes the growth, development, and physiological process of plants [147].

In this way, Sahoo et al. employed the biohydrothermal method to prepare ZnO-$ZnFe_2O_4$ nanocomposite using the extract of *Psidium guajava* and to employ an autoclave to carry out the reaction at 180 °C for 12 h. They obtained a crystalline material with a crystalline phase of ZnO and $ZnFe_2O_4$ with spherical morphology and a diameter range of 10–20 nm. With an efficient performance in the remotion of methylene blue dye. Furthermore, the evaluation of the effects of ZnO-$ZnFe_2O_4$ NPs overcrop shows an enhancement of plant immunity without any compromise in the plant growth [148].

1.3.2.3 Sol-gel method

The sol-gel process is the most popular and established process of the generation of colloidal nanoparticles from liquid phase processing of nanocomposite powders. The method's main advantage is that it requires low temperatures and less energy to carry out the synthesis [135], while offering outstanding control over the texture and surface properties. The process is described in five steps: (i) hydrolysis, (ii) polycondensation, (iii) aging, (iv) drying, and (v) thermal decomposition.

(i) The hydrolysis step takes place in water or alcohol. The method can be either aqueous or nonaqueous sol-gel, depending on whether water or an organic solvent is used as a reaction medium. Typically, compounds such as inorganic metal salt or a metal-organic species, like a metal alkoxide or acetylacetonate, are used as precursors. Furthermore, acidic or alkaline media could aid in conducting the hydrolysis step [149].

(ii) The condensation step eliminates the water/alcohol and polymeric networks growing on colloidal dimensions. Forming hydroxyl bridge is formed between metal centers, and an oxo bridge is formed between two metal centers. Furthermore, the solvent's viscosity is high due to the formation of porosity in the structure conserving the liquid phase, called gel. The alkoxide precursor and pH of the solution strongly affect the colloidal particles' size and cross-linking.

(iii) In the step of the aging process, the polycondensation continues with the changes of the structure and properties of the gel, decreasing the porosity and increasing the thickness between colloidal particles.

(iv) In the drying process, it is difficult to realize that the water and organic components are detached from the gel which disturbs its structure. The most used drying methods are atmospheric/thermal drying, supercritical drying, and freeze-drying, each of which has different modifications to the gel network structure, such as high porosity, pore volume, and surface area [150, 151].

(v) Moreover, the last step, thermal treatment, is realized to eliminate the residues and water molecules from the material. It is essential to establish and control the calcination temperature due to controlling the pore size and the density of the material [152].

The nanoparticles commonly synthesized by this technique are listed in Table 1.5. TiO_2 is a common material synthesis using sol-gel. Farag et al. synthesized TiO_2 nanoparticles using the isopropoxide of titanium (TTIP) as a precursor and a solution of isopropanol and water distilled. The mixture was stirred over 16 h, reporting 10–40 nm particle size and an agglomerate material. The photocatalytic and antimicrobial activities were evaluated over fabrics and compared with bacterial cellulose/TiO_2 hybrid nanocomposite synthesized through biosynthesis [153]. SiO_2 is another material generally synthesized by sol-gel technique. Imoisili et al. synthesized palm kernel shell ash as a silica precursor. The analysis shows that particle size between 50 and 98 nm, average pore diameters ranging from 2.2 to 6.3 nm, and surface area of 438 m^2/g exhibit better properties as commercial silica [154]. Palmqvist et al. synthesized Fe_2O_3 using $FeCl_2$-$4H_2O$ as a precursor and ethanol as a medium. The sol was dry at 100 °C. The results showed a crystal size of 3.8 nm and an agglomerate material. The Fe_2O_3 NPs were evaluated over plants transferred to pots irrigated with a nutrient solution with NPs. This exhibits an interaction between the enzymes in the plant with NPs [155].

In this way, Belhamel et al. obtained nanostructured alumina (NSA) by the sol-gel method using aluminum nitrate and butanediol; the gel was heated at 1,000 °C for 1 h. They obtained a polycrystalline and monophasic NSA material with an average crystallite size of NSA of 56.9 nm. NSA was evaluated as a seed protectant of the main seed-infesting insect pests. Furthermore, the effect of NSA on seed germination and plant growth was assessed. Moreover, to complete, the study has analyzed the presence of NSA in the leaves of bean plans germinated [156].

On the other hand, Herrera et al. synthesized α-Al_2O_3 nanoparticles employing the same precursor as Belhamel et al. [156]. However, the author realizes a heat treatment of 1,000 °C to obtain α-Al_2O_3 NPs. The crystal size was 58 nm, and an average surface area was 2.6 m^2/g. The NPs were evaluated over biomass matrix for adsorption of cadmium with an optimal value of pH 6 to obtain the maximum adsorption capacity [157].

The sol-gel technique allows the synthesis of composite materials, like Ag-TiO_2 NPs. Godrillo-Delgado et al. synthesized Ag-TiO_2 NPs employing TTIP and silver nitrate as precursors and methanol as a medium. Then, it was calcinated at different temperatures to grow the crystalline phases. The results exhibit a mix of anatase and rutile phases and increased crystallinity with the thermal treatment. The nanoparticles were applied over seeds as suspensions at different concentrations, which establish a dependence on the size of Ag-TiO_2 NPs with the beneficial effects over the physiological and morphological parameters related to plant growth [158].

Another hybrid nanocomposite synthesis for this technique is GRA quantum dot and iron codoped titanium dioxide (GQD-Fe-TiO_2). Khan et al. synthesized GQD-Fe-TiO_2 photocatalysts, employing TTIP, Fe, and GQD solution as precursors with acetic acid and water. The final materials were calcinated at 300 °C. They obtained an average particle size of 20 nm and a surface area of 179.79 m^2/g. The materials were evaluated over the decolorization of reactive black 5 dye. They studied the effects of overuse treated water with Fe-TiO_2, TiO_2, and GQD over the germination of *L. esculentum*. The use of GQD-Fe-TiO_2 exhibits higher seed germination [159].

1.3.2.4 Microwave method

Microwave-assisted synthesis allows a faster chemical reaction than traditional convection heating methods and has high yields and fewer secondary products. The excellent control over reaction mixing and holding up high temperatures and pressures demonstrate reproducibility between reactions. Microwave-assisted method is a popular method employed from biochemical processes to nanotechnology. Microwave-assisted heating could provide selectivity in activating the NM precursor materials [143].

Microwaves are electromagnetic waves with frequencies of 300 MHz to 300 GHz, which correspond to wavelengths between 1 m and 1 mm. The success of microwave heating when polar and charged molecules are exposed to extreme electromagnetic excitation creates dissipation of radiant energy and increases the internal temperature of the microwave-exposed object. The interaction of microwaves with matter creates an inverted temperature gradient profile compared to conventional heating. The transferred heat of the conventional treatments is from the surface to the inner solution or material, which makes it deficient. Heating with low homogeneity affects the stability of the product or reagent, decreasing reaction yield, generating undesired by-products, and making necessary the purification stage [160, 161].

The advantages of this technique are short reaction times, higher reaction yields, selective heating, enhanced crystallinity, and uniform morphology. The nature of the technique allows carrying out organic reactions and inorganic products and obtaining materials such as metal oxides, metallic NPs, polymers, or GRA-based materials (Table 1.5). One of the materials most synthesized by this technique is the TiO_2 powder. Younes et al. employed a microwave heat oven at 90 °C for 4 h; then it was washed and dried at 60 °C. The treatment was enough to obtain an anatase phase, with a particle diameter of 43 nm and a specific surface area of 14.4 m^2/g. Afterward, the TiO_2 NPs have assessed the performance of plant seedlings by applying a gel coating on different seeds, increasing the germination rate and showing better morphogenesis [145].

1.3.2.5 Sonochemistry method

The SQ technique, compared to other techniques, is a green process. It is induced by ultrasound under a liquid medium, producing the phenomenon of acoustic cavitation. Briefly, it involves the formation, growth, and implosion of bubbles, which generates hot spots with high temperatures up to 5,000 K, pressures up to 1,000 atmospheres, and cooling rates of over 1,010 K/s. These conditions allow the chemical interactions and enhance the chemical reactions, and the intensity of ultrasound increases the chemical reactions at high energy [162].

The sonochemical technique presents two processes; primary SQs employing volatile precursors were formed inside the bubble, and secondary SQs using nonvolatile precursors were formed out of the bubble, in which both processes have a synergistic impact. The reaction's parameters like temperature, sonication time, and power of ultrasonication are determined in the properties of NMs [163].

The SQ technique demonstrated the advantages for better control of morphology and thermal stability, higher surface area, phase purity, less hazardous use, and short synthesis times (Table 1.5) [164–166].

Shyla et al. synthesized through a solution of TiO_2 pellets on NaOH under ultrasonic irradiation at 20 kHz and 350 W. The NPs show a spherical shape ranging from 85 to 100 nm and anatase phase with high crystallinity. The NPs have evaluated the effects on seed treated with NP solutions [167].

1.4 Biosynthesis

There are diverse synthesis processes which have several advantages in processing but some disadvantages regarding the environment. That is why the "green chemistry" [168] was born using environmentally harmless materials as an alternative to obtain different materials, specifically NMs to address pollution and low-cost production [169].

Table 1.5: Synthesis methods of nanoparticles and their applications.

Materials	Synthesis method	Reagents and conditions	Size and surface area	Applications	References
ZnO	Hydrothermal	$ZnC_4H_6O_4$, $(C_6H_9NO)_n$ Autoclave: 70 °C, 50 kpsi, 6 h	7 nm 19.2 m²/g	Effects of the NPs as a coating over different seeds	[145]
ZnO-ZnFe₂O₄	Hydrothermal	$Zn(NO_3)_2 \cdot 6H_2O$ $Fe(NO_3)_3 9H_2O$, PVP Autoclave: 180 °C, 12 h	10–20 nm	Remotion of methylene blue dye Evaluation of immunity of the plant	[148]
Fe₃O₄	Hydrothermal	$FeCl_3 \cdot 6H_2O$, $NaHCO_3$ Autoclave: 150 °C, 4 h	7 nm	Comparative effects of phytotoxicity over cucumber plants	[146]
Fe₂O₃	Chemical precipitation	$FeCl_3$, $FeSO_4$, NH_4OH		Effects of toxicity over the soil	[142]
γ-Fe₂O₃-Fe₃O₄ γ-Fe₂O₃	Chemical precipitation	Solution (II, III) salts 100 °C Ultrasonic action Argon bubbling Oleic acid	–5 to 20 nm 92 m²/g –10 to 20 nm 52 m²/g –10 to 20 75 m²/g	Effects over lettuce seeds	[134]
Fe₂O₃	Sol-gel	$FeCl_2$-$4H_2O$, ethanol-water, dry at 100 °C	3.8 nm	Effects over *Brassica napus* (seeds and plants)	[155]
Ce₂O	Chemical precipitation	$Ce(NO_3)_3 \cdot 6H_2O$ Hexamethylenetetramine 3 h, 75 °C	7 nm 25 nm	Effects over cucumber plants	[141]
			7 nm	Phytotoxicity on lettuce seeds	[136]
CDs:Ce	Hydrothermal	$C_6H_8O_7$, urea, $C_6H_{14}CeO_4$ Autoclave: 12.5 h, 180 °C	—	Effects over wheat seeds	[147]
Al₂O₃	Sol-gel	$Al(NO_3)_3$, butanediol, water: ethanol Dry: 100 °C, 1 h	56.9 nm	Insecticidal effect over *P. vulgaris* and morphological effects	[156]
α-Al₂O₃	Sol-gel	$Al(NO_3)_3$, citric acid Calcinated 1,000 °C	58 nm 2.6 m²/g	Absorption of Cd for biomass	[157]

Table 1.5 (continued)

Materials	Synthesis method	Reagents and conditions	Size and surface area	Applications	References
TiO_2	Sol-gel	TTIP, isopropanol:water	10–40 nm	Photocatalytic activity and antimicrobial activity in textiles	[153]
	Microwave	TTIP, CH_3COOH, polyvinyl alcohol Microwave: 90 °C, 4 h.	43 nm 14.4 m²/g	Effects on eggplant, pepper, tomato over seeds	[145]
	Sonochemical	TiO_2 pellets, NaOH 20 kHz, 350 W	85–100 nm	Effects on seed treated with NP solution	[167]
Ag-TiO_2	Sol-gel	TTIP, Ag(NO_3), methanol medium 400, 500, 600, 700 °C, 2 h	9 nm, 15 nm, 35 nm, 43 nm	Morphology, phytotoxicity, and photosynthetic activity on spinach seeds	[158]
GOD-Fe-TiO_2	Sol-gel	TTIP, Fe, graphene quantum dot solution, CH_3COOH ethanol, water	20 nm 170.79 m²/g	Remotion of reactive black 5 and over germination of *L. esculentum* seeds	[159]

TTIP, isopropoxide of titanium.

The environmental-friendly synthesis routes are approaches that include fast, nontoxic, ecologically sustainable, and optimum pressure and temperature conditions compared to the traditional processes (Figure 1.7) [170, 171].

The green chemistry (GC) methods are also known as biosynthesis, which use natural substances from microbes and plant extracts, which work as reducing agents for crowning and stabilization medium, and eliminate the use of hazardous reagents such as sodium borohydride ($NaBH_4$) [172].

In a general way, the use of chemical compounds obtained from microbes or plant extracts is due to the considerable number of oxidoreductive proteins, alkaloids, flavonoids, saponins, steroids, tannins, and phenols, which act as reducing and stabilization agents [173].

The advantages of plant-mediated biosynthesis over other biomaterials include easy availability, safe to handle, cost-effective, and one-step simple process, composed of various metabolites that may aid in reduction, elimination of elaborate maintenance of cell cultures, rapid rate of synthesis, more environmental-friendly, more stable nanoparticles, and better control over size and shape [170, 173, 174].

Several factors affect the final properties of NMs obtained by biosynthesis, such as the type of precursors, reaction time, pH, temperature, extract concentration, radiation (if applied like light, MW, or SQ, for example), and synthesis protocol [174, 175].

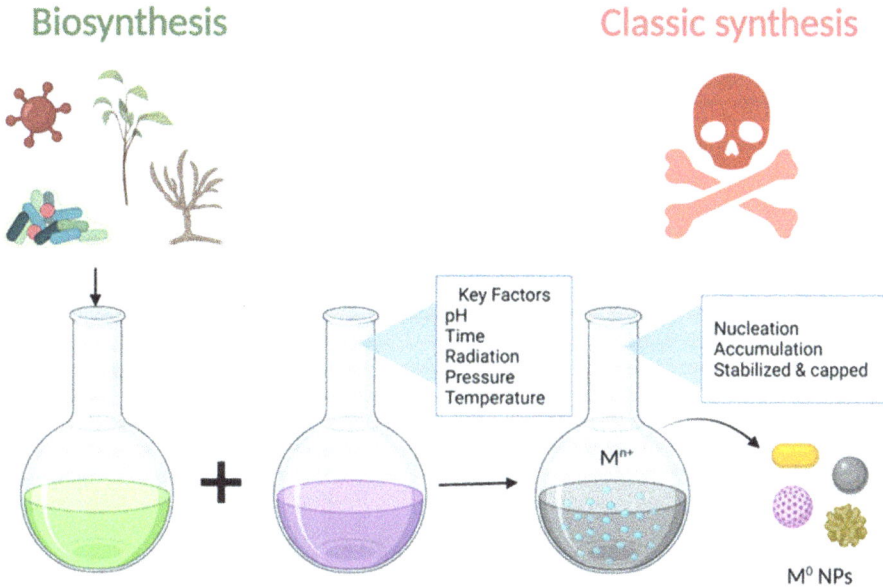

Figure 1.7: Experimental process of NP GC synthesis.

Among the microbes that help the biosynthesis, diverse types of bacteria, fungi, algae, virus, and yeast can be found [19, 176]. And for the plant extract, the diversity is bigger due to the only condition to apply these chemical compounds and its antioxidant activities [177]. Some of the most important microorganisms and plants according to the results and recent literature are presented in Table 1.6.

1.4.1 Metallic nanoparticles

As previously mentioned, the metallic nanoparticles (M^0-NPs) are considered the main NMs that have found several applications such as in optics, medicine, biology, agriculture, electronics, water, and air treatment, and specifically the gold and silver nanoparticles are the pioneers in their applications due to their excellent biocompatibility, physicochemical properties, and catalytic efficiency [292].

The metallic NPs through bioactive extracts involve three stages (as can be seen in Figure 1.8): (i) the reduction and nucleation of the metallic ion, known as activation stage [253], the growth phase, where NPs are close together to increase its size and an improvement in the thermodynamic stability of the NP. The parameters that must be controlled in this stage are temperature and pH. And (iii) the NPs take its final shape, which is determined by the composition of the herbal extract. The main functional groups involved are CO, OH, $R\text{-}NH_2$, and $R\text{-}OCH_3$ [170].

Table 1.6: Microbes and plant extracts used as bioreductive agents.

Biomass type	Scientific name	References
Bacteria	Sulfate-reducing bacteria	[178–181]
	Shewanella sp.	[182–186]
	Shewanella oneidensis	[187–190]
	Clostridium acetobutylicum	[191–193]
	Klebsiella pneumoniae	[194–196]
	Bacillus cereus	[197–199]
	Desulfovibrio alaskensis	[200–202]
	Morganella psychrotolerans	[203–205]
	Paracoccus haeundaensis	[206, 207]
	Rhodobacter capsulatus	[208]
Fungi	*Fusarium oxysporum*	[209–211]
	Pleurotus ostreatus	[212–214]
	Trichoderma harzianum	[215–217]
	Trametes trogii	[218, 219]
	Fusarium chlamydosporum	[19, 209, 220]
	Penicillium cyclopium	[221]
	Phomopsis liquidambaris	[19, 222]
	Macrophomina phaseolina	[223, 224]
	Ganoderma enigmaticum	[225, 226]
	Aspergillus niger	[227–230]
Alga	*Klebsormidium flaccidum*	[171, 231, 232]
	Tetraselmis kochinensis	[233, 234]
	Ulva faciata	[235–237]
	Pterocladia capillacea	[238, 239]
	Colpomenia sinusa	[240]
	Cystoseria baccatta	[241]
	Portieria hornemannii	[242, 243]
	Sargassum ilicifolium	[244–246]
	Gelidium corneum	[247]
Yeast	*Meyerozyma guilliermondii*	[248]
	Candida albicans	[249]
	Magnusiomyces ingens	[250]
	Yarrowia lipolytica	[251]
	Saccharomyces cerevisiae	[252, 253]
	Phanerochaete chrysosporium	[254, 255]
	Colletotrichum sp.	[256]

Table 1.6 (continued)

Biomass type	Scientific name	References
Plant	Acacia sp.	[257–259]
	Annona muricata	[260–262]
	Berberis vulgaris	[263–265]
	Cassia fistula	[266, 267]
	Cinnamomum camphora	[268–271]
	Eucalyptus sp.	[272–275]
	Ficus benghalensis	[276–278]
	Jatropha curcas	[279–281]
	Ginkgo biloba	[282–284]
	Olea europaea	[285, 286]
	Pistacia atlantica	[287–289]
	Theobroma cacao	[290, 291]

The difference of using herbal extracts and microbes that secrete compounds is that these last ones are rich in oxidoreductive proteins which help in the metallic ions' reduction, generating irresolvable metal protein nanocomplexes which finally form NPs [293, 294].

Several authors report two probable mechanisms related to the microorganisms that explain the metallic reduction. In the first one, there is the participation of

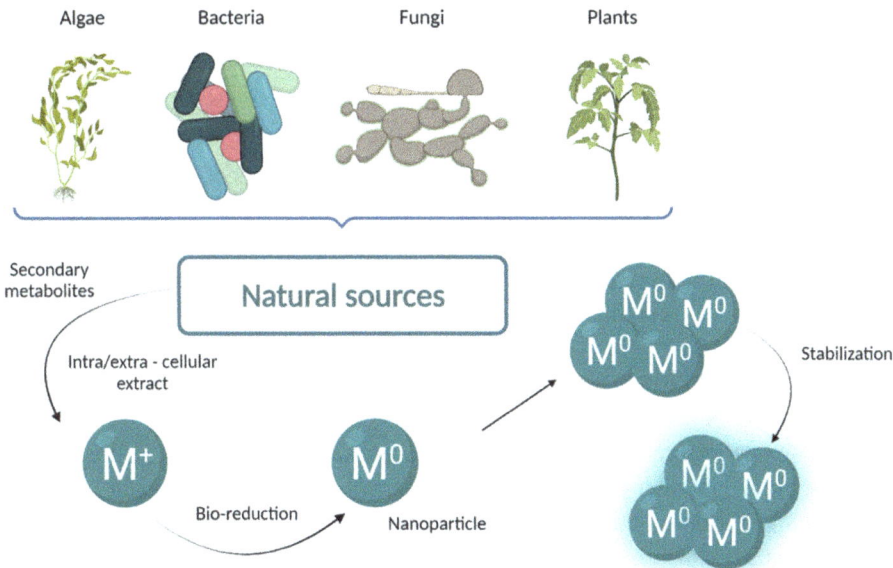

Figure 1.8: Stages of the biosynthesis of metallic NPs.

nicotinamide adenine dinucleotide [214] dependent enzymes and nitrate reductase that are secreted in the reaction medium. Figure 1.9(a) shows the general scheme of this mechanism, where NMs are formed by electron transfers from NADH resulting in reduction of the metallic species. The second probable mechanisms are the intracellular mechanisms, where both the fungal cell wall and proteins play a significant role in bioreduction as shown in Figure 1.9(b) [175].

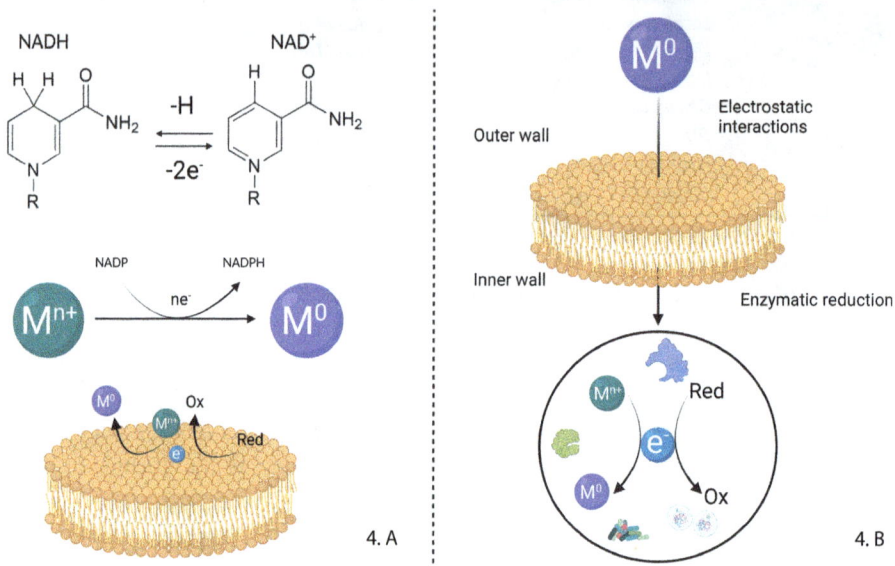

Figure 1.9: (a) Extracellular mechanism and (b) intracellular mechanisms proposed for NP biosynthesis.

As has been mentioned, the final properties and characteristics of the biosynthesized metallic NPs will depend on the herbal extract or microbes applied. Some of the main metallic nanoparticles generated by this type of synthesis are Ag, Au, and Cu, and its applications are also diverse (Table 1.7).

1.4.2 Metallic oxide nanomaterials

The metal oxide NMs are materials with a wide crystal structure types, that is, by the metal-oxygen bonds, their physicochemical properties are unique. As metallic nanoparticles, the applications of the metal oxide NMs are diverse such as in biomedical, medicine, materials chemistry, agriculture, electronics, catalysis, optical, environment, and energy [300].

The biosynthesis of this kind of materials has been established as a result of bioremediation of metal ion toxicity. Because most of the metal ions are toxic to microbes

Table 1.7: Metallic nanoparticles obtained through biosynthesis and their applications.

Materials	Extract	Size and shape	Characterization	Application	References
Ag-NPs	*Passiflora edulis* peel *Nypa fruticans* fruit husk *Malva parviflora* L. *Borassus flabellifer* fruit	7–50 nm spherical	Zeta potential, dynamic light scattering, FTIR, XRD, EDX, TGA, SEM, HR-TEM, SAED, microfluorescence, FE-SEM, TEM, UV-vis, XPS	Dye reduction, antibacterial and anticancer activity	[169, 292, 295, 296]
Au-NPs	*Passiflora edulis* peel *Borassus flabellifer* fruit Panchagavya				[169, 296, 297]
Cu-NPs	*Pseudomonas stutzeri* biomass Ginger root extract	1–60 nm roundabout shapes		Projection to produce molten salts for energy storage and antibacterial activity	[298, 299]

FTIR, Fourier-transform infrared; XRD, X-ray diffraction; SEM, scanning electron microscopy; EDX, energy-dispersive X-ray spectroscopy; TGA, thermogravimetric analysis; HR-TEM, high-resolution transmission electron microscopy; SAED, selected area electron diffraction; XPS, X-ray photoelectron spectroscopy.

and as a defense mechanism, those microbes (bacteria, alga, and fungi) transform these metal ions to less-toxic compounds, or in other words, the metal ions are oxidized or transformed to a more stable NM [301].

The main mechanism of how the metallic ions are oxidized by the microbes has been attributed to intracellular proteins, extracellular secreting enzymes, and polysaccharides such as alginate, laminarin, and fucoidan which contain carboxyl, hydroxides, and sulfhydryl groups, and secondary metabolites like flavonoids, monoterpenoids, triterpenoids, and tannins, specifically for alga [302]. For the plant extracts, the biosynthesis of metallic oxide NMs works under the same biomechanism as for the metallic nanoparticles [303]. Some of the metallic oxides obtained by biosynthesis are listed in Table 1.8.

Table 1.8: Metallic oxide nanomaterials obtained through biosynthesis and their applications.

Materials	Extract	Size and shape	Characterization	Applications	References
Au/Fe_3O_4	*Piper auritum*	35 nm spherical	HR-SEM, EDS, XRD, UV-vis, XPS, vibrating sample magnetometer, FTIR, FE-SEM, TEM, EDX, DLS, EDAX, Mössbauer spectroscopy, zeta potential, magnetic studies, TGA/DSC, HR-TEM, XRF, PL spectroscopy, diffuse reflectance, Raman spectroscopy	Catalytic degradation of dye	[304]
Ag-NP ZnO Ag-ZnO	*Pistacia atlantica*	<50 nm spherical <70 nm polyhedral shape		Antimicrobial activity	[289]
NiO	*Vitis vinifera* *Euphorbia heterophylla*	25 nm spherical 12–15 nm rhombohedral		Antibacterial activity and potential cytotoxicity against cancer cell lines	[305]
ZnO	*Eucalyptus globulus* *Bergenia ciliata* *Justicia spicigera*	10–100 nm spherical, semispherical, truncated octahedron		Antimicrobial activity and catalytic activity	[32, 306, 307]
$Fe_3O_4/$ $Fe_2O_3/$ FeO	*Moringa oleifera* *Syzygium aromaticum* *Rhodococcus erythropolis* ATCC 4277 free cells Green tea *Psidium guavaja–Moringa oleifera*	<60 nm		Biomedical application, and antibacterial and photocatalytic activities	[308–311] [312]
CuO CuO/C	*Enicostemma axillare* *Mimosa hamata* flower *Ficus religiosa*	6–45 nm sponge		Antibacterial activity and organic dye adsorption	[313–315]
$Pd-CeO_2$	*Salvadora persica*	10–20 nm		Toxicity effect on breast cancer cell line	[316]

Table 1.8 (continued)

Materials	Extract	Size and shape	Characterization	Applications	References
CeO$_2$	*Prosopis fracta* fruit	15–20 nm		Cytotoxic tests on cancer cell lines	[317]
TiO$_2$	*Monsonia burkeana* *Streptomyces* sp. *P. domestica* L./*P. persia* L./*A. deliciosa*	10–70 nm spherical		Antibacterial and photocatalytic potential	[168, 318, 319]
SiO$_2$	*Nerium oleander* leaves *Aspergillus tubingensis* F20	50 nm spherical		Absorbent	[320, 321]
Al$_2$O$_3$	*Prunus* × *yedoensis* leaves	50–100 nm hexagonal shape		Antibacterial and photocatalytic activities	[322]

XRD, X-ray diffraction; EDX, EDAX, energy-dispersive X-ray spectroscopy; DLS, dynamic light scattering; TGA, thermogravimetric analysis; DSC, differential scanning calorimetry; SEM, scanning electron microscopy, HR-TEM, high-resolution transmission electron microscopy; SAED, selected area electron diffraction; XRF, X-ray fluorescence; XPS, X-ray photoelectron spectroscopy.

1.4.3 Carbon-based materials

For these specific NMs, its synthesis via green chemistry is reduced only to obtain reduced graphene oxide (rGO), and is still under investigation due to the small amount of actual published papers. According to Ghosh et al. [323], it was possible to prepare high-quality reduced graphene oxide (rGO) using natural graphite as a raw material by utilizing various reducing agents such as starch, glucose, ascorbic acid, cloves (Syzygium aromaticum), and mint leaves (Mentha arvensis). These reducing agents were chosen due to their significant polyphenol content, which is known to possess chemically antioxidant properties. Consequently, they proved to be effective reducing agents for the rGO synthesis. Rattan et al. [324] reported a green method to obtain rGO using fresh ginger and garlic extracts due to the popular method of graphene oxide reduction by treatment with hydrazine hydrate which is not environment friendly.

Further research in this area should concentrate on deeper understanding of the reduction mechanism and controlled functionalization for obtaining a nondefective rGO.

As the metallic and metallic oxide NMs, the rGO obtained by green chemistry shows interesting results in antibacterial activity, anticancer performance, and electrochemical detector using green tea extract as a reducing agent, as reported by Vatandost et al. [325].

1.5 Physicochemical NM characterization (SEM, TEM, EDS, XRD, Raman, XPS, IR, UV-Vis, ZP, EQ)

Currently, the research involved in NMs applied in agriculture has increased exponentially. NM engineering is one of the most recent technological advances that display unique targeted characteristics with a definite function and properties not attainable in another chemical state or form of the identical material [326]. Nowadays, researchers can develop NMs with different physicochemical properties, such as size and shape [327]. Understanding the NM properties allows the development of new advanced novel materials and assesses their possible effects on living organisms [3]. Therefore, it is essential to become acquainted with different characterization techniques when working with NMs. This section revised different physicochemical characterization techniques and their applications to NMs.

1.5.1 Dynamic light scattering (DLS)

NMs' size and form are essential factors in determining their physicochemical properties. Dynamic light scattering (DLS) is a fast and nonintrusive approach for measuring the size of NMs in solution or suspension using a monochromatic wavelength light beam such as a laser [328, 329]. In simple terms, when a visible light source irradiates a particle, some of the light is transmitted through the sample, and the sample absorbs some light. When the particle is small enough in relation to the incident irradiation's wavelength ($<\lambda/20$), the irradiation is scattered in all directions without changing its energy or wavelength, a phenomenon known as Rayleigh scattering or elastic scattering of light [330].

In its most basic form, the DLS technique is based on monitoring fluctuations in Rayleigh scattering caused by the Brownian motion of nanoparticles much smaller than the wavelength of the incident light at a fixed angle. When a monochromatic light beam, such as a laser, shines through a solution containing randomly moving nanoparticles, the light is scattered with a different frequency proportional to the particle size. DLS assumes that nanoparticles suspended in a fluid move in Brownian motion. The particles' random collisions cause this movement with the solvent molecules, which causes the particles to diffuse through the medium [331].

DLS measurements are useful when evaluating the physicochemical properties of NMs. Recently, research in antimicrobial, antibacterial, absorbent, and cytotoxic applications of metal and metal oxide nanoparticles is increasing, especially when combined with a novel biosynthesis process [318, 320–322].

1.5.2 Scanning electron microscopy (SEM)

One of the most important aspects of NMs and their effects in different conditions is the particle size and size distribution, mainly due to the different synthesis methods yielding different morphologies with different sizes. Several methods can be applied to determine particle size and size distribution. One of the most utilized techniques is scanning electron microscopy (SEM), which operates with spatial resolution below 5 nm [331].

SEM is an imaging technique that generates micrographs of a sample by scanning it with a focused beam of accelerated electrons. It employs electromagnetic lenses to generate high-resolution images with up to 1 nm [328]. Incident electron beams interact with the surface of samples, producing signals that reflect the atomic composition and topography of the specimen.

In brief, the incident electrons cause the atoms on the sample surface to emit elastically backscattered electrons, secondary inelastically scattered electrons, and characteristic X-rays. The most common mode in SEM is secondary electron detection, in which the sample image is created by collecting the secondary electrons with a special detector [332]. The main drawback of the SEM technique is the requirement to operate in high vacuum and, thus, for many commercial instruments, SEM is applicable only to solid samples. Additionally, SEM provides information on conductive materials, limiting its application. However, this last drawback can be potentially overcome by, for example, sputtering a thin layer of a conductive material, such as silver or gold, on the sample. Nevertheless, this may lead to changes in NMs due to the added layers that increase the size of particles [329].

Despite limitations, SEM has been a fundamental technique for NMs since it can provide a much larger analysis area and provides more excellent reliability in terms of morphological and topological information of the sample. Moreover, it has a wide application in environmental resolution and reports concerning the interaction with crops. The main contribution is the material's morphology, sometimes composed of two or more systems that can be arranged randomly or ordered [331].

In addition to SEM analysis, most modern systems include the energy-dispersive X-ray spectroscopy (EDX) integrated with the equipment and are used to analyze the elements on the surface on the micrometer scale. The coupling application of SEM-EDX techniques allows acquiring data on the size, shape, and surface texture and the elements presented in a relatively short time [328].

One of the most important techniques for characterizing a material in purity and stoichiometry is EDX. It is a technique for determining the elements present in a sample. It is based on the interaction of an X-ray excitation source and a specimen. It operates on the premise that each unique atomic structure will have distinct X-ray emission spectrum peaks. Highly energetic particles, such as an electron or an X-ray beam, are allowed to interact with the specimen, exciting the electron of the inner energy level, causing a hole to form in it. The electron from the outer energy level can now move to the inner energy level by releasing excess energy in the form of X-ray photons. An energy-dispersive spectrometer measures the number and energy of these X-ray photons [333].

1.5.3 Transmission electron microscopy (TEM)

In addition to SEM, transmission electron microscopy (TEM) is an imaging technique that enables to study the particle size and morphology of NMs. TEM provides immediate images and chemical information about NMs with spatial resolution down to the atomic dimensions (<1 nm) [328]. An incident electron beam is transmitted through a fragile foil specimen to achieve this resolution. The incident electrons that interact with the specimen are changed to unscattered electrons, elastically scattered electrons, or inelastically scattered electrons. The scattered or unscattered electrons are focused by a series of electromagnetic lenses and then projected onto a screen to generate electron diffraction, amplitude-contrast image, a phase-contrast image, or a shadow image of varying darkness based on the density of unscattered electrons [332].

The high spatial resolution that TEM offers enhance the structural and morphological analysis of NMs. TEM and SEM can disclose the size and shape of NMs. However, TEM has a lead over SEM in terms of improved spatial resolution and capability of supplementary analytical measurements, especially when coupled to analytical techniques, such as electron energy loss spectroscopy and EDX, which can quantitatively assess the electronic structure and chemical composition of NMs. Nevertheless, certain drawbacks accompany TEM advantages, namely, a high vacuum and a narrow sample section are essential for electron-beam penetration. In general, high-resolution imaging enables examination of an extremely small fraction of the specimen and results in poor statistical sampling [329]. Also, TEM provides 2D imaging, leading to no depth sensitivity for 3D specimens. Another drawback is that samples must be narrow enough to transmit appropriate electrons to generate images. This can lead to extensive conditioning of narrow specimens, increasing the probability of modifying the sample structure and making the TEM analysis a very time-consuming process [332].

TEM applications have focused on structure determination at the atomical resolution to elucidate the accurate structure of NMs. The main applications of TEM are also concerned with the porous material characterization for promising applications as absorbents [320, 321] and characterization of metal nanoparticles for antimicrobial and antibacterial applications [296–299].

1.5.4 X-ray diffraction (XRD)

Characterization of NMs includes size and shape, polymorphism, and crystal structure. X-ray diffraction (XRD) is one of the most employed techniques to assess the polymorph, crystallographic structure, crystallinity, and crystallite size of NMs. When conducting an XRD test, the X-ray wavelength is similar to the distance among atoms [334]. The irradiated X-ray interacts with a crystalline sample, producing a diffraction arrangement displaying several sharp spots, called Bragg diffraction peaks. Figure 1.10 shows a graphic interpretation of diffraction in a crystal with corresponding atomic planes with interplanar distance d. The incidence angle of the two parallel rays is θ. The difference in path length among the scattered X-rays from the bottom and top planes is equal to ($2d \sin \theta$). When the variation in path length between the top and bottom X-rays is identical to the wavelength (i.e., $\lambda = 2d \sin \theta$), constructive wave interference and strong XRD occur [335].

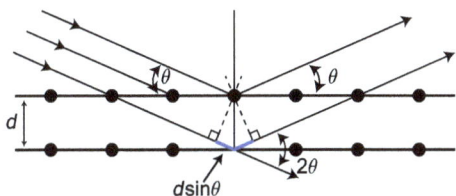

Figure 1.10: Bragg's law deduction diagram. X-rays with equal wavelength (λ) and phase are incident on parallel atomic planes with interplanar distance (d). The lower beam traverses an extra length $2d \sin \theta$.

The resulting diffractogram from NMs will be composed of its characteristic set of Bragg diffraction peaks. By analyzing the position and intensities of these peaks, the material's crystallographic structure can be determined [332]. Also, by fitting their peaks, positions, and the corresponding intensities to an established pattern of standard material, unknown NMs are detected. Nowadays, a novel biosynthesis process is gaining importance due to the use of nontoxic plant extract to oxidize or reduce ionic metals to form nanoparticles. In this context, the XRD technique helps identify and assess differences due to the synthesis methods [296, 297, 317, 319].

1.5.5 Fourier-transform infrared spectroscopy (FTIR)

Nowadays, nanotechnology and engineered materials allow researchers to modify and design NMs with specific functionalities. Due to their novel properties, dual composites and even ternary composite materials have recently attracted attention. In many cases, organic molecules are bound to the NM surface to achieve a specific objective. In this context, Fourier-transform infrared spectroscopy (FTIR) is commonly utilized to identify characteristic spectral bands that uncover surfactants, proteins, or other functional molecules attached to NM surfaces [336, 337]. The FTIR technique is used to assess the presence of IR-active functional groups or bonds in organic or inorganic samples through IR radiation in the range of 4,000 to 400 cm^{-1}. The required condition for a sample to show an infrared spectrum is the chance in the electric dipole moment of the functional group present in a molecule during the vibration caused by the absorption of the characteristic wavelength of the IR radiation [334]. Distinctive bands in FTIR spectroscopy correspond to "fingerprints" of NM molecule and/or organic molecule conjugation (e.g., proteins bound to nanoparticle surfaces or other types of substrates) [332]. Furthermore, the FTIR technique has been extended to study NMs synthesized by biosynthesis or green route, utilizing plant extracts to form metal nanoparticles and even metal oxide nanoparticles [32, 296, 297, 319].

1.5.6 Raman spectroscopy

Another scattering technique is Raman spectroscopy. Like FTIR, Raman scattering spectroscopy delivers a pattern of distinctive vibrations and rotations associated with the specimen. However, it is susceptible to molecular movements for which IR is less sensitive and vice versa. The primary idea behind this method is the observation of light scattered on crystalline lattice vibrations or oscillations of molecules. Raman spectroscopy is primarily used to investigate oscillations [334, 338].

The concept of Raman spectroscopy is to quantify the inelastic scattering produced by high-power laser light with a wavelength either in the visible or near-infrared regions incident on the sample [339]. The incident photons from the laser beam cause alternating polarization in the molecules, which in ambient conditions are in their ground-vibrational state [340]. Most of the incident photons are scattered from the sample through elastic scattering, the so-called Rayleigh scattering. The wavelength of the scattered photons is identical to the incident photon. A minute portion of the incident photons (~ 1 in 10 million) is scattered with a different frequency and generally smaller than the incident photons, the so-called Stokes Raman scattering. The energy shift of the inelastically scattered photons corresponds to distinctive frequencies correlated to the molecular vibrational states [334]. A smaller number of photons is scattered at higher energy, causing the anti-Stokes scattering. Figure 1.11 shows the different possibilities of light scattering.

Raman spectroscopy is a technique to acquire structural information such as crystalline phases, crystallite sizes, disorder degree, and strain state of bulk nanocrystallized catalysts. Recently, applications of Raman spectroscopy are being reported in food and food safety and its interactions with NMs [340].

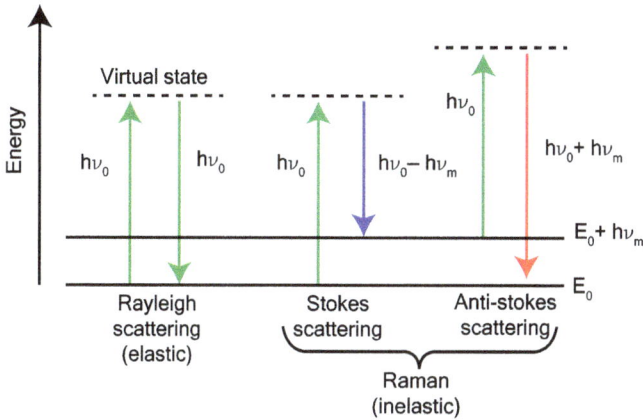

Figure 1.11: Jablonski energy diagram for the different scattering of light.

1.5.7 X-ray photoelectron spectroscopy (XPS)

The X-ray photoelectron spectroscopy (XPS) technique is considered a surface analysis technique that can detect and measure the chemical state, electronic state, elemental composition, and empirical formula of the elements contained in a material, with an atomic sensitivity in the range of 0.1–1%. Also, the elements' oxidation state can be evaluated with an average depth of approximately 5 nm [331]. Therefore, XPS is a formidable instrument for evaluating the surface chemical characteristics of NPs, nanostructured materials, surface layers, and thin films. XPS spectra are acquired by irradiating a sample surface with monoenergetic X-rays (often Al Kα or Mg) while determining the kinetic energy and number of photoelectrons released from the sample surface [341].

The elemental classification, the oxidation state, and the amount of a sensed element are defined by a photoelectron peak's binding energy and intensity. Since the work depth of XPS is about 5 nm, it is a crucial instrument for evaluating surface contamination or surface passivation. It offers a narrow surface analysis that techniques with a deeper analysis volume cannot provide. Nevertheless, the SEM-EDX technique is more appropriate if the bulk material properties are more important than the surface [334]. XPS technique has been essential in the characterization of nanostructures, focusing recently on the characterization of different types of NMs, including metal nanoparticles, namely, Au [296], Ag [297], and Cu [298]. Also, the characterization of metal oxide NMs is obtained by biosynthesis, such as ZnO [32], Fe_3O_4 [308], NiO [305], CeO_2 [317], and TiO_2 [319], with different applications in different science and technology fields.

1.5.8 UV-vis spectroscopy (UV-vis)

Nanostructured materials and nanoparticles commonly display optical properties responsive to concentration, size, and shape. UV-vis spectroscopy is frequently applied to quantitatively determine transition metal ions, organic, and biological macromolecules in solution. When conducting a UV-vis experiment, electromagnetic radiation between 190 and 800 nm is used to irradiate the sample. The UV-vis spectrometer quantifies the intensity of light before (I_0) and after passing (I) through a sample, which allows determining the absorbance A of the specimen [328]. Since the absorption of radiation, either ultraviolet or visible by a molecule, leads to an electronic transition among energy levels of the molecule, it is often called electronic spectroscopy. When coupled to other spectral data sources, the information acquired by this spectroscopy provides an evidence of valuable structural information of various molecules.

In UV-vis spectroscopy, the effects associated with the electric component of electromagnetic waves are essential due to the electronic transitions in the samples. In this context, the unique optical properties that NMs exhibit are of great interest when conducting a UV-vis measurement [342]. In the case of metal nanoparticles, the optical properties result from the collective oscillations of conduction electrons, which, when excited by electromagnetic radiation, are called surface plasmon polariton resonances [343]. The electron excitation influences the refractive index next to the surface of the nanoparticles; hence, it is possible to characterize by UV-vis spectroscopy. Their advantages of UV-vis spectroscopy include simplicity, sensitivity, selectivity to NMs, and short measurement time [344]. Therefore, this technique is increasingly used for NM characterization in many fields of science and industry, such as metal nanoparticles for dye remotion or antibacterial activity [292] and toxicity effect and cytotoxic test on cancer cell lines [316].

1.5.9 Zeta potential

Among the physicochemical properties of NMs, the surface charge is considered one of the most important. The charge on the surface alters the dispersion stability and aggregation states of NMs. Highly positive or negative charged particles tend to repulse each other and form a stable colloidal dispersion, while NMs with low surface charge often cluster and precipitate [345, 346]. The zeta potential technique determines the surface charge of nanoparticles in a colloidal solution [329].

The zeta potential technique is a powerful tool that allows estimating the surface charge of the nanoparticles. Hence, the measured magnitude can predict their colloidal stability in solution. Zeta potential can take values in the range of +100 to –100 mV. Generally, values higher than +30 mV or less than –30 mV are considered strongly cationic and anionic, respectively, and have high degrees of stability in colloidal solution. Values of zeta potential below +25 mV or higher than –25 mV have tendencies to agglomerate due to interactions between nanoparticles, such as van der Waals, and hydrophobic and hydrogen bonding interactions. Lastly, zeta potential values in the range of +10 and −10 mV are considered approximately neutral [329].

Nowadays, applications of zeta potential measurements include the assessment of coatings of NMs [347] and evaluation of metal nanoparticles. For example, My-Thao Nguyen et al. [169] evaluated the biosynthesis of metallic nanoparticles (Au and Ag); according to the zeta potential measurements, the nanoparticles exhibit a highly stable colloidal behavior.

1.6 Conclusion

Because of the social development, industrial growth, and pollution, it is necessary to propose new methodologies of application for materials for the growth, development, and monitoring of plants through agriculture. Nanotechnology through NMs offers endless new viable alternatives for implementing new technologies in the agricultural sector. However, as a new and emerging technology, it is necessary to ensure that the processes to obtain these materials with unique and specific properties are sustainable, do not pollute, and promote benefits concerning the technologies currently applied. NMs can have a beneficial impact in all research fields in which they are used; in the same way, it is essential to highlight the synthesis processes of these materials that they can carry out and their advantages over other methods. Each methodology used for the synthesis of NMs is different. Therefore, it is crucial to understand the physical and chemical principles that govern the processes and how it promotes the properties of each type of material to be used in different fields such as agriculture. Research carried out in agriculture can take advantage of this type of methodologies to produce materials that promote greater benefits, processes of greater reliability and durability, and prevent environmental pollution. Therefore, nanotechnology is crucial for new opportunities to solve problems that have not been solved until now.

References

[1] Prasad, R., Bhattacharyya, A., Nguyen, Q. D. Nanotechnology in sustainable agriculture: Recent developments, challenges, and perspectives. *Frontiers in Microbiology* 2017, 8(Jun), 1014–1014, doi: 10.3389/FMICB.2017.01014/BIBTEX.

[2] Usman, M., et al. Nanotechnology in agriculture: Current status, challenges and future opportunities. *Science of the Total Environment* 2020, 721, 137778–137778, doi: 10.1016/J. SCITOTENV.2020.137778.

[3] Shang, Y., Hasan, M. K., Ahammed, G. J., Li, M., Yin, H., Zhou, J. Applications of nanotechnology in plant growth and crop protection: A review. *Molecules* 2019, 24(14), 2558, doi: 10.3390/molecules24142558.

[4]   Dwivedi, S., Saquib, Q., Al-Khedhairy, A. A., Musarrat, J. Understanding the role of nanomaterials in agriculture. *Microbial Inoculants in Sustainable Agricultural Productivity: Vol. 2: Functional Applications* 2016, 271–288, doi: 10.1007/978-81-322-2644-4_17.

[5]   Lv, M., Liu, Y., Geng, J., Kou, X., Xin, Z., Yang, D. Engineering nanomaterials-based biosensors for food safety detection. *Biosensors Bioelectron* 2018, 106, 122–128, doi: 10.1016/J.BIOS.2018.01.049.

[6]   Mazari, S. A., et al. Nanomaterials: Applications, waste-handling, environmental toxicities, and future challenges – A review. *Journal of Environmental Chemical Engineering* 2021, 9(2), 105028–105028, doi: 10.1016/J.JECE.2021.105028.

[7]   Ramsden, J. Nanotechnology: An Introduction. William Andrew, 2011, 272–272.

[8]   Saleh, T. A. Nanomaterials: Classification, properties, and environmental toxicities. *Environmental Technology & Innovation* 2020, 20, 101067–101067, doi: 10.1016/J.ETI.2020.101067.

[9]   Kahan, D. M., Braman, D., Slovic, P., Gastil, J., Cohen, G. Cultural cognition of the risks and benefits of nanotechnology. *Nature Nanotechnology* 2008, 4(2), 87–90, doi: 10.1038/nnano.2008.341.

[10]   Mathur, P., Roy, S. Nanosilica facilitates silica uptake, growth and stress tolerance in plants. *Plant Physiology and Biochemistry* 2020, 157, 114–127, doi: 10.1016/J.PLAPHY.2020.10.011.

[11]   Dolez, P. I. Nanomaterials definitions, classifications, and applications. *Nanoengineering: Global Approaches to Health and Safety Issues* 2015, 3–40, doi: 10.1016/B978-0-444-62747-6.00001-4.

[12]   Sudha, P. N., Sangeetha, K., Vijayalakshmi, K., Barhoum, A. Nanomaterials history, classification, unique properties, production and market. *Emerging Applications of Nanoparticles and Architectural Nanostructures: Current Prospects and Future Trends* 2018, 341–384, doi: 10.1016/B978-0-323-51254-1.00012-9.

[13]   Pokropivny, V., Lohmus, R., Hussainova, I., Pokropivny, A., Vlassov, S. Introduction to Nanomaterials and Nanotechnology. Tartu University Press Ukraine, 2007.

[14]   Prasad, R. Synthesis of silver nanoparticles in photosynthetic plants. *Journal of Nanoparticles* 2014, 2014, 1–8, doi: 10.1155/2014/963961.

[15]   Aziz, N., et al. Facile algae-derived route to biogenic silver nanoparticles: Synthesis, antibacterial, and photocatalytic properties. *Langmuir* 2015, 31(42), 11605–11612, doi: 10.1021/ACS. LANGMUIR.5B03081/SUPPL_FILE/LA5B03081_SI_001.PDF.

[16]   Prasad, R., Pandey, R., Barman, I. Engineering tailored nanoparticles with microbes: Quo vadis?. *Wiley Interdisciplinary Reviews: Nanomedicine and Nanobiotechnology* 2016, 8(2), 316–330, doi: 10.1002/ WNAN.1363.

[17]   Sagadevan, S., Periasamy, M. Recent trends in nanobiosensors and their applications-a review. *Reviews on Advanced Materials Science* 2014, 36(2014), 62–69.

[18]   Kulkarni, N., Muddapur, U. Biosynthesis of metal nanoparticles: A review. *Journal of Nanotechnology* 2014, 2014, doi: 10.1155/2014/510246.

[19]   Saravanan, A., et al. A review on biosynthesis of metal nanoparticles and its environmental applications. *Chemosphere* 2021, 264, 128580, doi: https://doi.org/10.1016/j. chemosphere.2020.128580.

[20]   Salavati-Niasari, M., Davar, F., Mir, N. Synthesis and characterization of metallic copper nanoparticles via thermal decomposition. *Polyhedron* 2008, 17(27), 3514–3518, doi: 10.1016/J. POLY.2008.08.020.

[21]   Wisniewska, J., Guesmi, H., Ziolek, M., Tielens, F. Stability of nanostructured silver-platinum alloys. *Journal of Alloys and Compounds* 2019, 770, 934–941, doi: 10.1016/J.JALLCOM.2018.08.208.

[22]   Huang, X., El-Sayed, M. A. Gold Nanoparticles: Optical Properties and Implementations in Cancer Diagnosis and Photothermal Therapy, Vol. 1, Elsevier, 2010, 13–28.

[23]   Jung, H. Y., Yeo, I. S., Kim, T. U., Ki, H. C., Gu, H. B. Surface plasmon resonance effect of silver nanoparticles on a TiO2 electrode for dye-sensitized solar cells. *Applied Surface Science* 2018, 432, 266–271, doi: 10.1016/j.apsusc.2017.04.237.

[24] Ealias, A. M., Saravanakumar, M. P. A review on the classification, characterisation, synthesis of nanoparticles and their application. *IOP Conference Series: Materials Science and Engineering* 2017, 263(3), 032019–032019, doi: 10.1088/1757-899X/263/3/032019.

[25] Schröfel, A., Kratošová, G., Šafařík, I., Šafaříková, M., Raška, I., Shor, L. M. Applications of biosynthesized metallic nanoparticles – A review. *Acta Biomaterialia* 2014, 10(10), 4023–4042, doi: 10.1016/J.ACTBIO.2014.05.022.

[26] Shah, Z., et al. Synthesis of high surface area AgNPs from Dodonaea viscosa plant for the removal of pathogenic microbes and persistent organic pollutants. *Materials Science and Engineering: B* 2021, 263, 114770–114770, doi: 10.1016/J.MSEB.2020.114770.

[27] Vanti, G. L., et al. Synthesis of Gossypium hirsutum-derived silver nanoparticles and their antibacterial efficacy against plant pathogens. *Applied Organometallic Chemistry* 2019, 33(1), e4630–e4630, doi: 10.1002/AOC.4630.

[28] Alafeef, M., Moitra, P., Pan, D. Nano-enabled sensing approaches for pathogenic bacterial detection. *Biosensors Bioelectron* 2020, 165, 112276–112276, doi: 10.1016/J.BIOS.2020.112276.

[29] Rastogi, A., et al. Impact of metal and metal oxide nanoparticles on plant: A critical review. *Frontiers in Chemistry* 2017, 5, 78–78, doi: 10.3389/FCHEM.2017.00078/BIBTEX.

[30] Sanchez-Dominguez, M., Boutonnet, M., Solans, C. A novel approach to metal and metal oxide nanoparticle synthesis: The oil-in-water microemulsion reaction method. *Journal of Nanoparticle Research* 2009, 11(7), 1823–1829, doi: 10.1007/S11051-009-9660-8/FIGURES/3.

[31] Findik, F. Nanomaterials and their applications. *Periodicals of Engineering and Natural Sciences* 2021, 9(3), 62–75, doi: 10.21533/PEN.V9I3.1837.

[32] Soto-Robles, C. A., et al. Biosynthesis, characterization and photocatalytic activity of ZnO nanoparticles using extracts of Justicia spicigera for the degradation of methylene blue. *Journal of Molecular Structure* 2021, 1225, 129101, doi: 10.1016/j.molstruc.2020.129101.

[33] Shafei, A., Sheibani, S. Visible light photocatalytic activity of Cu doped TiO2-CNT nanocomposite powder prepared by sol–gel method. *Materials Research Bulletin* 2019, 110, 198–206, doi: 10.1016/j.materresbull.2018.10.035.

[34] Olvera-Rodríguez, I., et al. TiO2/Au/TiO2 multilayer thin-film photoanodes synthesized by pulsed laser deposition for photoelectrochemical degradation of organic pollutants. *Separation and Purification Technology* 2019, 224, 189–198, doi: 10.1016/J.SEPPUR.2019.05.020.

[35] Ziental, D., et al. Titanium dioxide nanoparticles: Prospects and applications in medicine. *Nanomaterials* 2020, 10(2), doi: 10.3390/NANO10020387.

[36] Jiang, B. P., Zhou, B., Lin, Z., Liang, H., Shen, X. C. Recent advances in carbon nanomaterials for cancer phototherapy. *Chemistry – A European Journal* 2019, 25(16), 3993–4004, doi: 10.1002/CHEM.201804383.

[37] Fu, X., Xu, L., Li, J., Sun, X., Peng, H. Flexible solar cells based on carbon nanomaterials. *Carbon* 2018, 139, 1063–1073, doi: 10.1016/J.CARBON.2018.08.017.

[38] Russier, J., et al. Oxidative biodegradation of single- and multi-walled carbon nanotubes. *Nanoscale* 2011, 3(3), 893–896, doi: 10.1039/C0NR00779J.

[39] Gupta, S., Murthy, C. N., Prabha, C. R. Recent advances in carbon nanotube based electrochemical biosensors. *International Journal of Biological Macromolecules* 2018, 108, 687–703, doi: 10.1016/J.IJBIOMAC.2017.12.038.

[40] Luo, Q., et al. All-carbon-electrode-based endurable flexible perovskite solar cells. *Advanced Functional Materials* 2018, 28(11), 1706777–1706777, doi: 10.1002/ADFM.201706777.

[41] Mehra, N. K., Jain, A. K., Nahar, M. Carbon nanomaterials in oncology: An expanding horizon. *Drug Discovery Today: Technologies* 2018, 23(5), 1016–1025, doi: 10.1016/J.DRUDIS.2017.09.013.

[42] Mukherjee, A., Majumdar, S., Servin, A. D., Pagano, L., Dhankher, O. P., White, J. C. Carbon nanomaterials in agriculture: A critical review. *Frontiers in Plant Science* 2016, 7(2016), 172, doi: 10.3389/FPLS.2016.00172/BIBTEX.

[43] Zhang, Y., et al. Flexible and stretchable lithium-ion batteries and supercapacitors based on electrically conducting carbon nanotube fiber springs. *Angewandte Chemie* 2014, 126(52), 14792–14796, doi: 10.1002/ANGE.201409366.

[44] Yi, H., et al. Selective prepared carbon nanomaterials for advanced photocatalytic application in environmental pollutant treatment and hydrogen production. *Applied Catalysis B Environmental* 2018, 239, 408–424, doi: 10.1016/J.APCATB.2018.07.068.

[45] Hu, S., Hu, C. Carbon nanotube-based electrochemical sensors: Principles and applications in biomedical systems. *Journal of Sensors* 2009, 2009, doi: 10.1155/2009/187615.

[46] Di Bartolomeo, A. Graphene Schottky diodes: An experimental review of the rectifying graphene/semiconductor heterojunction. *Physics Reports* 2016, 606, 1–58, doi: 10.1016/J.PHYSREP.2015.10.003.

[47] Ali, I., et al. Graphene based adsorbents for remediation of noxious pollutants from wastewater. *Environment International* 2019, 127, 160–180, doi: 10.1016/J.ENVINT.2019.03.029.

[48] Meyer, J. C., Geim, A. K., Katsnelson, M. I., Novoselov, K. S., Booth, T. J., Roth, S. The structure of suspended graphene sheets. *Nature* 2007, 446(7131), 60–63, doi: 10.1038/nature05545.

[49] Li, X., Wang, X., Zhang, L., Lee, S., Dai, H. Chemically derived, ultrasmooth graphene nanoribbon semiconductors. *Science* 2008, 319(5867), 1229–1232, doi: 10.1126/SCIENCE.1150878/SUPPL_FILE/PAP.PDF.

[50] Mohajeri, M., Behnam, B., Sahebkar, A. Biomedical applications of carbon nanomaterials: Drug and gene delivery potentials. *Journal of Cellular Physiology* 2019, 234(1), 298–319, doi: 10.1002/JCP.26899.

[51] Novoselov, K. S., et al. Electric field in atomically thin carbon films. *Science* 2004, 306(5696), 666–669, doi: 10.1126/SCIENCE.1102896/SUPPL_FILE/NOVOSELOV.SOM.PDF.

[52] Wu, J., Hong, Y., Wang, B. The applications of carbon nanomaterials in fiber-shaped energy storage devices *. *Journal of Semiconductors* 2018, 39(1), 011004, doi: 10.1088/1674-4926/39/1/011004.

[53] Amer, M. S., El-ashry, M. M. Selective adsorption from methanol/water mixtures by C60 fullerene nanospheres. *Chemical Physics Letters* 2006, 430(4–6), 323–325, doi: 10.1016/J.CPLETT.2006.08.120.

[54] Yan, W., Seifermann, S. M., Pierrat, P., Bräse, S. Synthesis of highly functionalized C60 fullerene derivatives and their applications in material and life sciences. *Organic and Biomolecular Chemistry* 2014, 13(1), 25–54, doi: 10.1039/C4OB01663G.

[55] Avanasi, R., Jackson, W. A., Sherwin, B., Mudge, J. F., Anderson, T. A. C60 fullerene soil sorption, biodegradation, and plant uptake. *Environmental Science Technology* 2014, 48(5), 2792–2797, doi: 10.1021/ES405306W.

[56] Liu, Y., Li, X., Fan, L., Li, S., Maleki Kheimeh Sari, H., Qin, J. A review of carbon-based materials for safe lithium metal anodes. *Frontiers in Chemistry* 2019, 7, 721–721, doi: 10.3389/FCHEM.2019.00721/BIBTEX.

[57] Arie, A. A., Vovk, O. M., Song, J. O., Cho, B. W., Lee, J. K. Carbon film covering originated from fullerene C60 on the surface of lithium metal anode for lithium secondary batteries. *Journal of Electroceramics* 2009, 23(2–4), 248–253, doi: 10.1007/S10832-008-9413-6/FIGURES/9.

[58] Baker, S. N., Baker, G. A. Luminescent carbon nanodots: Emergent nanolights. *Angewandte Chemie* 2010, 49(38), 6726–6744, doi: 10.1002/ANIE.200906623.

[59] Liu, J.-H., Yang, S.-T., Chen, -X.-X., Wang, H. Fluorescent carbon dots and nanodiamonds for biological imaging: Preparation, application, pharmacokinetics and toxicity. *Current Drug Metabolism* 2012, 13(8), 1046–1056, doi: 10.2174/138920012802850083.

[60] Zuo, P., Lu, X., Sun, Z., Guo, Y., He, H. A review on syntheses, properties, characterization and bioanalytical applications of fluorescent carbon dots. *Microchimica Acta* 2015, 183(2), 519–542, doi: 10.1007/S00604-015-1705-3.

[61] Li, L., Wu, G., Yang, G., Peng, J., Zhao, J., Zhu, J. J. Focusing on luminescent graphene quantum dots: Current status and future perspectives. *Nanoscale* 2013, 5(10), 4015–4039, doi: 10.1039/C3NR33849E.

[62] Nie, S. Understanding and overcoming major barriers in cancer nanomedicine. *Nanomedicine (London, England)* 2010, 5(4), 523–528, doi: 10.2217/NNM.10.23.

[63] Hola, K., Zhang, Y., Wang, Y., Giannelis, E. P., Zboril, R., Rogach, A. L. Carbon dots – Emerging light emitters for bioimaging, cancer therapy and optoelectronics. *Nano Today* 2014, 9(5), 590–603, doi: 10.1016/J.NANTOD.2014.09.004.

[64] Zheng, M., et al. Integrating oxaliplatin with highly luminescent carbon dots: An unprecedented theranostic agent for personalized medicine. *Advanced Materials* 2014, 26(21), 3554–3560, doi: 10.1002/ADMA.201306192.

[65] Niu, J., Gao, H. Synthesis and drug detection performance of nitrogen-doped carbon dots. *Journal of Luminescence* 2014, 149, 159–162, doi: 10.1016/J.JLUMIN.2014.01.026.

[66] Peng, Z., et al. Advances in the application, toxicity and degradation of carbon nanomaterials in environment: A review. *Environment International* 2020, 134, 105298–105298, doi: 10.1016/J.ENVINT.2019.105298.

[67] Tulinski, M., Jurczyk, M. Nanomaterials synthesis methods. *Metrology and Standardization of Nanotechnology* 2017, 75–98, doi: 10.1002/9783527800308.CH4.

[68] Deng, Y., Chen, W., Li, B., Wang, C., Kuang, T., Li, Y. Physical vapor deposition technology for coated cutting tools: A review. *Ceramics International* 2020, 46(11), 18373–18390, doi: 10.1016/J.CERAMINT.2020.04.168.

[69] Virji, M. A., Stefaniak, A. B. A review of engineered nanomaterial manufacturing processes and associated exposures. *Comprehensive Materials Processing* 2014, 8, 103–125, doi: 10.1016/B978-0-08-096532-1.00811-6.

[70] Lukaszkowicz, K. Review of nanocomposite thin films and coatings deposited by PVD and CVD technology. *Nanomaterials* 2011, doi: 10.5772/25799.

[71] Asanithi, P., Chaiyakun, S., Limsuwan, P. Growth of silver nanoparticles by DC magnetron sputtering. *Journal of Nanomaterials* 2012, 2012, doi: 10.1155/2012/963609.

[72] Ayyub, P., Chandra, R., Taneja, P., Sharma, A. K., Pinto, R. Synthesis of nanocrystalline material by sputtering and laser ablation at low temperatures. *Applied Physics A* 2001, 73(1), 67–73, doi: 10.1007/S003390100833.

[73] Sportelli, M. C., et al. The pros and cons of the use of laser ablation synthesis for the production of silver nano-antimicrobials. *Antibiotics* 2018, 7(3), 67–67, doi: 10.3390/ANTIBIOTICS7030067.

[74] Karatutlu, A., Barhoum, A., Sapelkin, A. Liquid-phase synthesis of nanoparticles and nanostructured materials. *Emerging Applications of Nanoparticles and Architectural Nanostructures: Current Prospects and Future Trends* 2018, 1–28, doi: 10.1016/B978-0-323-51254-1.00001-4.

[75] Frishberg, I. V. Gas-phase method of metal powder production. *Handbook of Non-Ferrous Metal Powders* 2009, 143–153, doi: 10.1016/B978-1-85617-422-0.00006-9.

[76] Shang, S. M., Zeng, W. Conductive nanofibres and nanocoatings for smart textiles. *Multidisciplinary Know-How for Smart-Textiles Developers* 2013, 92–128, doi: 10.1533/9780857093530.1.92.

[77] Wei, R., Vajo, J. J., Matossian, J. N., Gardos, M. N. Aspects of plasma-enhanced magnetron-sputtered deposition of hard coatings on cutting tools. *Surface and Coatings Technology* 2002, 158–159, 465–472, doi: 10.1016/S0257-8972(02)00289-X.

[78] Mehran, Q. M., Fazal, M. A., Bushroa, A. R., Rubaiee, S. A critical review on physical vapor deposition coatings applied on different engine components 2017, 43(2), doi: 10.1080/10408436.2017.1320648.

[79] Ravi-Kumar, S., Lies, B., Lyu, H., Qin, H. Laser ablation of polymers: A review. *Procedia Manufacturing* 2019, 34, 316–327, doi: 10.1016/J.PROMFG.2019.06.155.

[80] Brown, M. S., Arnold, C. B. Fundamentals of laser-material interaction and application to multiscale surface modification. *Springer Series in Materials Science* 2010, 135, 91–120, doi: 10.1007/978-3-642-10523-4_4.

[81] Ahmed, N., Darwish, S., Alahmari, A. M. Laser ablation and laser-hybrid ablation processes: A review 2016, 31(9), doi: 10.1080/10426914.2015.1048359.

[82] Von Der Linde, D., Sokolowski-Tinten, K. The physical mechanisms of short-pulse laser ablation. *Applied Surface Science* 2000, 154–155, 1–10, doi: 10.1016/S0169-4332(99)00440-7.

[83] Hoffman, J. The effect of recoil pressure in the ablation of polycrystalline graphite by a nanosecond laser pulse. *Journal of Physics D: Applied Physics* 2015, 48(23), 235201–235201, doi: 10.1088/0022-3727/48/23/235201.

[84] Singh, S., Argument, M., Tsui, Y. Y., Fedosejevs, R. Effect of ambient air pressure on debris redeposition during laser ablation of glass. *Journal of Applied Physics* 2005, 98(11), 113520–113520, doi: 10.1063/1.2138800.

[85] Tangwarodomnukun, V., Likhitangsuwat, P., Tevinpibanphan, O., Dumkum, C. Laser ablation of titanium alloy under a thin and flowing water layer. *International Journal of Machine Tools and Manufacture* 2015, 89, 14–28, doi: 10.1016/J.IJMACHTOOLS.2014.10.013.

[86] Sanchette, F., Ducros, C., Schmitt, T., Steyer, P., Billard, A. Nanostructured hard coatings deposited by cathodic arc deposition: From concepts to applications. *Surface and Coatings Technology* 2011, 205(23–24), 5444–5453, doi: 10.1016/J.SURFCOAT.2011.06.015.

[87] Brown, I. G. Cathodic Arc Deposition of Films 2003, 28(1), doi: 10.1146/ANNUREV.MATSCI.28.1.243.

[88] Yamamoto, K., Kujime, S., Takahara, K. Properties of nano-multilayered hard coatings deposited by a new hybrid coating process: Combined cathodic arc and unbalanced magnetron sputtering. *Surface and Coatings Technology* 2005, 200(1–4), 435–439, doi: 10.1016/J.SURFCOAT.2005.02.175.

[89] Hoseinzadeh, S., Ghasemiasl, R., Bahari, A., Ramezani, A. H. The injection of Ag nanoparticles on surface of WO3 thin film: Enhanced electrochromic coloration efficiency and switching response. *Journal of Materials Science: Materials in Electronics* 2017, 28(19), 14855–14863, doi: 10.1007/S10854-017-7357-9/FIGURES/12.

[90] Jang, G. S., Kim, D. Y., Hwang, N. M. The effect of charged Ag nanoparticles on thin film growth during DC magnetron sputtering. *Coatings* 2020, 10(8), 736–736, doi: 10.3390/COATINGS10080736.

[91] Wei, Z., Zhou, M., Qiao, H., Zhu, L., Yang, H., Xia, T. Particle size and pore structure characterization of silver nanoparticles prepared by confined arc plasma. *Journal of Nanomaterials* 2009, 2009, doi: 10.1155/2009/968058.

[92] Takahashi, S., et al. Oxygen reduction reaction activity and structural stability of Pt–Au nanoparticles prepared by arc-plasma deposition. *Physical Chemistry Chemical Physics* 2015, 17(28), 18638–18644, doi: 10.1039/C5CP02048D.

[93] Lung, J. K., et al. Preparation of gold nanoparticles by arc discharge in water. *Journal of Alloys and Compounds* 2007, 434–435(SPEC. ISS), 655–658, doi: 10.1016/J.JALLCOM.2006.08.213.

[94] Hussain, S., et al. Heat-treatment effects on the ORR activity of Pt nanoparticles deposited on multi-walled carbon nanotubes using magnetron sputtering technique. *International Journal of Hydrogen Energy* 2017, 42(9), 5958–5970, doi: 10.1016/J.IJHYDENE.2016.11.164.

[95] Mukherjee, S., et al. Ultrafine sputter-deposited Pt nanoparticles for triiodide reduction in dye-sensitized solar cells: Impact of nanoparticle size, crystallinity and surface coverage on catalytic activity. *Nanotechnology* 2012, 23(48), 485405–485405, doi: 10.1088/0957-4484/23/48/485405.

[96] Saito, M., Yasukawa, K., Umeda, T., Aoi, Y. Copper nanoparticles fabricated by laser ablation in polysiloxane. *Optical Materials* 2008, 30(7), 1201–1204, doi: 10.1016/J.OPTMAT.2007.05.049.

[97] Qiao, H., Wei, Z., Zhou, M., He, Z. Preparation and particle size characterization of Cu nanoparticles prepared by anodic arc plasma. *Advanced Materials Research* 2010, 92, 163–169, doi: 10.4028/WWW.SCIENTIFIC.NET/AMR.92.163.

[98] Chau, Y. T. R., Deng, L., Nguyen, M. T., Yonezawa, T. Monitor the growth and oxidation of Cu-nanoparticles in PEG after sputtering. *MRS Advances* 2019, 4(5–6), 305–309, doi: 10.1557/ADV.2019.55.

[99] Sanaee, M. R., Chaitoglou, S., Aguiló-Aguayo, N., Bertran, E. Size control of carbon encapsulated iron nanoparticles by arc discharge plasma method. *Applied Sciences* 2016, 7(1), 26–26, doi: 10.3390/APP7010026.

[100] Vitta, Y., Piscitelli, V., Fernandez, A., Gonzalez-Jimenez, F., Castillo, J. α-Fe nanoparticles produced by laser ablation: Optical and magnetic properties. *Chemical Physics Letters* 2011, 512(1–3), 96–98, doi: 10.1016/J.CPLETT.2011.07.020.

[101] Zhao, J., et al. Formation mechanism of fe nanocubes by magnetron sputtering inert gas condensation. *ACS Nano* 2016, 10(4), 4684–4694, doi: 10.1021/ACSNANO.6B01024/SUPPL_FILE/ NN6B01024_SI_002.AVI.

[102] Carette, X., et al. On the sputtering of titanium and silver onto liquids, discussing the formation of nanoparticles. *Journal of Physical Chemistry C* 2018, 122(46), 26605–26612, doi: 10.1021/ACS. JPCC.8B06987/SUPPL_FILE/JP8B06987_SI_001.PDF.

[103] Kwon, J. H., Kim, D. Y., Hwang, N. M. Generation of charged Ti nanoparticles and their deposition behavior with a substrate bias during RF magnetron sputtering. *Coatings* 2020, 10(5), 443–443, doi: 10.3390/COATINGS10050443.

[104] Dolgaev, S. I., Simakin, A. V., Voronov, V. V., Shafeev, G. A., Bozon-Verduraz, F. Nanoparticles produced by laser ablation of solids in liquid environment. *Applied Surface Science* 2002, 186(1–4), 546–551, doi: 10.1016/S0169-4332(01)00634-1.

[105] Escobar-Alarcón, L., Solis-Casados, D. A., Romero, S., Haro-Poniatowski, E. Thin films prepared by a hybrid deposition configuration combining two laser ablation plasmas with one sputtering plasma. *Applied Physics A* 2019, 126(1), 1–8, doi: 10.1007/S00339-019-3200-X.

[106] Martínez-Chávez, L. A., Esquivel, K., Solis-Casados, D. A., Velázquez-Castillo, R., Haro-Poniatowski, E., Escobar-Alarcón, L. Nanocomposite Bi/TiO2 multilayer thin films deposited by a crossed beam laser ablation configuration. *Applied Physics A* 2021, 127(11), 808, doi: 10.1007/s00339-021-04957-0.

[107] Martínez-Chávez, L. A., Rivera-Muñoz, E. M., Velázquez-Castillo, R. R., Escobar-Alarcón, L., Esquivel, K. Au-Ag/TiO2 thin films preparation by laser ablation and sputtering plasmas for its potential use as photoanodes in Electrochemical Advanced Oxidation Processes (EAOP). *Catalysts* 2021, 11(11), 1406–1406, doi: 10.3390/CATAL11111406.

[108] Bachari, E. M., Baud, G., Ben Amor, S., Jacquet, M. Structural and optical properties of sputtered ZnO films. *Thin Solid Films* 1999, 348(1–2), 165–172, doi: 10.1016/S0040-6090(99)00060-7.

[109] Karamat, S., et al. Nitrogen doping in pulsed laser deposited ZnO thin films using dense plasma focus. *Applied Surface Science* 2011, 257(6), 1979–1985, doi: 10.1016/J.APSUSC.2010.09.038.

[110] Minami, T., Ida, S., Miyata, T., Minamino, Y. Transparent conducting ZnO thin films deposited by vacuum arc plasma evaporation. *Thin Solid Films* 2003, 445(2), 268–273, doi: 10.1016/S0040-6090(03) 01159-3.

[111] Yang, X., Liang, S., Wang, X., Xiao, P., Fan, Z. Effect of WC and CeO2 on microstructure and properties of W–Cu electrical contact material. *International Journal of Refractory Metals & Hard Materials* 2010, 28(2), 305–311, doi: 10.1016/J.IJRMHM.2009.11.009.

[112] Usmanov, R. A., et al. Diffuse vacuum arc with heated cathode made of ceramic (CeO2) and metal (Cr) mixture. *Plasma Sources Science and Technology* 2020, 29(1), 015004–015004, doi: 10.1088/1361-6595/AB5F33.

[113] Ma, R., Jahurul Islam, M., Amaranatha Reddy, D., Kim, T. K. Transformation of CeO2 into a mixed phase CeO2/Ce2O3 nanohybrid by liquid phase pulsed laser ablation for enhanced photocatalytic activity through Z-scheme pattern. *Ceramics International* 2016, 42(16), 18495–18502, doi: 10.1016/J. CERAMINT.2016.08.186.

[114] Paramês, M. L., et al. Magnetic properties of Fe3O4 thin films grown on different substrates by laser ablation. *Applied Surface Science* 2007, 253(19), 8201–8205, doi: 10.1016/J.APSUSC.2007.02.134.

[115] Wu, X. J., Zhang, Z. Z., Liang, Q. S., Meng, J. Evolution from (110) Fe to (111) Fe3O4 thin films grown by magnetron sputtering using Fe2O3 target. *Journal of Crystal Growth* 2012, 340(1), 74–77, doi: 10.1016/J.JCRYSGRO.2011.12.041.

[116] Liu, X. F., Li, P., Jin, C., Bai, H. L. Enhancement of the magnetization in the Fe3O4/BiFeO3 epitaxial heterostructures fabricated by magnetron sputtering. *Applied Physics Letters* 2011, 99(18), 182511–182511, doi: 10.1063/1.3659483.

[117] Jeong, S. H., Kim, J. K., Kim, B. S., Shim, S. H., Lee, B. T. Characterization of SiO2 and TiO2 films prepared using rf magnetron sputtering and their application to anti-reflection coating. *Vacuum* 2004, 76(4), 507–515, doi: 10.1016/J.VACUUM.2004.06.003.

[118] Xu, S., Qiu, J., Jia, T., Li, C., Sun, H., Xu, Z. Femtosecond laser ablation of crystals SiO2 and YAG. *Optics Communications* 2007, 274(1), 163–166, doi: 10.1016/J.OPTCOM.2007.01.079.

[119] Zaikovskii, A. V. Silicon nanowhisker formation during SiO2 evaporation by arc discharge. *Journal of Physics: Conference Series* 2019, 1382(1), 012172–012172, doi: 10.1088/1742-6596/1382/1/012172.

[120] Acharya, J., Wilt, J., Liu, B., Wu, J. Probing the dielectric properties of ultrathin Al/Al2O3/Al trilayers fabricated using in situ sputtering and atomic layer deposition. *ACS Applied Materials and Interfaces* 2018, 10(3), 3112–3120, doi: 10.1021/ACSAMI.7B16506/SUPPL_FILE/AM7B16506_SI_001.PDF.

[121] Elfaham, M. M., Okil, M., Mostafa, A. M. Effects of post-laser irradiation on the optical and structure properties of Al2O3 nanoparticles produced by laser ablation. *Journal of Applied Physics* 2020, 128(15), 153104–153104, doi: 10.1063/5.0022554.

[122] Gavrilov, N. V., Kamenetskikh, A. S., Tretnikov, P. V., Chukin, A. V. Synthesis of α-Al2O3 coatings by reactive anodic arc evaporation under high-density low-energy ion assistance. *Journal of Physics: Conference Series* 2019, 1281(1), 012020–012020, doi: 10.1088/1742-6596/1281/1/012020.

[123] Arora, N., Sharma, N. N. Arc discharge synthesis of carbon nanotubes: Comprehensive review. *Diamond and Related Materials* 2014, 50, 135–150, doi: 10.1016/J.DIAMOND.2014.10.001.

[124] U. Rather, S. Preparation, characterization and hydrogen storage studies of carbon nanotubes and their composites: A review. *International Journal of Hydrogen Energy* 2020, 45(7), 4653–4672, doi: 10.1016/J.IJHYDENE.2019.12.055.

[125] Rao, R., et al. Carbon nanotubes and related nanomaterials: Critical advances and challenges for synthesis toward mainstream commercial applications. *ACS Nano* 2018, 12(12), 11756–11784, doi: 10.1021/ACSNANO.8B06511.

[126] Raphey, V. R., Henna, T. K., Nivitha, K. P., Mufeedha, P., Sabu, C., Pramod, K. Advanced biomedical applications of carbon nanotube. *Materials Science and Engineering: C* 2019, 100, 616–630, doi: 10.1016/J.MSEC.2019.03.043.

[127] Churilov, G. N. Synthesis of fullerenes and other nanomaterials in arc discharge 2010, 16(5–6), doi: 10.1080/15363830802281641.

[128] Bunshah, R. F., et al. Fullerene formation in sputtering and electron beam evaporation processes. *The Journal of Physical Chemistry* 2002, 96(17), 6866–6869, doi: 10.1021/J100196A005.

[129] Su, Y., et al. Facile synthesis and photoelectric properties of carbon dots with upconversion fluorescence using arc-synthesized carbon by-products. *RSC Advances* 2014, 4(10), 4839–4842, doi: 10.1039/C3RA45453C.

[130] Sajid, M. I., Jamshaid, U., Jamshaid, T., Zafar, N., Fessi, H., Elaissari, A. Carbon nanotubes from synthesis to in vivo biomedical applications. *International Journal of Pharmaceutics* 2016, 501(1–2), 278–299, doi: 10.1016/J.IJPHARM.2016.01.064.

[131] Dhar, S., et al. A new route to graphene layers by selective laser ablation. *AIP Advances* 2011, 1(2), 022109–022109, doi: 10.1063/1.3584204.

[132] Herbig, C., Michely, T. Graphene: The ultimately thin sputtering shield. *2D Materials* 2016, 3(2), 025032–025032, doi: 10.1088/2053-1583/3/2/025032.

[133] Li, Y., et al. Chemical vapor deposition-grown carbon nanotubes/graphene hybrids for electrochemical energy storage and conversion. *FlatChem* 2019, 15, 100091–100091, doi: 10.1016/J. FLATC.2019.100091.

[134] Olga, A. S., et al. Aqueous chemical co-precipitation of iron oxide magnetic nanoparticles for use in agricultural technologies. *Letters in Applied NanoBioScience* 2020, 10(2), 2215–2239, doi: 10.33263/LIANBS102.22152239.

[135] Majid, A., Bibi, M. Wet chemical synthesis methods. In Cadmium based II-VI Semiconducting Nanomaterials: Synthesis Routes and Strategies, Majid, A., Bibi, M. (Eds.). Cham, Springer International Publishing, 2018, 43–101.

[136] Cui, D., et al. Effect of cerium oxide nanoparticles on asparagus lettuce cultured in an agar medium. *Environmental Science Nano* 2014, 1(5), 459–465, doi: 10.1039/C4EN00025K.

[137] Akpomie, K. G., Ghosh, S., Gryzenhout, M., Conradie, J. One-pot synthesis of zinc oxide nanoparticles via chemical precipitation for bromophenol blue adsorption and the antifungal activity against filamentous fungi. *Scientific Reports* 2021, 11(1), 8305, doi: 10.1038/s41598-021-87819-2.

[138] Biesuz, M., Spiridigliozzi, L., Dell'Agli, G., Bortolotti, M., Sglavo, V. M. Synthesis and sintering of (Mg, Co, Ni, Cu, Zn)O entropy-stabilized oxides obtained by wet chemical methods. *Journal of Materials Science* 2018, 53(11), 8074–8085, doi: 10.1007/s10853-018-2168-9.

[139] Oosthuizen, D. N., Motaung, D. E., Swart, H. C. Gas sensors based on CeO_2 nanoparticles prepared by chemical precipitation method and their temperature-dependent selectivity towards H_2S and NO_2 gases. *Applied Surface Science* 2020, 505, 144356, doi: 10.1016/j.apsusc.2019.144356.

[140] Buraso, W., Lachom, V., Siriya, P., Laokul, P. Synthesis of TiO_2 nanoparticles via a simple precipitation method and photocatalytic performance. *Materials Research Express* 2018, 5(11), 115003, doi: 10.1088/2053-1591/aadbf0.

[141] Zhang, Z., et al. Uptake and distribution of ceria nanoparticles in cucumber plants†. *Metallomics* 2011, 3(8), 816–822, doi: 10.1039/c1mt00049g.

[142] Mahanty, B., Jesudas, S., Padmaprabha, A. Toxicity of surface functionalized iron oxide nanoparticles toward pure suspension culture and soil microcosm. *Environmental Nanotechnology, Monitoring and Management* 2019, 12, 100235, doi: 10.1016/j.enmm.2019.100235.

[143] Rane, A. V., Kanny, K., Abitha, V. K., Thomas, S. Methods for synthesis of nanoparticles and fabrication of nanocomposites. *Synthesis of Inorganic Nanomaterials* 2018, 121–139, doi: 10.1016/B978-0-08-101975-7.00005-1.

[144] Mustafa, M. K., Iqbal, Y., Majeed, U., Sahdan, M. Z. Effect of precursor's concentration on structure and morphology of ZnO nanorods synthesized through hydrothermal method on gold surface. *AIP Conference Proceedings* 2017, 1788(1), 030120, doi: 10.1063/1.4968373.

[145] Younes, N. A., Hassan, H. S., Elkady, M. F., Hamed, A. M., Dawood, M. F. A. Impact of synthesized metal oxide nanomaterials on seedlings production of three Solanaceae crops. *Heliyon* 2020, 6(1), e03188, doi: 10.1016/j.heliyon.2020.e03188.

[146] Konate, A., et al. Comparative effects of nano and bulk-Fe_3O_4 on the growth of cucumber (Cucumis sativus). In Ecotoxicology and Environmental Safety, Vol. 165, 2018, 547–554, doi: 10.1016/j.ecoenv.2018.09.053.

[147] Gong, Y., Dong, Z. Transfer, transportation, and accumulation of cerium-doped carbon quantum dots: Promoting growth and development in wheat. *Ecotoxicology & Environmental Safety* 2021, 226, 112852, doi: 10.1016/j.ecoenv.2021.112852.

[148] Sahoo, S. K., Panigrahi, G. K., Sahoo, A., Pradhan, A. K., Dalbehera, A. Bio-hydrothermal synthesis of ZnO–$ZnFe_2O_4$ nanoparticles using Psidium guajava leaf extract: Role in waste water remediation and plant immunity. *Journal of Cleaner Production* 2021, 318, 128522, doi: 10.1016/j.jclepro.2021.128522.

[149] Kajihara, K. Recent advances in sol–gel synthesis of monolithic silica and silica-based glasses. *Journal of Asian Ceramic Societies* 2013, 1(2), 121–133, doi: 10.1016/j.jascer.2013.04.002.

[150] Abdollahi, A., Roghani-Mamaqani, H., Salami-Kalajahi, M., Mousavi, A., Razavi, B., Shahi, S. Preparation of organic-inorganic hybrid nanocomposites from chemically modified epoxy and novolac resins and silica-attached carbon nanotubes by sol-gel process: Investigation of thermal

degradation and stability. *Progress in Organic Coatings* 2018, 117, 154–165, doi: 10.1016/j. porgcoat.2018.01.001.

[151] Velázquez, J. J., et al. Novel sol-gel SiO2-NaGdF4 transparent nano-glass-ceramics. *Journal of Non-Crystalline Solids* 2019, 520, 119447, doi: 10.1016/j.jnoncrysol.2019.05.023.

[152] Parashar, M., Shukla, V. K., Singh, R. Metal oxides nanoparticles via sol–gel method: A review on synthesis, characterization and applications. *Journal of Materials Science: Materials in Electronics* 2020, 31(5), 3729–3749, doi: 10.1007/s10854-020-02994-8.

[153] Farag, S., Amr, A., El-Shafei, A., Asker, M. S., Ibrahim, H. M. Green synthesis of titanium dioxide nanoparticles via bacterial cellulose (BC) produced from agricultural wastes. *Cellulose* 2021, 28(12), 7619–7632, doi: 10.1007/s10570-021-04011-5.

[154] Imoisili, P. E., Ukoba, K. O., Jen, T.-C. Green technology extraction and characterisation of silica nanoparticles from palm kernel shell ash via sol–gel. *Jouranal of Materials Research and Technology* 2020, 9(1), 307–313, doi: 10.1016/j.jmrt.2019.10.059.

[155] Palmqvist, N. G. M., Seisenbaeva, G. A., Svedlindh, P., Kessler, V. G. Maghemite Nanoparticles Acts as Nanozymes, Improving Growth and Abiotic Stress Tolerance in Brassica napus. *Nanoscale Research Letters* 2017, 12(1), 631, doi: 10.1186/s11671-017-2404-2.

[156] Belhamel, C., et al. Nanostructured alumina as seed protectant against three stored-product insect pests. *Journal of Stored Products Research* 2020, 87, 101607, doi: 10.1016/j.jspr.2020.101607.

[157] Herrera, A., Tejada-Tovar, C., González-Delgado, Á. D. Enhancement of cadmium adsorption capacities of agricultural residues and industrial fruit byproducts by the incorporation of Al2O3 nanoparticles. *ACS Omega* 2020, 5(37), 23645–23653, doi: 10.1021/acsomega.0c02298.

[158] Gordillo-Delgado, F., Zuluaga-Acosta, J., Restrepo-Guerrero, G. Effect of the suspension of Ag-incorporated TiO2 nanoparticles (Ag-TiO2 NPs) on certain growth, physiology and phytotoxicity parameters in spinach seedlings. *PLOS ONE* 2020, 15(12), e0244511, doi: 10.1371/journal. pone.0244511.

[159] Khan, M. S., et al. Graphene quantum dot and iron co-doped TiO2 photocatalysts: Synthesis, performance evaluation and phytotoxicity studies. *Ecotoxicology & Environmental Safety* 2021, 226, 112855, doi: 10.1016/j.ecoenv.2021.112855.

[160] Yang, G., Park, S.-J. Conventional and microwave hydrothermal synthesis and application of functional materials: A review. *Materials* 2019, 12(7), 1177, doi: 10.3390/ma12071177.

[161] Díaz de Greñu, B., De Los Reyes, R., Costero, A. M., Amorós, P., Ros-Lis, J. V. Recent progress of microwave-assisted synthesis of silica materials. *Nanomaterials* 2020, 10(6), 1092, doi: 10.3390/ nano10061092.

[162] Yadav, V. K., et al. Synthesis and characterization of amorphous iron oxide nanoparticles by the sonochemical method and their application for the remediation of heavy metals from wastewater. *Nanomaterials* 2020, 10(8), 1551, doi: 10.3390/nano10081551.

[163] Xu, H., Zeiger, B. W., Suslick, K. S. Sonochemical synthesis of nanomaterials. *Chemical Society Reviews* 2013, 42(7), 2555–2567, doi: 10.1039/C2CS35282F.

[164] Gedanken, A. Using sonochemistry for the fabrication of nanomaterials. *Ultrasonics Sonochemistry* 2004, 11(2), 47–55, doi: 10.1016/j.ultsonch.2004.01.037.

[165] Kinzyabaeva, Z. S., Sabirov, D. S. Sonochemical synthesis of novel C60 fullerene 1,4-oxathiane derivative through the intermediate fullerene radical anion. *Ultrasonics Sonochemistry* 2020, 67, 105169, doi: 10.1016/j.ultsonch.2020.105169.

[166] Kumar, R., Kumar, V. B., Gedanken, A. Sonochemical synthesis of carbon dots, mechanism, effect of parameters, and catalytic, energy, biomedical and tissue engineering applications. *Ultrasonics Sonochemistry* 2020, 64, 105009, doi: 10.1016/j.ultsonch.2020.105009.

[167] Natarajan, K. K. S. Customizing zinc oxide, silver and titanium dioxide nanoparticles for enhancing groundnut seed quality. *Indian Journal of Science and Technology* 2014, 7(9), 1376, doi: 10.17485/ijst/ 2014/v7i9.29.

[168] Ağçeli, G. K., Hammachi, H., Kodal, S. P., Cihangir, N., Aksu, Z. A novel approach to synthesize TiO2 nanoparticles: Biosynthesis by using streptomyces sp. HC1. *Journal of Inorganic and Organometallic Polymers and Materials* 2020, 30(8), 3221–3229, doi: 10.1007/s10904-020-01486-w.

[169] My-Thao Nguyen, T., et al. Biosynthesis of metallic nanoparticles from waste Passiflora edulis peels for their antibacterial effect and catalytic activity. *Arabian Journal of Chemistry* 2021, 14(4), 103096, doi: https://doi.org/10.1016/j.arabjc.2021.103096.

[170] Bhati, M., Biogenic synthesis of metallic nanoparticles: Principles and applications, *Materials Today: Proceedings*, 2021, doi: https://doi.org/10.1016/j.matpr.2021.04.272.

[171] Khanna, P., Kaur, A., Goyal, D. Algae-based metallic nanoparticles: Synthesis, characterization and applications. *Journal of Microbiological Methods* 2019, 163, 105656, doi: https://doi.org/10.1016/j.mimet.2019.105656.

[172] Kumar, J. A., et al. A focus to green synthesis of metal/metal based oxide nanoparticles: Various mechanisms and applications towards ecological approach. *Journal of Cleaner Production* 2021, 324, 129198, doi: https://doi.org/10.1016/j.jclepro.2021.129198.

[173] Vijayaraghavan, K., Ashokkumar, T. Plant-mediated biosynthesis of metallic nanoparticles: A review of literature, factors affecting synthesis, characterization techniques and applications. *Journal of Environmental Chemical Engineering* 2017, 5(5), 4866–4883, doi: https://doi.org/10.1016/j.jece.2017.09.026.

[174] Cuong, H. N., et al. New frontiers in the plant extract mediated biosynthesis of copper oxide (CuO) nanoparticles and their potential applications: A review. *Environmental Research* 2022, 203, 111858, doi: https://doi.org/10.1016/j.envres.2021.111858.

[175] De Jesus, R. A., et al. Environmental remediation potentialities of metal and metal oxide nanoparticles: Mechanistic biosynthesis, influencing factors, and application standpoint. *Environmental Technology & Innovation* 2021, 24, 101851, doi: https://doi.org/10.1016/j.eti.2021.101851.

[176] Salem, M. Z. M., et al. Plants-derived bioactives: Novel utilization as antimicrobial, antioxidant and phytoreducing agents for the biosynthesis of metallic nanoparticles. *Microbial Pathogenesis* 2021, 158, 105107, doi: https://doi.org/10.1016/j.micpath.2021.105107.

[177] Brar, K. K., et al. Green route for recycling of low-cost waste resources for the biosynthesis of nanoparticles (NPs) and nanomaterials (NMs)-A review. *Environmental Research* 2021, 112202, doi: https://doi.org/10.1016/j.envres.2021.112202.

[178] Blumenberg, M., et al. Biosynthesis of hopanoids by sulfate-reducing bacteria (genus Desulfovibrio). *Environmental Microbiology* 2006, 8(7), 1220–1227, doi: 10.1111/j.1462-2920.2006.01014.x.

[179] Qi, P., Zhang, D., Zeng, Y., Wan, Y. Biosynthesis of CdS nanoparticles: A fluorescent sensor for sulfate-reducing bacteria detection. *Talanta* 2016, 147, 142–146, doi: https://doi.org/10.1016/j.talanta.2015.09.046.

[180] Yue, L., Wu, Y., Liu, X., Xin, B., Chen, S. Controllable extracellular biosynthesis of bismuth sulfide nanostructure by sulfate-reducing bacteria in water-oil two-phase system. *Biotechnology Progress* Jul-Aug 2014, 30(4), 960–966, doi: 10.1002/btpr.1894.

[181] Qi, S., Yang, S., Chen, J., Niu, T., Yang, Y., Xin, B. High-yield extracellular biosynthesis of ZnS quantum dots through a unique molecular mediation mechanism by the peculiar extracellular proteins secreted by a mixed sulfate reducing bacteria. *ACS Applied Materials and Interfaces* 2019, 11(11), 10442–10451, doi: 10.1021/acsami.8b18574.

[182] Zhou, N.-Q., et al. Extracellular biosynthesis of copper sulfide nanoparticles by Shewanella oneidensis MR-1 as a photothermal agent. *Enzyme & Microbial Technology* 2016, 95, 230–235, doi: https://doi.org/10.1016/j.enzmictec.2016.04.002.

[183] Song, X., Shi, X. Biosynthesis of Ag/reduced graphene oxide nanocomposites using Shewanella oneidensis MR-1 and their antibacterial and catalytic applications. *Applied Surface Science* 2019, 491, 682–689, doi: https://doi.org/10.1016/j.apsusc.2019.06.154.

[184] Rajput, V. D., et al. Insights into the biosynthesis of nanoparticles by the genus *Shewanella*. *ApplEnvironMicrobiol* 2021, 87(22), e01390–21, doi: 10.1128/AEM.01390-21.

[185] Kim, T.-Y., Kim, M. G., Lee, J.-H., Hur, H.-G. Biosynthesis of nanomaterials by shewanella species for application in lithium ion batteries. *Frontiers in Microbiology* Mini Review 2018, 9(2817), doi: 10.3389/fmicb.2018.02817.

[186] Mondal, A. H., Yadav, D., Mitra, S., Mukhopadhyay, K. Biosynthesis of silver nanoparticles using culture supernatant of shewanella sp. ARY1 and their antibacterial activity. *International Journal of Nanomedicine* 2020, 15, 8295–8310, doi: 10.2147/IJN.S274535.

[187] Lin, T., Ding, W., Sun, L., Wang, L., Liu, C.-G., Song, H. Engineered Shewanella oneidensis-reduced graphene oxide biohybrid with enhanced biosynthesis and transport of flavins enabled a highest bioelectricity output in microbial fuel cells. *Nano Energy* 2018, 50, 639–648, doi: https://doi.org/10.1016/j.nanoen.2018.05.072.

[188] Min, D., et al. Enhancing extracellular electron transfer of shewanella oneidensis MR-1 through coupling improved flavin synthesis and metal-reducing conduit for pollutant degradation. *Environmental Science Technology* 2017, 51(9), 5082–5089, doi: 10.1021/acs.est.6b04640.

[189] Xu, H., et al. Microbial synthesis of Pd-Pt alloy nanoparticles using Shewanella oneidensis MR-1 with enhanced catalytic activity for nitrophenol and azo dyes reduction. *Nanotechnology* 2019, 30(6), 065607, doi: 10.1088/1361-6528/aaf2a6.

[190] Kimber, R. L., et al. Biosynthesis and characterization of copper nanoparticles using shewanella oneidensis: Application for click chemistry. *Small* Mar 2018, 14(10), doi: 10.1002/smll.201703145.

[191] Fahmy, H. M., et al. Review of green methods of iron nanoparticles synthesis and applications. *BioNanoScience* 2018, 8(2), 491–503, doi: 10.1007/s12668-018-0516-5.

[192] Venil, C. K., Usha, R., Devi, P. R. Chapter 12 – Green synthesis of nanoparticles from microbes and their prospective applications. In Nanomaterials, Kumar, R. P., Bharathiraja, B. (Eds.). Academic Press, 2021, 283–298.

[193] Mondal, P., Anweshan, A., Purkait, M. K. Green synthesis and environmental application of iron-based nanomaterials and nanocomposite: A review. *Chemosphere* 2020, 259, 127509, doi: https://doi.org/10.1016/j.chemosphere.2020.127509.

[194] Prema, P., Iniya, P. A., Immanuel, G. Microbial mediated synthesis, characterization, antibacterial and synergistic effect of gold nanoparticles using Klebsiella pneumoniae (MTCC-4030). *RSC Advances* 2016, 6(6), 4601–4607, doi: 10.1039/C5RA23982F.

[195] Srinath, B. S., Ravishankar Rai, V. Biosynthesis of highly monodispersed, spherical gold nanoparticles of size 4–10 nm from spent cultures of Klebsiella pneumoniae. *3 Biotech* 2015, 5(5), 671–676, doi: 10.1007/s13205-014-0265-2.

[196] Müller, A., Behsnilian, D., Walz, E., Gräf, V., Hogekamp, L., Greiner, R. Effect of culture medium on the extracellular synthesis of silver nanoparticles using Klebsiella pneumoniae, Escherichia coli and Pseudomonas jessinii. *Biocatalysis & Agricultural Biotechnology* 2016, 6, 107–115, doi: https://doi.org/10.1016/j.bcab.2016.02.012.

[197] Fatemi, M., Mollania, N., Momeni-Moghaddam, M., Sadeghifar, F. Extracellular biosynthesis of magnetic iron oxide nanoparticles by Bacillus cereus strain HMH1: Characterization and in vitro cytotoxicity analysis on MCF-7 and 3T3 cell lines. *Journal of Biotechnology* 2018, 270, 1–11, doi: https://doi.org/10.1016/j.jbiotec.2018.01.021.

[198] Pourali, P., Badiee, S. H., Manafi, S., Noorani, T., Rezaei, A., Yahyaei, B. Biosynthesis of gold nanoparticles by two bacterial and fungal strains, Bacillus cereus and Fusarium oxysporum, and assessment and comparison of their nanotoxicity in vitro by direct and indirect assays. *Electronic Journal of Biotechnology* 2017, 29, 86–93, doi: https://doi.org/10.1016/j.ejbt.2017.07.005.

[199] Mujaddidi, N., et al. Pharmacological properties of biogenically synthesized silver nanoparticles using endophyte Bacillus cereus extract of Berberis lyceum against oxidative stress and pathogenic

multidrug-resistant bacteria. *Saudi Journal of Biological Sciences* 2021, 28(11), 6432–6440, doi: https://doi.org/10.1016/j.sjbs.2021.07.009.

[200] Gong, J., Song, X., Gao, Y., Gong, S., Wang, Y., Han, J. Microbiological synthesis of zinc sulfide nanoparticles using Desulfovibrio desulfuricans. *Inorganic and Nano-Metal Chemistry* 2018, 48(2), 96–102, doi: 10.1080/15533174.2016.1216451.

[201] Era, Y., Dennis, J. A., Wallace, S., Horsfall, L. E. Micellar catalysis of the Suzuki Miyaura reaction using biogenic Pd nanoparticles from Desulfovibrio alaskensis. *Green Chemistry* 2021, 23(22), 8886–8890, doi: 10.1039/D1GC02392F.

[202] Gomez-Bolivar, J., Mikheenko, I. P., Macaskie, L. E., Merroun, M. L. Characterization of palladium nanoparticles produced by healthy and microwave-injured cells of desulfovibrio desulfuricans and escherichia coli. *Nanomaterials* 2019, 9(6), 857, doi: https://www.mdpi.com/2079-4991/9/6/857.

[203] Pantidos, N., Horsfall, L. Understanding the role of SilE in the production of metal nanoparticles by Morganella psychrotolerans using MicroScale Thermophoresis. *New Biotechnology* 2020, 55, 1–4, doi: https://doi.org/10.1016/j.nbt.2019.09.002.

[204] Marooufpour, N., Alizadeh, M., Hatami, M., Asgari Lajayer, B. Biological synthesis of nanoparticles by different groups of bacteria. In *Microbial Nanobionics: Volume 1, State-of-the-Art*, Prasad, R. (Ed.). Cham, Springer International Publishing, 2019, 63–85.

[205] Gahlawat, G., Choudhury, A. R. A review on the biosynthesis of metal and metal salt nanoparticles by microbes. *RSC Advances* 2019, 9(23), 12944–12967, doi: 10.1039/C8RA10483B.

[206] Patil, M. P., et al. Extracellular synthesis of gold nanoparticles using the marine bacterium Paracoccus haeundaensis BC74171T and evaluation of their antioxidant activity and antiproliferative effect on normal and cancer cell lines. *Colloids and Surfaces B: Biointerfaces* 2019, 183, 110455, doi: https://doi.org/10.1016/j.colsurfb.2019.110455.

[207] Rónavári, A., et al. Green silver and gold nanoparticles: Biological synthesis approaches and potentials for biomedical applications. *Molecules* 2021, 26(4), 844, doi: https://www.mdpi.com/1420-3049/26/4/844.

[208] Borghese, R., et al. Reprint of "Extracellular production of tellurium nanoparticles by the photosynthetic bacterium Rhodobacter capsulatus". *Journal of Hazardous Materials* 2017, 324, 31–38, doi: https://doi.org/10.1016/j.jhazmat.2016.11.002.

[209] Rai, M., et al. Fusarium as a novel fungus for the synthesis of nanoparticles: Mechanism and applications. *Journal of Fungi* 2021, 7(2), 139, doi: https://www.mdpi.com/2309-608X/7/2/139.

[210] Gupta, K., Chundawat, T. S. Bio-inspired synthesis of platinum nanoparticles from fungus Fusarium oxysporum: Its characteristics, potential antimicrobial, antioxidant and photocatalytic activities. *Materials Research Express* 2019, 6(10), 1050d6, doi: 10.1088/2053-1591/ab4219.

[211] Almeida, É. S., De Oliveira, D., Hotza, D. Characterization of silver nanoparticles produced by biosynthesis mediated by Fusarium oxysporum under different processing conditions. *Bioprocess and Biosystems Engineering* 2017, 40(9), 1291–1303, doi: 10.1007/s00449-017-1788-9.

[212] Bhardwaj, K., et al. Pleurotus macrofungi-assisted nanoparticle synthesis and its potential applications: A review. *Journal of Fungi* 2020, 6(4), 351, doi: https://www.mdpi.com/2309-608X/6/4/351.

[213] Martínez-Flores, H. E., Contreras-Chávez, R., Garnica-Romo, M. G. Effect of extraction processes on bioactive compounds from pleurotus ostreatus and pleurotus djamor: Their applications in the synthesis of silver nanoparticles. *Journal of Inorganic and Organometallic Polymers and Materials* 2021, 31(3), 1406–1418, doi: 10.1007/s10904-020-01820-2.

[214] Owaid, M. N. Green synthesis of silver nanoparticles by Pleurotus (oyster mushroom) and their bioactivity: Review. *Environmental Nanotechnology, Monitoring and Management* 2019, 12, 100256, doi: https://doi.org/10.1016/j.enmm.2019.100256.

[215] Konappa, N., et al. Ameliorated Antibacterial and Antioxidant Properties by Trichoderma harzianum Mediated Green Synthesis of Silver Nanoparticles. *Biomolecules* 2021, 11(4), 535, doi: https://www.mdpi.com/2218-273X/11/4/535.

[216] Consolo, V. F., Torres-Nicolini, A., Alvarez, V. A. Mycosinthetized Ag, CuO and ZnO nanoparticles from a promising Trichoderma harzianum strain and their antifungal potential against important phytopathogens. *Scientific Reports* 2020/11/24 2020, 10(1), 20499, doi: 10.1038/s41598-020-77294-6.

[217] Guilger-Casagrande, M., Germano-Costa, T., Pasquoto-Stigliani, T., Fraceto, L. F., D. Lima, R. Biosynthesis of silver nanoparticles employing Trichoderma harzianum with enzymatic stimulation for the control of Sclerotinia sclerotiorum. *Scientific Reports* 2019/10/04 2019, 9(1), 14351, doi: 10.1038/s41598-019-50871-0.

[218] Kobashigawa, J. M., Robles, C. A., Martínez Ricci, M. L., Carmarán, C. C. Influence of strong bases on the synthesis of silver nanoparticles (AgNPs) using the ligninolytic fungi Trametes trogii. *Saudi Journal of Biological Sciences* 2019, 26(7), 1331–1337, doi: https://doi.org/10.1016/j.sjbs.2018.09.006.

[219] Annamalai, J., Murugan, K., Shanmugam, J., Boopathy, U. Chapter 14 – Mycosynthesis of silver nanoparticles: Mechanism and applications. In Green Synthesis of Silver Nanomaterials, Abd-Elsalam, K. A. (Ed.). Elsevier, 2022, 391–411.

[220] Khalil, N. M., Abd El-Ghany, M. N., Rodríguez-Couto, S. Antifungal and anti-mycotoxin efficacy of biogenic silver nanoparticles produced by Fusarium chlamydosporum and Penicillium chrysogenum at non-cytotoxic doses. *Chemosphere* 2019, 218, 477–486, doi: https://doi.org/10.1016/j.chemosphere.2018.11.129.

[221] Wanarska, E., Maliszewska, I. The possible mechanism of the formation of silver nanoparticles by Penicillium cyclopium. *Bioorganic Chemistry* 2019, 93, 102803, doi: https://doi.org/10.1016/j.bioorg.2019.02.028.

[222] Seetharaman, P. K., et al. Antimicrobial and larvicidal activity of eco-friendly silver nanoparticles synthesized from endophytic fungi Phomopsis liquidambaris. *Biocatalysis & Agricultural Biotechnology* 2018, 16, 22–30, doi: https://doi.org/10.1016/j.bcab.2018.07.006.

[223] Spagnoletti, F. N., Spedalieri, C., Kronberg, F., Giacometti, R. Extracellular biosynthesis of bactericidal Ag/AgCl nanoparticles for crop protection using the fungus Macrophomina phaseolina. *Journal of Environmental Management* 2019, 231, 457–466, doi: https://doi.org/10.1016/j.jenvman.2018.10.081.

[224] Sreedharan, S. M., Gupta, S., Saxena, A. K., Singh, R. Macrophomina phaseolina: Microbased biorefinery for gold nanoparticle production. *Annals of Microbiology* 2019/04/01 2019, 69(4), 435–445, doi: 10.1007/s13213-018-1434-z.

[225] Krishna, G., Srileka, V., Singara Charya, M. A., Abu Serea, E. S., Shalan, A. E. Biogenic synthesis and cytotoxic effects of silver nanoparticles mediated by white rot fungi. *Heliyon* 2021, 7(3), e06470, doi: https://doi.org/10.1016/j.heliyon.2021.e06470.

[226] Chhipa, H. Chapter 5 – Mycosynthesis of nanoparticles for smart agricultural practice: A green and eco-friendly approach. In Green Synthesis, Characterization and Applications of Nanoparticles, Shukla, A. K., Iravani, S. (Eds.). Elsevier, 2019, 87–109.

[227] Zomorodian, K., et al. Biosynthesis and characterization of silver nanoparticles by *Aspergillus* species. *BioMed Research International* 2016, 2016, 5435397, doi: 10.1155/2016/5435397.

[228] Kalpana, V. N., Kataru, B. A. S., Sravani, N., Vigneshwari, T., Panneerselvam, A., Devi Rajeswari, V. Biosynthesis of zinc oxide nanoparticles using culture filtrates of Aspergillus niger: Antimicrobial textiles and dye degradation studies. *OpenNano* 2018, 3, 48–55, doi: https://doi.org/10.1016/j.onano.2018.06.001.

[229] Farrag, H. M. M., Mostafa, F. A. A. M., Mohamed, M. E., Huseein, E. A. M. Green biosynthesis of silver nanoparticles by Aspergillus niger and its antiamoebic effect against Allovahlkampfia spelaea trophozoite and cyst. *Experimental Parasitology* 2020, 219, 108031, doi: https://doi.org/10.1016/j.exppara.2020.108031.

[230] Elegbede, J. A., et al. Biofabrication of gold nanoparticles using xylanases through valorization of corncob by Aspergillus niger and Trichoderma longibrachiatum: Antimicrobial, antioxidant,

anticoagulant and thrombolytic activities. *Waste Biomass Valorization* 2020, 11(3), 781–791, doi: 10.1007/s12649-018-0540-2.

[231] Moraes, L. C., et al. High diversity of microalgae as a tool for the synthesis of different silver nanoparticles: A species-specific green synthesis. *Colloid and Interface Science Communications* 2021, 42, 100420, doi: https://doi.org/10.1016/j.colcom.2021.100420.

[232] Rahman, A., Kumar, S., Nawaz, T. Chapter 17 – Biosynthesis of nanomaterials using algae. In Microalgae Cultivation for Biofuels Production, Yousuf, A. (Ed.). Academic Press, 2020, 265–279.

[233] Khan, A. U., Khan, M., Malik, N., Cho, M. H., Khan, M. M. Recent progress of algae and blue–green algae-assisted synthesis of gold nanoparticles for various applications. *Bioprocess and Biosystems Engineering* 2019, 42(1), 1–15, doi: 10.1007/s00449-018-2012-2.

[234] Kalimuthu, K., Cha, B. S., Kim, S., Park, K. S. Eco-friendly synthesis and biomedical applications of gold nanoparticles: A review. *Microchemical Journal* 2020, 152, 104296, doi: https://doi.org/10.1016/j.microc.2019.104296.

[235] Abd El Raouf, N., Hozyen, W., Abd El Neem, M., Ibraheem, I. Potentiality of silver nanoparticles prepared by ulva fasciata as anti-nephrotoxicity in albino-rats. *Egyptian Journal of Botany* 2017, 57(3), 479–494, doi: 10.21608/ejbo.2017.913.1070.

[236] Alsaggaf, M. S., Diab, A. M., ElSaied, B. E. F., Tayel, A. A., Moussa, S. H. Application of ZnO nanoparticles phycosynthesized with ulva fasciata extract for preserving peeled shrimp quality. *Nanomaterials* 2021, 11(2), 385, doi: https://www.mdpi.com/2079-4991/11/2/385.

[237] Massironi, A., et al. Ulvan as novel reducing and stabilizing agent from renewable algal biomass: Application to green synthesis of silver nanoparticles. *Carbohydrate Polymers* 2019, 203, 310–321, doi: https://doi.org/10.1016/j.carbpol.2018.09.066.

[238] Salem, D. M. S. A., Ismail, M. M., Aly-Eldeen, M. A. Biogenic synthesis and antimicrobial potency of iron oxide (Fe3O4) nanoparticles using algae harvested from the Mediterranean Sea, Egypt. *The Egyptian Journal of Aquatic Research* 2019, 45(3), 197–204, doi: https://doi.org/10.1016/j.ejar.2019.07.002.

[239] El-malek, F. A., Rofeal, M., Farag, A., Omar, S., Khairy, H. Polyhydroxyalkanoate nanoparticles produced by marine bacteria cultivated on cost effective Mediterranean algal hydrolysate media. *Journal of Biotechnology* 2021, 328, 95–105, doi: https://doi.org/10.1016/j.jbiotec.2021.01.008.

[240] El-Sheekh, M. M., El-Kassas, H. Y., Shams El-Din, N. G., Eissa, D. I., El-Sherbiny, B. A. Green synthesis, characterization applications of iron oxide nanoparticles for antialgal and wastewater bioremediation using three brown algae. *International Journal of Phytoremediation* 2021, 23(14), 1538–1552, doi: 10.1080/15226514.2021.1915957.

[241] Machado, S., et al. Toxicity in vitro and in Zebrafish embryonic development of gold nanoparticles biosynthesized using cystoseira macroalgae extracts. *International Journal of Nanomedicine* 2021, 16, 5017–5036, doi: 10.2147/IJN.S300674.

[242] Fatima, R., Priya, M., Indurthi, L., Radhakrishnan, V., Sudhakaran, R. Biosynthesis of silver nanoparticles using red algae Portieria hornemannii and its antibacterial activity against fish pathogens. *Microbial Pathogenesis* 2020, 138, 103780, doi: https://doi.org/10.1016/j.micpath.2019.103780.

[243] Karthic, A., Wagh, N. S., Lakkakula, J. R. Chapter 10 – Algae-assisted synthesis of nanoparticles. In Nanobiotechnology, Ghosh, S., Webster, T. J. (Eds.). Elsevier, 2021, 145–165.

[244] Koopi, H., Buazar, F. A novel one-pot biosynthesis of pure alpha aluminum oxide nanoparticles using the macroalgae Sargassum ilicifolium: A green marine approach. *Ceramics International* 2018, 44(8), 8940–8945, doi: https://doi.org/10.1016/j.ceramint.2018.02.091.

[245] Koopi, H., Buazar, F. Synthesis and characterization of aluminum oxide nano particles using macro algae sargassum ilicifolium. *Journal of Marine Science and Technology* 2018, 17(3), 58–64, doi: 10.22113/jmst.2016.40996.

[246] AlNadhari, S., Al-Enazi, N. M., Alshehrei, F., Ameen, F. A review on biogenic synthesis of metal nanoparticles using marine algae and its applications. *Environmental Research* 2021, 194, 110672, doi: https://doi.org/10.1016/j.envres.2020.110672.

[247] Yılmaz Öztürk, B., Yenice Gürsu, B., Dağ, İ. Antibiofilm and antimicrobial activities of green synthesized silver nanoparticles using marine red algae Gelidium corneum. *Process Biochemical* 2020, 89, 208–219, doi: https://doi.org/10.1016/j.procbio.2019.10.027.

[248] Saad, A. M. A., Mohamed, *, H., Nivien, A. N., Mahmoud, A. S., Ali, M. A. Controllable biogenic synthesis of intracellular silver/silver chloride nanoparticles by Meyerozyma guilliermondii KX008616. *Journal of Microbiology and Biotechnology* 2018, 28(6), 917–930, doi: 10.4014/jmb.1802.02010.

[249] Shamsuzzaman, A. M., Khanam, H., Aljawfi, R. N. Biological synthesis of ZnO nanoparticles using C. albicans and studying their catalytic performance in the synthesis of steroidal pyrazolines. *Arabian Journal of Chemistry* 2017, 10, S1530–S1536, doi: https://doi.org/10.1016/j.arabjc.2013.05.004.

[250] Lian, S., et al. Characterization of biogenic selenium nanoparticles derived from cell-free extracts of a novel yeast Magnusiomyces ingens. *3 Biotech* 2019, 9(6), 221, doi: 10.1007/s13205-019-1748-y.

[251] Bolbanabad, E. M., Ashengroph, M., Darvishi, F. Development and evaluation of different strategies for the clean synthesis of silver nanoparticles using Yarrowia lipolytica and their antibacterial activity. *Process Biochemical* 2020, 94, 319–328, doi: https://doi.org/10.1016/j.procbio.2020.03.024.

[252] Faramarzi, S., Anzabi, Y., Jafarizadeh-Malmiri, H. Nanobiotechnology approach in intracellular selenium nanoparticle synthesis using Saccharomyces cerevisiae – Fabrication and characterization. *Archives of Microbiology* 2020, 202(5), 1203–1209, doi: 10.1007/s00203-020-01831-0.

[253] Asghari-Paskiabi, F., Imani, M., Rafii-Tabar, H., Razzaghi-Abyaneh, M. Physicochemical properties, antifungal activity and cytotoxicity of selenium sulfide nanoparticles green synthesized by Saccharomyces cerevisiae. *Biochemical and Biophysical Research Communications* 2019, 516(4), 1078–1084, doi: https://doi.org/10.1016/j.bbrc.2019.07.007.

[254] Tarver, S., Gray, D., Loponov, K., Das, D. B., Sun, T., Sotenko, M. Biomineralization of Pd nanoparticles using Phanerochaete chrysosporium as a sustainable approach to turn platinum group metals (PGMs) wastes into catalysts. *International Biodeterioration & Biodegradation* 2019, 143, 104724, doi: https://doi.org/10.1016/j.ibiod.2019.104724.

[255] Laxmi Sharma, J., Dhayal, V., Kumar Sharma, R., Antibacterial effect of glycerol assisted ZnO nanoparticles synthesized by white rot fungus Phanerochaete chrysosporium, *Materials Today: Proceedings* 2021, 43, 2855–2860, doi: https://doi.org/10.1016/j.matpr.2021.01.075.

[256] Suryavanshi, P., Pandit, R., Gade, A., Derita, M., Zachino, S., Rai, M. Colletotrichum sp.- mediated synthesis of sulphur and aluminium oxide nanoparticles and its in vitro activity against selected food-borne pathogens. *LWT – Food Science and Technology* 2017, 81, 188–194, doi: https://doi.org/10.1016/j.lwt.2017.03.038.

[257] Al-Ansari, M. M., Al-Dahmash, N. D., Ranjitsingh, A. J. A. Synthesis of silver nanoparticles using gum Arabic: Evaluation of its inhibitory action on Streptococcus mutans causing dental caries and endocarditis. *Journal of Infection and Public Health* 2021, 14(3), 324–330, doi: https://doi.org/10.1016/j.jiph.2020.12.016.

[258] Escárcega-González, C. E., et al. In vivo antimicrobial activity of silver nanoparticles produced via a green chemistry synthesis using Acacia rigidula as a reducing and capping agent. *International Journal of Nanomedicine* 2018, 13, 2349–2363, doi: 10.2147/IJN.S160605.

[259] Riaz, M., et al. Exceptional antibacterial and cytotoxic potency of monodisperse greener AgNPs prepared under optimized pH and temperature. *Scientific Reports* 2021, 11(1), 2866, doi: 10.1038/s41598-021-82555-z.

[260] Folorunso, A., et al. Biosynthesis, characterization and antimicrobial activity of gold nanoparticles from leaf extracts of Annona muricata. *Journal of Nanostructure in Chemistry* 2019, 9(2), 111–117, doi: 10.1007/s40097-019-0301-1.

[261] Gavamukulya, Y., et al. Green synthesis and characterization of highly stable silver nanoparticles from ethanolic extracts of fruits of annona muricata. *Journal of Inorganic and Organometallic Polymers and Materials* 2020, 30(4), 1231–1242, doi: 10.1007/s10904-019-01262-5.

[262] Jabir, M. S., et al. Green synthesis of silver nanoparticles using annona muricata extract as an inducer of apoptosis in cancer cells and inhibitor for NLRP3 inflammasome via enhanced autophagy. *Nanomaterials* 2021, 11(2), 384, doi: https://www.mdpi.com/2079-4991/11/2/384.

[263] Behravan, M., Hossein Panahi, A., Naghizadeh, A., Ziaee, M., Mahdavi, R., Mirzapour, A. Facile green synthesis of silver nanoparticles using Berberis vulgaris leaf and root aqueous extract and its antibacterial activity. *International Journal of Biological Macromolecules* 2019, 124, 148–154, doi: https://doi.org/10.1016/j.ijbiomac.2018.11.101.

[264] Anzabi, Y. Biosynthesis of ZnO nanoparticles using barberry (Berberis vulgaris) extract and assessment of their physico-chemical properties and antibacterial activities. *Green Processing and Synthesis* 2018, 7(2), 114–121, doi: 10.1515/gps-2017-0014.

[265] Safipour Afshar, A., Saeid Nematpour, F. Evaluation of the cytotoxic activity of biosynthesized silver nanoparticles using berberis vulgaris leaf extract. *Jentashapir Journal of Cellular and Molecular Biology* 2021, 12(1), e112437, doi: 10.5812/jjcmb.112437.

[266] Naseer, M., Aslam, U., Khalid, B., Chen, B. Green route to synthesize zinc oxide nanoparticles using leaf extracts of cassia fistula and melia azadarach and their antibacterial potential. *Scientific Reports* 2020, 10(1), 9055, doi: 10.1038/s41598-020-65949-3.

[267] Jyoti, K., Arora, D., Fekete, G., Lendvai, L., Dogossy, G., Singh, T. Antibacterial and anti-inflammatory activities of Cassia fistula fungal broth-capped silver nanoparticles. *Materials Technology* 2021, 36(14), 883–893, doi: 10.1080/10667857.2020.1802841.

[268] Ramya, A. V., Joseph, N., Balachandran, M. Facile synthesis of few-layer graphene oxide from cinnamomum camphora. *Nanobiotechnology Reports* 2021, 16(2), 183–187, doi: 10.1134/S2635167621020130.

[269] Li, W., et al. Antimicrobial activity of sliver nanoparticles synthesized by the leaf extract of Cinnamomum camphora. *Biochemical Engineering Journal* 2021, 172, 108050, doi: https://doi.org/10.1016/j.bej.2021.108050.

[270] Zhu, W., et al. Green synthesis of zinc oxide nanoparticles using Cinnamomum camphora (L.) Presl leaf extracts and its antifungal activity. *Journal of Environmental Chemical Engineering* 2021, 9(6), 106659, doi: https://doi.org/10.1016/j.jece.2021.106659.

[271] Aref, M. S., Salem, S. S. Bio-callus synthesis of silver nanoparticles, characterization, and antibacterial activities via Cinnamomum camphora callus culture. *Biocatalysis & Agricultural Biotechnology* 2020, 27, 101689, doi: https://doi.org/10.1016/j.bcab.2020.101689.

[272] Salgado, P., Mártire, D. O., Vidal, G. Eucalyptus extracts-mediated synthesis of metallic and metal oxide nanoparticles: Current status and perspectives. *Materials Research Express* 2019, 6(8), 082006, doi: 10.1088/2053-1591/ab254c.

[273] Sawalha, H., et al. Toward a better understanding of metal nanoparticles, a novel strategy from eucalyptus plants. *Plants* 2021, 10(5), 929, doi: https://www.mdpi.com/2223-7747/10/5/929.

[274] Syukri, D. M., et al. Antibacterial-coated silk surgical sutures by ex situ deposition of silver nanoparticles synthesized with Eucalyptus camaldulensis eradicates infections. *Journal of Microbiological Methods* 2020, 174, 105955, doi: https://doi.org/10.1016/j.mimet.2020.105955.

[275] Munir, H., et al. Eucalyptus camaldulensis gum as a green matrix to fabrication of zinc and silver nanoparticles: Characterization and novel prospects as antimicrobial and dye-degrading agents. *Jouranal of Materials Research and Technology* 2020, 9(6), 15513–15524, doi: https://doi.org/10.1016/j.jmrt.2020.11.026.

[276] Lagashetty, A., Ganiger, S. K., K, P. R., Reddy, S., Pari, M. Microwave-assisted green synthesis, characterization and adsorption studies on metal oxide nanoparticles synthesized using Ficus

Benghalensis plant leaf extracts. *New Journal of Chemistry* 2020, 44(33), 14095–14102, doi: 10.1039/D0NJ01759K.

[277] Yadav, S., Rani, N., Saini, K., Green synthesis of ZnO and CuO NPs using Ficus benghalensis leaf extract and their comparative study for electrode materials for high performance supercapacitor application, *Materials Today: Proceedings* 2021, doi: https://doi.org/10.1016/j.matpr.2021.08.323.

[278] Yadav, P. K., et al. Synthesis of green fluorescent carbon quantum dots from the latex of Ficus benghalensis for the detection of tyrosine and fabrication of Schottky barrier diode. *New Journal of Chemistry* 2021, 45(28), 12549–12556, doi: 10.1039/D1NJ01655E.

[279] Goutam, S. P., Saxena, G., Singh, V., Yadav, A. K., Bharagava, R. N., Thapa, K. B. Green synthesis of TiO2 nanoparticles using leaf extract of Jatropha curcas L. for photocatalytic degradation of tannery wastewater. *Chemical Engineering Journal* 2018, 336, 386–396, doi: https://doi.org/10.1016/j.cej.2017.12.029.

[280] Nayak, S., Sajankila, P., Rao, C. V., Hegde, A. R., Mutalik, S. Biogenic synthesis of silver nanoparticles using Jatropha curcas seed cake extract and characterization: Evaluation of its antibacterial activity. *Energy Sources, Part A: Recovery, Utilization and Environmental Effects* 2021, 43(24), 3415–3423, doi: 10.1080/15567036.2019.1632394.

[281] Magudieshwaran, R., et al. Green and chemical synthesized CeO2 nanoparticles for photocatalytic indoor air pollutant degradation. *Materials Letters* 2019, 239, 40–44, doi: https://doi.org/10.1016/j.matlet.2018.11.172.

[282] Huang, L., Sun, Y., Mahmud, S., Liu, H. Biological and environmental applications of silver nanoparticles synthesized using the aqueous extract of ginkgo biloba leaf. *Journal of Inorganic and Organometallic Polymers and Materials* 2020, 30(5), 1653–1668, doi: 10.1007/s10904-019-01313-x.

[283] Wang, F., Zhang, W., Tan, X., Wang, Z., Li, Y., Li, W. Extract of Ginkgo biloba leaves mediated biosynthesis of catalytically active and recyclable silver nanoparticles. *Colloids and Surfaces A: Physicochemical and Engineering Aspects* 2019, 563, 31–36, doi: https://doi.org/10.1016/j.colsurfa.2018.11.054.

[284] Thatikayala, D., Min, B. Ginkgo leaves extract-assisted synthesis of ZnO/CuO nanocrystals for efficient UV-induced photodegradation of organic dyes and antibacterial activity. *Journal of Materials Science: Materials in Electronics* 2021, 32(13), 17154–17169, doi: 10.1007/s10854-021-06169-x.

[285] De Matteis, V., et al. Cultivar-dependent anticancer and antibacterial properties of silver nanoparticles synthesized using leaves of different olea europaea trees. *Nanomaterials* 2019, 9(11), 1544, doi: https://www.mdpi.com/2079-4991/9/11/1544.

[286] Issam, N., Naceur, D., Nechi, G., Maatalah, S., Zribi, K., Mhadhbi, H. Green synthesised ZnO nanoparticles mediated by Olea europaea leaf extract and their antifungal activity against Botrytis cinerea infecting faba bean plants. *Archives of Phytopathology & Plant Protection* 2021, 54(15–16), 1083–1105, doi: 10.1080/03235408.2021.1889859.

[287] Golabiazar, R., Othman, K. I., Khalid, K. M., Maruf, D. H., Aulla, S. M., Yusif, P. A. Green synthesis, characterization, and investigation antibacterial activity of silver nanoparticles using pistacia atlantica leaf extract. *BioNanoScience* 2019, 9(2), 323–333, doi: 10.1007/s12668-019-0606-z.

[288] Veisi, H., Kavian, M., Hekmati, M., Hemmati, S. Biosynthesis of the silver nanoparticles on the graphene oxide's surface using Pistacia atlantica leaves extract and its antibacterial activity against some human pathogens. *Polyhedron* 2019, 161, 338–345, doi: https://doi.org/10.1016/j.poly.2019.01.034.

[289] Jomehzadeh, N., Koolivand, Z., Dahdouh, E., Akbari, A., Zahedi, A., Chamkouri, N. Investigating in-vitro antimicrobial activity, biosynthesis, and characterization of silver nanoparticles, zinc oxide nanoparticles, and silver-zinc oxide nanocomposites using Pistacia Atlantica Resin. *Materials Today Communications* 2021, 27, 102457, doi: https://doi.org/10.1016/j.mtcomm.2021.102457.

[290] Thatikayala, D., Jayarambabu, N., Banothu, V., Ballipalli, C. B., Park, J., Rao, K. V. Biogenic synthesis of silver nanoparticles mediated by Theobroma cacao extract: Enhanced antibacterial and

photocatalytic activities. *Journal of Materials Science: Materials in Electronics* 2019, 30(18), 17303–17313, doi: 10.1007/s10854-019-02077-3.

[291] Mellinas, C., Jiménez, A., D. C. Garrigós, M. Microwave-assisted green synthesis and antioxidant activity of selenium nanoparticles using theobroma cacao L. Bean Shell extract. *Molecules* 2019, 24(22), 4048, doi: https://www.mdpi.com/1420-3049/24/22/4048.

[292] Doan, V.-D., Phung, M.-T., Nguyen, T. L.-H., Mai, T.-C., Nguyen, T.-D. Noble metallic nanoparticles from waste Nypa fruticans fruit husk: Biosynthesis, characterization, antibacterial activity and recyclable catalysis. *Arabian Journal of Chemistry* 2020, 13(10), 7490–7503, doi: https://doi.org/10.1016/j.arabjc.2020.08.024.

[293] Bao, Z., Lan, C. Q. Advances in biosynthesis of noble metal nanoparticles mediated by photosynthetic organisms – A review. *Colloids and Surfaces B: Biointerfaces* 2019, 184, 110519, doi: https://doi.org/10.1016/j.colsurfb.2019.110519.

[294] Nasrollahzadeh, M., Sajjadi, M., Iravani, S., Varma, R. S. Green-synthesized nanocatalysts and nanomaterials for water treatment: Current challenges and future perspectives. *Journal of Hazardous Materials* 2021, 401, 123401, doi: https://doi.org/10.1016/j.jhazmat.2020.123401.

[295] Al-Otibi, F., et al. Biosynthesis of silver nanoparticles using Malva parviflora and their antifungal activity. *Saudi Journal of Biological Sciences* 2021, 28(4), 2229–2235, doi: https://doi.org/10.1016/j.sjbs.2021.01.012.

[296] Vandarkuzhali, S. A. A., Karthikeyan, G., Pachamuthu, M. P. Microwave assisted biosynthesis of Borassus flabellifer fruit mediated silver and gold nanoparticles for dye reduction, antibacterial and anticancer activity. *Journal of Environmental Chemical Engineering* 2021, 9(6), 106411, doi: https://doi.org/10.1016/j.jece.2021.106411.

[297] Sathiyaraj, S., et al. Biosynthesis, characterization, and antibacterial activity of gold nanoparticles. *Journal of Infection and Public Health* 2021, doi: https://doi.org/10.1016/j.jiph.2021.10.007.

[298] Wong-Pinto, L.-S., Mercado, A., Chong, G., Salazar, P., Ordóñez, J. I. Biosynthesis of copper nanoparticles from copper tailings ore – An approach to the 'Bionanomining'. *Journal of Cleaner Production* 2021, 315, 128107, doi: https://doi.org/10.1016/j.jclepro.2021.128107.

[299] Abbas, A. H., Fairouz, N. Y., Characterization, biosynthesis of copper nanoparticles using ginger roots extract and investigation of its antibacterial activity, *Materials Today: Proceedings* 2021, doi: https://doi.org/10.1016/j.matpr.2021.09.551.

[300] Zikalala, N., Matshetshe, K., Parani, S., Oluwafemi, O. S. Biosynthesis protocols for colloidal metal oxide nanoparticles. *Nano-Structures and Nano-Objects* 2018, 16, 288–299, doi: https://doi.org/10.1016/j.nanoso.2018.07.010.

[301] Gebre, S. H., Sendeku, M. G. New frontiers in the biosynthesis of metal oxide nanoparticles and their environmental applications: An overview. *SN Applied Sciences* 2019, 1(8), 928, doi: 10.1007/s42452-019-0931-4.

[302] Taghizadeh, S.-M., Morowvat, M. H., Negahdaripour, M., Ebrahiminezhad, A., Ghasemi, Y. Biosynthesis of metals and metal oxide nanoparticles through microalgal nanobiotechnology: Quality control aspects. *BioNanoScience* 2021, 11(1), 209–226, doi: 10.1007/s12668-020-00805-2.

[303] Gupta, V., Chandra, N. Biosynthesis and antibacterial activity of metal oxide nanoparticles using brassica oleracea subsp. botrytis (L.) leaves, an agricultural waste. *Proceedings of the National Academy of Sciences, India, Section B: Biological Sciences* 2020, 90(5), 1093–1100, doi: 10.1007/s40011-020-01184-0.

[304] D. J. Ruíz-baltazar, Á. Sonochemical activation-assisted biosynthesis of Au/Fe3O4 nanoparticles and sonocatalytic degradation of methyl orange. *Ultrasonics Sonochemistry* 2021, 73, 105521, doi: https://doi.org/10.1016/j.ultsonch.2021.105521.

[305] Hussein, B. Y., Mohammed, A. M. Biosynthesis and characterization of nickel oxide nanoparticles by using aqueous grape extract and evaluation of their biological applications. *Results in Chemistry* 2021, 3, 100142, doi: https://doi.org/10.1016/j.rechem.2021.100142.

[306] Obeizi, Z., Benbouzid, H., Ouchenane, S., Yılmaz, D., Culha, M., Bououdina, M. Biosynthesis of Zinc oxide nanoparticles from essential oil of Eucalyptus globulus with antimicrobial and anti-biofilm activities. *Materials Today Communications* 2020, 25, 101553, doi: https://doi.org/10.1016/j.mtcomm.2020.101553.

[307] Ahmed Rather, G., et al. Biosynthesis of Zinc oxide nanoparticles using Bergenia ciliate aqueous extract and evaluation of their photocatalytic and antioxidant potential. *Inorganic Chemistry Communications* 2021, 134, 109020, doi: https://doi.org/10.1016/j.inoche.2021.109020.

[308] Laid, T. M., Abdelhamid, K., Eddine, L. S., Abderrhmane, B. Optimizing the biosynthesis parameters of iron oxide nanoparticles using central composite design. *Journal of Molecular Structure* 2021, 1229, 129497, doi: https://doi.org/10.1016/j.molstruc.2020.129497.

[309] Maass, D., Valério, A., Lourenço, L. A., De Oliveira, D., Hotza, D. Biosynthesis of iron oxide nanoparticles from mineral coal tailings in a stirred tank reactor. *Hydrometallurgy* 2019, 184, 199–205, doi: https://doi.org/10.1016/j.hydromet.2019.01.010.

[310] T, S. J. K., V.r, A., M, V., Muthu, A. Biosynthesis of multiphase iron nanoparticles using Syzygium aromaticum and their magnetic properties. *Colloids and Surfaces A: Physicochemical and Engineering Aspects* 2020, 603, 125241, doi: https://doi.org/10.1016/j.colsurfa.2020.125241.

[311] Shahrashoub, M., Bakhtiari, S., Afroosheh, F., Googheri, M. S. Recovery of iron from direct reduction iron sludge and biosynthesis of magnetite nanoparticles using green tea extract. *Colloids and Surfaces A: Physicochemical and Engineering Aspects* 2021, 622, 126675, doi: https://doi.org/10.1016/j.colsurfa.2021.126675.

[312] Madubuonu, N., et al. Biosynthesis of iron oxide nanoparticles via a composite of Psidium guavaja-Moringa oleifera and their antibacterial and photocatalytic study. *Journal of Photochemistry and Photobiology B: Biology* 2019, 199, 111601, doi: https://doi.org/10.1016/j.jphotobiol.2019.111601.

[313] Chand Mali, S., Raj, S., Trivedi, R. Biosynthesis of copper oxide nanoparticles using Enicostemma axillare (Lam.) leaf extract. *Biochemistry and Biophysics Reports* 2019, 20, 100699, doi: https://doi.org/10.1016/j.bbrep.2019.100699.

[314] Sackey, J. et al., Biosynthesis of CuO nanoparticles using Mimosa hamata extracts, *Materials Today: Proceedings* 2021, 36, 540–548, doi: https://doi.org/10.1016/j.matpr.2020.05.325.

[315] G, B. P., S, X. T. A critical green biosynthesis of novel CuO/C porous nanocomposite via the aqueous leaf extract of Ficus religiosa and their antimicrobial, antioxidant, and adsorption properties. *Chemical Engineering Journal Advances* 2021, 8, 100152, doi: https://doi.org/10.1016/j.ceja.2021.100152.

[316] Hamidian, K., Saberian, M. R., Miri, A., Sharifi, F., Sarani, M. Doped and un-doped cerium oxide nanoparticles: Biosynthesis, characterization, and cytotoxic study. *Ceramics International* 2021, 47(10, Part A), 13895–13902, doi: https://doi.org/10.1016/j.ceramint.2021.01.256.

[317] Nazaripour, E., et al. Biosynthesis of lead oxide and cerium oxide nanoparticles and their cytotoxic activities against colon cancer cell line. *Inorganic Chemistry Communications* 2021, 131, 108800, doi: https://doi.org/10.1016/j.inoche.2021.108800.

[318] Ngoepe, N. M., Mathipa, M. M., Hintsho-Mbita, N. C. Biosynthesis of titanium dioxide nanoparticles for the photodegradation of dyes and removal of bacteria. *Optik* 2020, 224, 165728, doi: https://doi.org/10.1016/j.ijleo.2020.165728.

[319] Ajmal, N., Saraswat, K., Bakht, M. A., Riadi, Y., Ahsan, M. J., Noushad, M. Cost-effective and eco-friendly synthesis of titanium dioxide (TiO2) nanoparticles using fruit's peel agro-waste extracts: Characterization, in vitro antibacterial, antioxidant activities. *Green Chemistry Letters and Reviews* 2019, 12(3), 244–254, doi: 10.1080/17518253.2019.1629641.

[320] El Messaoudi, N., et al. Biosynthesis of SiO2 nanoparticles using extract of Nerium oleander leaves for the removal of tetracycline antibiotic. *Chemosphere* 2022, 287, 132453, doi: https://doi.org/10.1016/j.chemosphere.2021.132453.

[321] Abd-elmohsen, S. A., Mohmed, S. A., Daigham, G. E., Hoballah, E. M., Sidkey, N. M. Green synthesis, optimization and characterization of SiO2 nanoparticles using Aspergillus tubingensis F20 isolated

from drinking water. *Novel Research in Microbiology Journal* 2019, 3(6), 546–557, doi: 10.21608/nrmj.2019.66747.

[322] Manikandan, V., Jayanthi, P., Priyadharsan, A., Vijayaprathap, E., Anbarasan, P. M., Velmurugan, P. Green synthesis of pH-responsive Al2O3 nanoparticles: Application to rapid removal of nitrate ions with enhanced antibacterial activity. *Journal of Photochemistry and Photobiology A: Chemistry* 2019, 371, 205–215, doi: https://doi.org/10.1016/j.jphotochem.2018.11.009.

[323] Ghosh, T. K., Sadhukhan, S., Rana, D., Bhattacharyya, A., Chattopadhyay, D., Chakraborty, M. Green approaches to synthesize reduced graphene oxide and assessment of its electrical properties. *Nano-Structures and Nano-Objects* 2019, 19, 100362, doi: https://doi.org/10.1016/j.nanoso.2019.100362.

[324] Rattan, S., Kumar, S., Goswamy, J. K., Graphene oxide reduction using green chemistry, *Materials Today: Proceedings* 2020, 26, 3327–3331, doi: https://doi.org/10.1016/j.matpr.2019.09.168.

[325] Vatandost, E., Ghorbani-hasansaraei, A., Chekin, F., Naghizadeh Raeisi, S., Shahidi, S.-A. Green tea extract assisted green synthesis of reduced graphene oxide: Application for highly sensitive electrochemical detection of sunset yellow in food products. *Food Chemistry* 2020, 6, 100085–100085, doi: 10.1016/j.fochx.2020.100085.

[326] Saleem, H., Zaidi, S. J. Recent developments in the application of nanomaterials in agroecosystems. *Nanomaterials* 2020, 10(12), 2411, doi: 10.3390/nano10122411.

[327] Liu, C., Zhou, H., Zhou, J. The applications of nanotechnology in crop production. *Molecules* 2021, 26(23), 7070, doi: 10.3390/molecules26237070.

[328] Barhoum, A., Luisa García-Betancourt, M. Chapter 10 – Physicochemical characterization of nanomaterials: Size, morphology, optical, magnetic, and electrical properties. In Emerging Applications of Nanoparticles and Architecture Nanostructures, Barhoum, A., Makhlouf, A. S. H. (Eds.). Micro and Nano Technologies: Elsevier, 2018, 279–304.

[329] Kaliva, M., Vamvakaki, M. Chapter 17 – Nanomaterials characterization. In Polymer Science and Nanotechnology, Narain, R. (Ed.). Elsevier, 2020, 401–433.

[330] Bhattacharjee, S. DLS and zeta potential – What they are and what they are not?. *Journal of Controlled Release* 2016, 235, 337–351, doi: 10.1016/j.jconrel.2016.06.017.

[331] Rasmussen, K., Rauscher, H., Mech, A. Chapter 2 – Physicochemical characterization. In Adverse Effects of Engineered Nanomaterials (Second Edition), Fadeel, B., Pietroiusti, A., Shvedova, A. A. (Eds.). Academic Press, 2017, 15–49.

[332] Lin, P.-C., Lin, S., Wang, P. C., Sridhar, R. Techniques for physicochemical characterization of nanomaterials. *Biotechnology Advances* 2014, 32(4), 711–726, doi: 10.1016/j.biotechadv.2013.11.006.

[333] Khatua, L., Das, S. K., Energy dispersive X-ray spectroscopy study of compound semiconductor zinc orthotitanate prepared by solid state reaction method, *Materials Today: Proceedings* 2020, 33, 5628–5631, doi: 10.1016/j.matpr.2020.03.794.

[334] Barhoum, A., García-Betancourt, M. L., Rahier, H., Van Assche, G. Chapter 9 – Physicochemical characterization of nanomaterials: Polymorph, composition, wettability, and thermal stability. In Emerging Applications of Nanoparticles and Architecture Nanostructures, Barhoum, A., Makhlouf, A. S. H. (Eds.). Micro and Nano Technologies: Elsevier, 2018, 255–278.

[335] Lamas, D. G., De Oliveira Neto, M., Kellermann, G., Craievich, A. F. 5 – X-ray diffraction and scattering by nanomaterials. In Nanocharacterization Techniques, Da Róz, A. L., Ferreira, M., De Lima Leite, F., Oliveira, O. N. (Eds.). Micro and Nano Technologies: William Andrew Publishing, 2017, 111–182.

[336] Dendisová, M., Jeništová, A., Parchaňská-Kokaislová, A., Matějka, P., Prokopec, V., Švecová, M. The use of infrared spectroscopic techniques to characterize nanomaterials and nanostructures: A review. *Analytica Chimica Acta* 2018, 1031, 1–14, doi: 10.1016/j.aca.2018.05.046.

[337] Kumar, A., Khandelwal, M., Gupta, S. K., Kumar, V., Rani, R. Chapter 6 – Fourier transform infrared spectroscopy: Data interpretation and applications in structure elucidation and analysis of small

molecules and nanostructures. In Data Processing Handbook for Complex Biological Data Sources, Misra, G. (Ed.). Academic Press, 2019, 77–96.

[338] Loridant, S. Chapter 2 – Raman spectroscopy of nanomaterials: Applications to heterogeneous catalysis. In Characterization of Nanomaterials, Bhagyaraj, S. M., Oluwafemi, O. S., Kalarikkal, N., Thomas, S. (Eds.). Woodhead Publishing, 2018, 37–59.

[339] Szybowicz, M., Nowicka, A. B., Dychalska, A. Chapter 1 – Characterization of carbon nanomaterials by raman spectroscopy. In Characterization of Nanomaterials, Bhagyaraj, S. M., Oluwafemi, O. S., Kalarikkal, N., Thomas, S. (Eds.). Micro and Nano Technologies: Woodhead Publishing, 2018, 1–36Micro and Nano Technologies:.

[340] Li, Y.-S., Church, J. S. Raman spectroscopy in the analysis of food and pharmaceutical nanomaterials. *Journal of Food and Drug Analysis* 2014, 22(1), 29–48, doi: 10.1016/j.jfda.2014.01.003.

[341] Shard, A. G. Chapter 4.3.1 – X-ray photoelectron spectroscopy. In Characterization of Nanoparticles, Hodoroaba, V.-D., Unger, W. E. S., Shard, A. G. (Eds.). Micro and Nano Technologies: Elsevier, 2020, 349–371.

[342] Kaur, G., Singh, H., Singh, J. Chapter 2 – UV-vis spectrophotometry for environmental and industrial analysis. In Green Sustainable Process for Chemical and Environmental Engineering and Science, Inamuddin, R. B., Asiri, A. M. (Eds.). Elsevier, 2021, 49–68.

[343] George, G., Wilson, R., Joy, J. Chapter 3 – Ultraviolet spectroscopy: A facile approach for the characterization of nanomaterials. In Spectroscopic Methods for Nanomaterials Characterization, Thomas, S., Thomas, R., Zachariah, A. K., Mishra, R. K. (Eds.). Micro and Nano Technologies: Elsevier, 2017, 55–72.

[344] Kafle, B. P. Chapter 6 – Introduction to nanomaterials and application of UV–Visible spectroscopy for their characterization. In Chemical Analysis and Material Characterization by Spectrophotometry, Kafle, B. P. (Ed.). Elsevier, 2020, 147–198.

[345] Dukhin, A. S., Xu, R. Chapter 3.2.5 – Zeta-potential measurements. In Characterization of Nanoparticles, Hodoroaba, V.-D., Unger, W. E. S., Shard, A. G. (Eds.). Micro and Nano Technologies: Elsevier, 2020, 213–224.

[346] Feng, Y., Kilker, S. R., Lee, Y. Chapter Seven – Surface charge (zeta-potential) of nanoencapsulated food ingredients. In Characterization of Nanoencapsulated Food Ingredients, Vol. 4, Jafari, S. M. (Ed.). Nanoencapsulation in the Food Industry: Academic Press, 2020, 213–241.

[347] Mariam, J., Dongre, P. M., Kothari, D. C. Study of Interaction of Silver Nanoparticles with Bovine Serum Albumin Using Fluorescence Spectroscopy. *Journal of Fluorescence* 2011, 21(6), 2193, doi: 10.1007/s10895-011-0922-3.

Muhittin Kulak*, Canan Gulmez Samsa

Chapter 2
Nanoparticles and medicinal plants: a visualized analysis of the core and theme content of the reports in the period of 2018–2022

Abstract: Corresponding to the increases in global population and either anthropogenic or natural mediated environmental perturbations, the demand with respect to crop improvement and agricultural productivity necessitated and led to the critical searches. For this purpose, in the last decades, a great interest has been devoted to engineer and develop nanomaterials in multiple disciplines and accordingly, remarkable findings/improvements have been noted owing to the superior physicochemical properties of nanomaterials. Herein, the chapter comprises two major subgroups. The first one was oriented on the compilation of the literatures available through a traditional descriptive approach. The second group was based on the analysis of the relevant reports through a dimension reduction approach. For dimension reduction, we used R-studio-based Bibliometrix software. In this regard, we extracted 2,000 documents in the period 2018–2022 using "nanomaterial (s)" and "medicinal plants" within the field of "agriculture" from SCOPUS database. The retrieved documents were analyzed for country and author productivity, co-occurrences of keyword and Keyword Plus as well as thematic maps. The strongly and weakly developed topics were presented. The motor themes of the documents were related antioxidant behavior of the nanomaterials, whereas niche themes were observed as toxicity of the nanomaterials. Corresponding to the documents considered in the period of last five years, gene expression like molecular analysis was categorized in emerging and declining themes as marginal or weakly devel-

Acknowledgments: We both author declare that we have no competing interest. The Software R-studio based Bibliometrix and VOSviewer are free of uses. Each program can easily be approached. The figures we presented were automatically generated by the Software. So, any permission is not kindly required with respect to the copy-right. The relevant figures were based on the words we used for data search on SCOPUS database. We would like to express our gratitude to software developers for their great contribution in reducing a huge amount of data.

*Corresponding author: Muhittin Kulak,** Department of Herbal and Animal Production, Vocational School of Technical Sciences, Igdir University, Igdir 76000, Türkiye, e-mail: muhyttynx@gmail.com; muhittin.kulak@igdir.edu.tr
Canan Gulmez Samsa, Department of Pharmacy Services, Tuzluca Vocational High School, Igdir University, Igdir 76000, Türkiye

https://doi.org/10.1515/9781501523229-002

oped topics. On the other hand, transversal topics as basic themes were linked to the green synthesis and silver nanoparticles, which are shared in other disciplines, as well.

Keywords: antioxidant, bibliometric analysis, green synthesis, nanomaterials, nanotechnology, metallic nanoparticles, oxidative stress

2.1 Introduction

2.1.1 First part of the chapter: a descriptive approach to the nanomaterials and medicinal plants

2.1.1.1 (A)biotic stress and increasing global population: critical menaces to the productivity of plants

Due to their sessile natures, plants are constantly challenged by an array of suboptimal environmental conditions, either in abiotic or biotic nature. These environmental perturbations impose critical menaces to the growth and survival of the plants through modifications at morphological, physiological, biochemical, cellular and molecular levels [1, 2]. The impaired responses of the plants are translated into the critical losses/reductions in yield. In addition to the suboptimal conditions, significant rises in global populations have necessitated and led to the critical searches linked to the sustainable solutions for agricultural productivity through crop improvement. Considering the predictable or unpredictable climate scenarios; such crop improvement studies are considered as the *rungs of a long ladder* to meet the demands of ever-increasing population and ever-changing environmental conditions including either biotic or abiotic conditions [3]. Not only being confined to the studies of crop improvement; higher crop yield or resilience might be attained through adopting/introducing the suitable genotypes and management practices [4]. Of the management practices, attempts such as enrichment soil with organic matters [5, 6], organic/inorganic soil mulching [7], polymer-mediated soil conditioning [8], and minimum soil tillage [9, 10] are the widely adopted strategies to conserve and improve the soil structure. In addition to the buffering practices for degraded soils; a plethora of molecules (mostly phenolic nature compounds) have been assayed through either priming or foliar in plants against either stress or non-stress conditions [11–15]. In the recent decades, nano-engineered materials gained a bewildering interest in agricultural researches. The following subsections of the chapter will address definition, class, and the synthesis of nanomaterials.

2.1.1.2 Astonishing nano-engineered particles: pathway from nano-scale chemical structures to the macro-scale of plants

Nanotechnology and nanoscience are the cornerstones of the age of science and technology and have marked a significant revolution in almost every field of science. These areas are concerned with the production and applications of nanoparticles for use in a variety of applications. On the other hand, the nanoparticles and its green synthesis with plants have paved the way for the formation of nanomedicine field and are an important step for humanity and the environment. Innovative fields where nanotechnology is used are generally biomedical applications, diagnostic and health care, drug manufacturing and improvement, environmental cleanup, cosmetics, and food and agriculture industries [16, 17].

Nanomaterials are materials that have been engineered on a nanoscale, meaning they have at least one dimension that measures 100 nanometers or less. These materials have unique properties that make them different from bulk materials, and they are often used in a variety of applications, such as in electronics, energy production, medicine, and agriculture. Some common types of nanomaterials include nanoparticles, nanotubes, and nanofibers [18, 19].

There are three types in nanotechnology: wet nanotechnology associated with living organisms and their components (enzyme, hormone, tissue), dry nanotechnology associated with inorganic compounds (silicon, carbon), and computational nanotechnology associated with the simulation of nanometer-sized components [20, 21]. There are three main topics regarding the classification of nanomaterials, i.e. (i) nanomaterials based on the number of dimensions, (ii) nanomaterials based on their properties, (iii) nanomaterials based on source of origin [22]. According to the most general classification, nanoparticles are evaluated as carbon-based (fullerenes and carbon nanotubes), metallic (copper, gold, silver, etc.), lipid-based, polymeric, ceramics, and semiconductor nanoparticles [23].

Methods for nanoparticle synthesis are classified into two chief groups, i.e. (i) Bottom-up synthesis and (ii) Top-down synthesis. These two groups are categorized into three subgroups as biological, physical, and chemical. Bottom-up synthesis, bulk state metals are separated into much smaller particles and thus metal nanoparticles of desired dimensions are obtained. This approach includes some physical techniques such as pyrolysis, laser ablation grinding, physical vapor deposition, inert gas condensation. In the other approach, nanoparticles are formed from moderately smaller and simpler molecules, and synthesis occurs assembly small substances into nanoparticles by biological and chemical methods [22, 24].

The combination of nanotechnology and green chemistry has been key to the future of nanotechnology. The physical and chemical methods used in nanoparticle synthesis have some limitations. These methods include disadvantages such as unwanted and harmful by-products, toxicity, not being eco-friendly, and expensive and may not meet desired shape, size, quality, and quantity. In addition, nanoparticles synthesized by the traditional physical method require high temperature, pressure, and expensive tools, which involve high energy consumption [25–27].

In the last 10 years, green chemistry or biological techniques for synthesizing metal nanoparticles from plant extracts in catalysis, sensing, electronics, and photonics have drawn the attention of many researches (Figure 2.1). In particular, biological methods are very popular because they prevent the harmful effects of chemicals and energy loss, and also offer the advantage of synthesizing nanoparticles using natural reduction, capping and stabilizing agents [28]. Biological processes primarily performed via medicinal plants offer advantages over chemical and physical methods. Green synthesis of nanoparticles is emphasized to overcome or minimize the current potential risks and offers numerous advantages, including energy efficiency, simple procedure, rapid and large-scale production, low toxicity, high yields, cost effective, zero/low contamination, eco-friendly and clean analytic methodologies, and ready availability. In green synthesis, organic and natural resources are utilized and biological assets such as bacteria, algae, plants, actinomycetes, and fungi are used [28, 29].

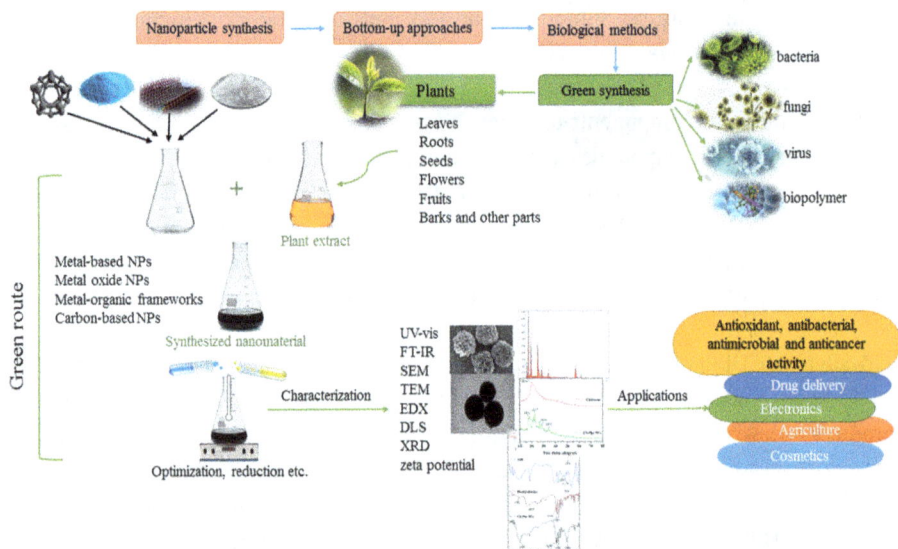

Figure 2.1: Green and chemical route of nanomaterial synthesis.

2.1.1.3 The characterization of nanomaterials

Researchers have been generally focused on the evaluating and testing of plants for the synthesis of metallic nanoparticles, especially silver, copper and gold nanoparticles, but important properties such as reducing agents, capping agents, solvents, metal salts, nucleation, growth, aggregation, stabilization, and characterization, which concern the synthesis mechanism, have attracted less attention [18, 20].

The chemical features like structure, morphology, surface energy, composition and charge, and molecular behavior of nanomaterials are elucidated by a series of analytical methods. Some of the frequently used microscopy and spectroscopy techniques are UV-Visible Spectroscopy (UV-vis), Fourier Transform Infrared Spectroscopy (FT-IR), Atomic Force Microscopy (AFM), Scanning Electron Microscopy (SEM), Transmission Electron Microscopy (TEM), Energy Dispersive Spectroscopy (EDX), Dynamic Light Scattering (DLS), Powder X-ray Diffraction (XRD), High Performance Liquid Chromatography (HPLC), Raman Spectroscopy, and zeta potential (ζ pot) (Table 2.1) [30].

Table 2.1: Structure, source, characterization techniques, and morphological properties of some metallic nanomaterials.

Nanomaterial	Natural source	Characterization	Size (nm)	Shapes	Reference
CuO NPs/NCs	*Duchsnea indica* fruits	UV–vis, FT-IR, XRD, EDX and TEM	2–31	Cubic, crystalline	Bhatia et al. [40]
Ag NPs	*Lonicera japonica* leaves	UV–vis, XRD, SEM and TEM	50 ± 5	Cubic, crystalline	Rajivgandhi et al. [41]
AgNPs	*H. perforatum L.* (aerial parts)	UV–vis, DLS, ζ-potential, FTIR, XRD, SEM, EDXS, and TEM	20–40	Spherical	Alahmad et al. [42]
AgNPs	*Eugenia roxburghii* leaves	UV–vis, XRD, HR-TEM, and ζ-potential	19–39	Crystalline	Giri et al. [43]
AuNPs	Parkia biglobosa leaf	UV–vis, XRD, FT-IR and TEM	1–35	Truncated, pentagonal, spherical, triangular	Davids et al. [44]
AuNP	*Cinnamon* bark	UV–vis and TEM	35	Spherical	ElMitwalli et al. [45]
ZnNPs	*Cayratia pedata* leaves	UV–vis, FESEM, XRD, FT-IR and EDX	52.24	Spherical	Jayachandran et al. [46]
ZnNPs	*Eucalyptus globules* leaves	UV-Vis, FT-IR and TEM	52–70	Spherical, elongated	Ahmad et al. [47]

Table 2.1 (continued)

Nanomaterial	Natural source	Characterization	Size (nm)	Shapes	Reference
TiO$_2$ NPs	*Pouteria campechiana* leaves	UV–vis, EDX, XRD, FT-IR and SEM	73–140	Spherical	Narayanan et al. [48]
TiO$_2$ NPs	*Syzygium cumini* leaves	HR-TEM, HR-SEM, DLS, BET, XRD, FTIR and EDX	22	Spherical round	Sethy et al. [49]
PdNP	*Coleus amboinicus* leaves	UV–vis, XRD, FT-IR and SEM	40–50	Spherical	Bathula et al. [50]
Pt NPs, Pd NPs, and Pt–Pd NPs	*Peganum harmala* L. seeds	UV–vis, ζ-potential, UV–vis, XRD, FT-IR and TEM	20.3 ± 1.9, 22.5 ± 5.7 and 33.5 ± 5.4	Spherical	Fahmy et al. [51]

UV-vis spectroscopy can easily show the formation of nanoparticles and is a critical tool for assessing nanoparticle growth and stability. As the basic structure clarification tool, UV-vis provides information about the structure, size, stability, concentration, and aggregation of nanoparticles [31]. TEM is a microscopy technique in which a beam of electrons is transmitted through a sample to create an image. It is one of the most popular and effective methods for characterizing the size, shape, and density of nanoparticles. The technique inserts images that allow to understand the morphological and chemical status of the material. SEM is an electron microscope that determines the size, structure, and surface fractures of the nanoparticle. It produces images of a sample by scanning the surface with a focused beam of electrons. These electrons interact with atoms to produce various signals containing information about the surface topography and composition of the sample [30, 32]. XRD is based on the principle that each crystal refracts X-rays in a characteristic pattern, depending on the unique atomic arrangement of the phase. This technique is considered as an important key instrument due to its function of determining the size and crystallinity of the synthesized nanoparticles [33]. EDX is an analytical technique for elemental analysis or chemical characterization of a sample and is based on the interaction of a sample with X-ray excitation sources. Each element has a unique, distinctive, and characteristic atomic structure in the electromagnetic emission spectrum, making it a frequently applied technique in nanoparticle characterization [34]. For example, Au-nanoparticles synthesized using Licorice root extract [35] or Pt-nanoparticles from Anbara fruits [36] have been confirmed Pt and Au atoms with optical signals in EDX analysis. A widely used technique, called FTIR, is used to obtain the infrared absorption or emission spectrum of a solid, liquid, or gas. FTIR measures, over a wide spectral range and simultaneously, the absorption of radiation

(IR) by each bond in the molecule, giving a spectrum commonly known as percent transmittance versus wavenumber [37]. The IR region between 4,000 and 670 cm^{-1} provides information about the functional groups, side chains, and cross-links of nanoparticles and thus helps to elucidate the chemical structure and stabilization mechanism of them. Usually, O–H, C = O, and C–N are ascribed to the peaks in nanoparticle synthesis using natural products [38]. Zeta potential is the electro kinetic potential at the surface of a particle. It is a commonly used technique for nanometer-sized liquids. It is widely known that particles dispersions with a ζ pot < −30 mV or > +30 mV are stable dispersions. Zeta potential is a measurable indicator of the stability of colloidal dispersions and investigates using a zeta sizer nano [39].

2.1.1.4 Medicinal plants: minor royal of plant kingdom

Plant kingdom is characterized with an estimated 500,000 plant taxa. Amid the taxa, a number of plants between 50,000 and 80,000 flowering plants are considered to be used for a plethora of medicinal purposes, especially in developing countries [52]. The World Health Organization (WHO) declared that approximately 60% of the world populations and about 80% of the population in developing countries are dependent on plant-based medicine for their primary health care needs [26]. Since time immemorial, sine qua non roles of the plants have been passed from generation through either *heuristic approach* by our prehistoric ancestors or contemporary experimental evidences [53, 54]. The beneficial uses of the plants considered are crucially linked to specialized metabolites available, a great number of reports say [55–58]. In the following parts of the chapter, we will address our comments on specialized metabolites of the plant, their classifications and potential functions under suboptimal growth conditions. In addition, we will further extend our direction to the interactions of secondary metabolites and nanomaterials.

2.1.1.5 Plant specialized metabolites (plant secondary metabolites)

There are two metabolic pathways in the complex system of the plant. The distinction of those pathways is linked to their essentiality for proper metabolism of the plants. The first one is primary metabolism, which is essential for growth and development. The second one is the specialized metabolism (traditionally defined as secondary metabolism), which are not essential for such purposes [59]. Harborne (1993) [60] defines the difference as the functional rather than structural terms. In the current chapter, we will further address the main interest on the specialized metabolism. Being confined to the plants, in general; a fascinated number of specialized metabolites (approximately equivalent ~ 200,000) is synthesized via a series of metabolic pathways, but this bewildering number of chemical structures is considered to be over one million. Recent studies have revealed that 15–25% of plant genes play a role in secondary metabolite formation.

Secondary metabolites are divided into important groups as follows: (1) C5-type terpenoids (isoprenoids) from isopentenyl diphosphate (IPP) – average 35,000 structures; (2) nitrogen-containing secondary metabolites – over 12,000; (3) phenolic compounds formed by shikimate or malonate-acetate – over 8,000 structures; and (4) minor secondary metabolites – an average of 10,000 structures [61].

These metabolites are organic compounds, as primary metabolites. As noted above, these metabolites are not involved in the primary metabolic processes of growth, development, and reproduction, but a plethora of functions in plants have been attributed to secondary metabolites. The attributed functions are definitely linked to the huge structural diversity, and these "specialized" metabolites are confined to certain plant taxa, exhibiting a high degree of interspecific and intraspecific variability [59]. Being considered for their plausible functions with respect to the plant-specific adaptation to the challenging environmental conditions, decision of carbon fluxes between primary metabolism and secondary metabolism results in trade-off as an energy-drainage from growth toward those specialized metabolites for defense [62]. To reveal/understand response and subsequently improve the resilience of the plants, a very large number of researches have been addressed on accumulation of these small molecules and their antioxidative functions on reactive oxygen species (ROS) triggered by stress factors [63, 64].

2.1.1.6 Roles of novel nano-engineered particles in the war of abiotic stress and medicinal plants

The medicinal properties of the plant are linked to the secondary metabolite present. As the other side of coin, the plants synthesize and accumulate those compounds as a response against abiotic stresses. The response is, in turn, manifested as a lower accumulation of reactive oxygen species. The success of the plants is considered to be partially linked to polyphenol chemistry (i.e., oxidation, condensation). In addition to critical modifications in chemistry of the compounds, the main question links whether the accumulation of the compounds in pools of compounds or de novo synthesis of the constitutive or defense compounds (well-reviewed by [62]. In relation to the physiological and agronomic attributes, the number of the reports with respect to the stress and nanoparticle interaction is very low. For that reason, we also included the findings obtained under optimal growth conditions. For instance, silver nanoparticles (AgNPs) increased biomass accumulation and elicited total phenolics and flavonoids, as well individual compounds such as flavonols, hydroxycinnamic and hydroxybenzoic acids in *Cucumis anguria*. The increases were translated to the enhanced activities such as antioxidant, antimicrobial and anticancer [65]. The phenolic compound and then antioxidant activities of *Capsicum annuum* L. at the germination stage were critically enhanced with zinc oxide nanoparticles [66]. In a concentration and time dependent, gallic acid, cinnamic acid, and rutin increased in callus of *Dracocephalum kotschyi* hairy roots elicited by SiO_2 nanoparticles [67]. CuO NPs treated plants exhibited high content of phenolics and corresponding antioxidant activity of *Withania somnifera L.* [68]. Those reports clearly suggested that nanoparticles, as novel abiotic elicitors, induce bioactive compounds of plants through altering the expression levels of the genes responsible for secondary metabolites synthesis [67, 69–71]. Considering the stress and NPs interaction, no significant changes in total phenolic content of shoot and root as well as total flavonoid of shoot were reported in *Dracocephalum moldavica* L. treated with iron oxide NPs under salt stress conditions. Flavonoid content of root was highly accumulated [72]. Corresponding to treatment of drought stress and TiO_2 nanoparticles on Moldavian dragonhead, rosmarinic acid, luteolin 7-O-glucoside, p-cumaric acid and gentisic, ellagitannin, chlorogenic acid, caffeic acid, and geraniol were significantly responsive to the treatments [73] In the reports, the cost of tolerance level and growth *vs.* defense trade-off dilemma have been discussed but still depth-mechanistic works are solidly required to report common postulations with respect to the interactive effects of stress and nanoparticles.

2.1.2 Second part of the chapter: a reductive approach to the nanomaterials and medicinal plants

2.1.2.1 Bibliometrix-aided analysis for dimension reduction of published documents

Corresponding to dynamic nature of environmental conditions; the number of disseminated reports for the ever-changing environmental conditions increases in parallel. In any certain study area, it is possible to see the bewildering number of reputed review articles. Those reputed review articles pioneered the researchers to new avenues by pointing the way-forward as research agenda or pointing the gaps in the literature. Also, state of the art for each topic was also suggested. As of the best literature survey done by the authors of this chapter, a plethora of satisfying and remarkable descriptive review articles has been disseminated as the case of topic of the current chapter [74–77]. However; in comparison to the descriptive or narrative review articles, lower number of bibliometric analysis compilation studies are available. For that reason, we constructed the theoretical frame of section of this chapter on bibliometric analysis, aiming at collection of the documents and then following a dimension-reduction approach. In this regard, a series of analysis such as author keywords, Keyword Plus, title, and source were carried out. In brief, we aimed to reveal the core content and trending topic in the case of nanomaterials and medicinal plants. We hypothesized that the present study would be fundamental of the ahead potential related studies. To realize the analysis considered, we used R-studio based Bibliometrix, an online free-software (https://www.bibliometrix.org/home/) after optimization of running conditions of the program and construction of Bib-Tex file version of SCOPUS documents. The relevant software was already applied in a quite number of documents [78–80].

2.1.2.2 Search strategy and collection of the data

Since the topic is quite broad with an estimated 892,938 documents according to SCO-PUS database, in this chapter we need to narrow our horizon with respect to the nanomaterials and agriculture, in general and medicinal and aromatic plants, in particular. In this context, we performed a search using the following criteria as (TITLE-ABS-KEY (nanomaterial OR nanomaterials OR nanoparticle OR nanoparticles)) AND (medicinal AND plants OR medicinal AND plant) AND (LIMIT-TO (SUBJAREA, "AGRI")) AND (LIMIT-TO (PUBYEAR, 2022) OR LIMIT-TO (PUBYEAR, 2021) OR LIMIT-TO (PUBYEAR, 2020) OR LIMIT-TO (PUBYEAR, 2019) OR LIMIT-TO (PUBYEAR, 2018)), we recorded 2,000 documents and used for bibliometric analysis.

2.2 Results and comments

2.2.1 Descriptive analysis

Table 2.2 represents the main information of the extracted documents by bibliometric analysis. A total of 2,000 articles were recorded within the time span of 2018–2022 according to SCOPUS database. Bibliometrix software provides opportunity of analyzing the extracted documents of SCOPUS, Web of Science, or PubMed. However, we preferred the SCOPUS, as we did in our previous works [81–83]. The SCOPUS indexes a plethora number of journals (N = 44,034) across multiple scientific disciplines, also being considered to be of the most comprehensive and reliable bibliographic sources [84]. Present bibliometric analysis revealed that a total of 8,140 authors published in that period, reporting a total of 5,434 keywords, and producing a mean of 10.27 citations per document. Critically, annual growth rate of the documents was observed as 47.54%, suggesting us that there is a growing interest with respect to the uses of nano-engineered materials in agricultural and biological sciences to enhance/improve the crop productivity of the plants, under either optimal or suboptimal growth conditions. In the following subsections of the chapter, the relevant analysis of keywords will be core interest of our study. In this regard, we did a series of analysis including the sources and their impact, productivity of countries, co-occurrences of keywords, thematic map of keywords, word cloud of keywords/abstracts/titles, conceptual structure map, and topic dendogram of the keywords, etc.

2.2.2 Authors

Figure 2.2 depicts the top 10 authors about nanomaterials and their application in agricultural systems, as the case of medicinal plants with the most published papers. Amid the authors, "Rajeshkumar, S" takes the first place with a contribution of 34 articles,

Table 2.2: Main information of the retrieved documents.

Description	Results
Time-span	2018–2022
Sources (journals, books, etc.)	461
Documents	2,000
Annual growth rate %	47,54
Document average age	1,39
Average citation per doc	10,27
Keyword Plus	8,591
Author keywords	5,434
Authors	8,140
Authors of single-authored documents	63
Articles	1,373
Book	7
Book chapter	270
Conference paper	3
Review	341

even "Rajeshkumar, S" did not report an article after 2020 according to the analysis of *Authors' Production over Time* (Figure 2.3). The author is from India. We also analyzed the scientific production of country, observing that India topped at the list with a huge number of documents (N = 1,277 out of the 2,000 documents considered for analysis) (Figure 2.3). Then, the author was followed by two Chinese researchers (Li, Y: N = 17; Wang, Y: N = 15). Both Chinese authors kept a steady of publishing the relevant field, analysis of *Authors' Production over Time* says that (Figure 2.2). According the analysis of country scientific production, China placed at the first place, whereas Iran and Egypt ranked at third and fourth places, respectively. Considering the number of citations received, within the Top 3, India topped with 4,094 citations, followed by China ($N_{citation}$ = 4016) and Iran $N_{citation}$ = 2,998) in 2nd and 3rd positions, respectively (Figure 2.4).

Of the first 10 authors considered for analysis, 8 out of 10 was, except Ghorbanpour, M (N = 15) and Jafari, SM (N = 14), India- and China-based authors. Corresponding to the author analysis, Lotka's law was applied to the documents extracted. Figure 2.5 exhibits the article numbers to which each author has contributed. For the present analysis, 85.3% of the authors (N = 6,944) contributed with at least one document, whereas 9.3% (N = 759) with at least two documents. To be more specific, the authors were assessed for their productivity/impact through their H-index. H-index is linked to the citation number received in other researches. Higher H-index value might be a significant indicator of originality of authors' studies. However, self-citation might be critical disadvantage in this assessment. Exclusion of self-citation might provide real assessment. In Figure 2.6, different groups of researchers were observed than Figure 2.2, which were related to the number of documents generated by the researchers. "Jafari, SM" ranked at the first with its H-index (11), then followed by "Li, Y (H-index = 8)" and "Wang, L (H-index = 8)," respectively. Such an observed variation in generated documents and H-indexes of authors might be linked to the journals (for instance; local impact and access options).

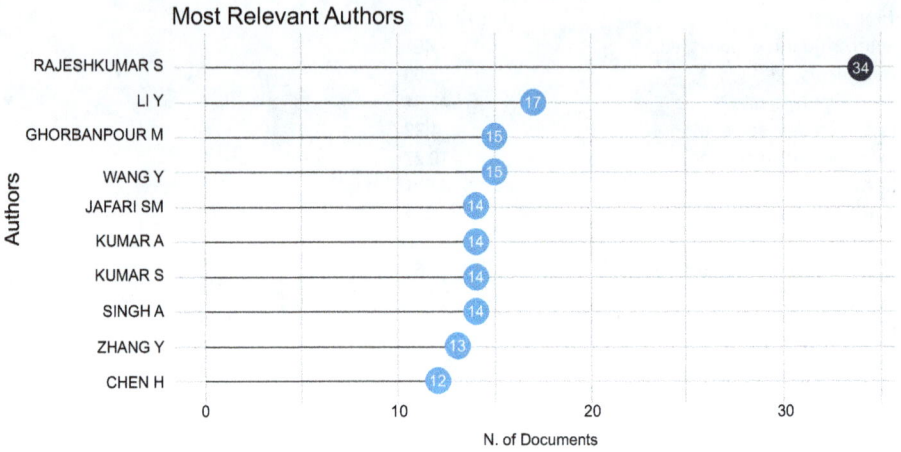

Figure 2.2: Most relevant authors.

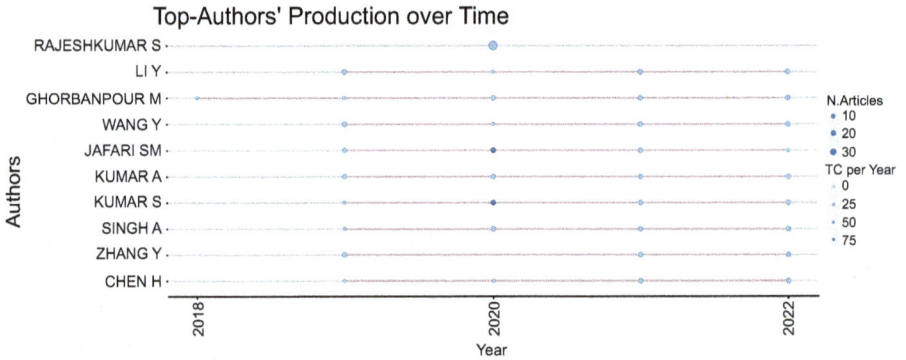

Figure 2.3: Authors' production over time.

2.2.3 Corresponding author's country

Figure 2.7 shows the countries of the corresponding authors. That analysis also provides to get insight linked to the number of publications carried out individually or in a collaboration manner. The relevant collaboration assessment is based on the Intra-country (SCP) and inter-country (MCP) collaboration during period of 2018–2022. The figure depicts the first 20 countries. India ranks at the first by leading the largest number of documents with a total of 462, which includes 375 SCP and 87 MCP, followed in

Most Cited Countries

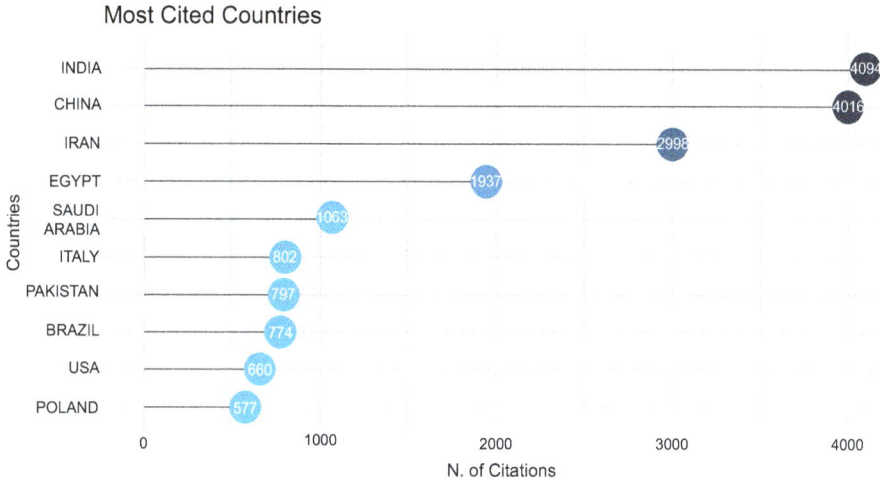

Figure 2.4: Most cited countries.

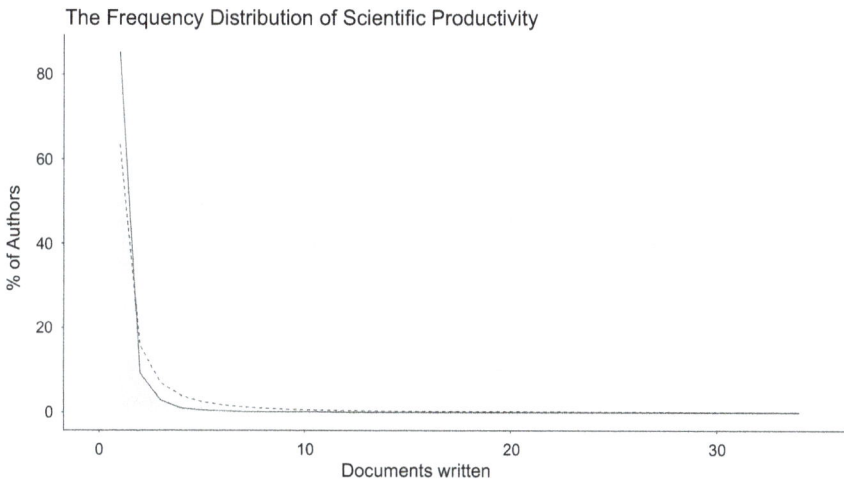

Figure 2.5: Author productivity through Lotka's Law.

the 2nd position by Iran with 237 documents (SCP:192; MCP:45) and China ranks the 3rd position with 217 documents (SCP: 217; MCP: 68). Considering the MCP ratio, the top three countries are Saudi Arabia (MCP$_{ratio}$: 0.654; document: 81; SCP: 28; MCP:53), Korea (MCP$_{ratio}$: 0.633; document: 30; SCP: 11; MCP:19) and Spain (MCP$_{ratio}$: 0.480; document: 25; SCP: 13; MCP:12). Amid the first 20 countries considered for analysis, Turkey has the lowest MCP$_{ratio}$: 0.029, with a total of 34 documents (SCP:33; MCP:1), suggesting that Turkey-based researchers did not have much international collaborations during the period of 2018–2022.

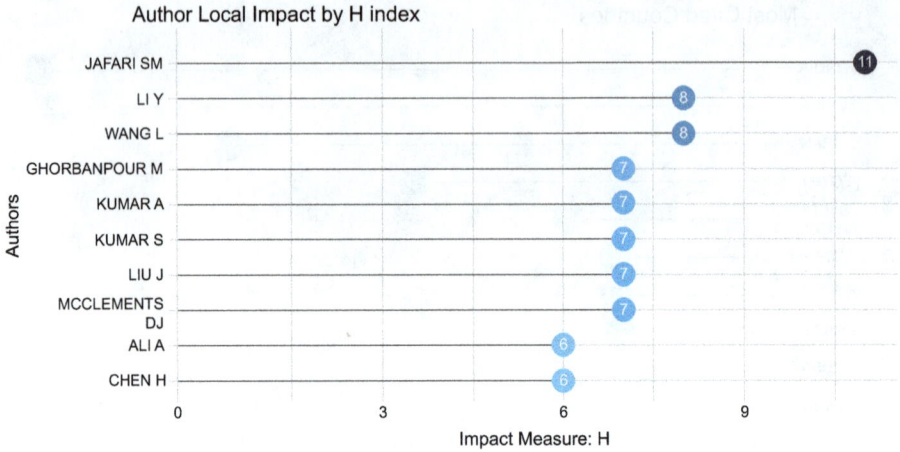

Figure 2.6: Author local impact.

2.2.4 Sources

Figure 2.8 depicts the most relevant publication sources. The source "Nanotechnology in the Life Sciences" topped at the list with a 115-number of documents. The source is a compilation of book series (Springer). Then, the list is followed by peer-reviewed journals as "Plants (N = 66; MDPI)," "Biocatalysis and Agricultural Biotechnology (N = 62; Elsevier)," "Saudi Journal of Biological Sciences (N = 52; Elsevier)", "Food Chemistry (N = 51; Elsevier)," etc. As we previously noted in descriptive analysis (Table 2.2), the number of sources was 461. We furthermore clustered those journals using Bradford's Law to reveal the core sources (Figure 2.9). Corresponding to the 461 sources, we obtained three zones. The first zone (cumulative frequency: 661) comprised 15 sources. Figure 2.9 shows the 10 out of 15 sources and the rest five sources were as "South African Journal of Botany, Plant Archives, Trends in Food Science and Technology, Nutrients, Plant Cell, Tissue and Organ Culture, and International Journal of Environmental Analytical Chemistry." The second zone (cumulative frequency: 1,341) and third zone (cumulative frequency: 2,000) included 71 and 375 sources, respectively.

2.2.5 Source local impact

Also, we analyzed the local impact of sources. Figure 2.10 shows the source local impact by H-index. Food Chemistry ranked at the first with a H index of 25 (Table 2.3). The second and third ranks were occupied by "Biocatalysis and Agricultural Biotechnology (H-index: 22)" and "Saudi Journal of Biological Sciences (H-index: 22)" respectively. As noted by Virú-Vásquez et al. [85], such high H-index might be cause of consideration of

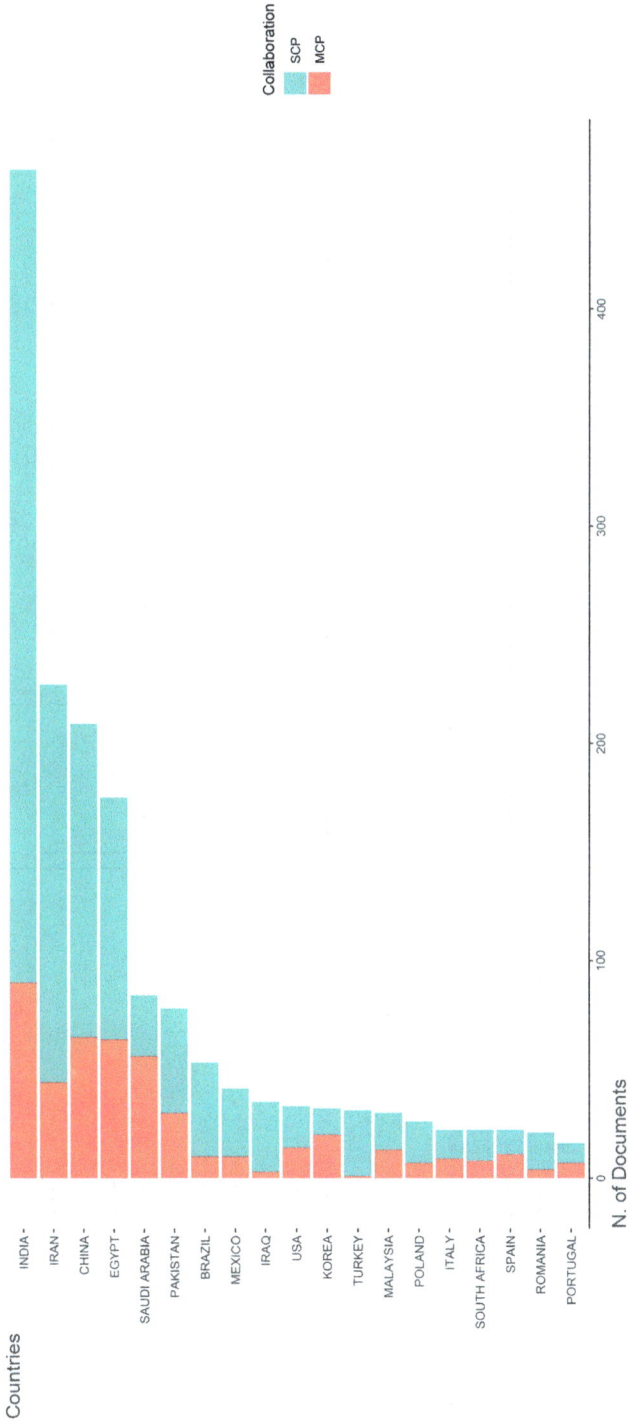

Figure 2.7: Corresponding author's country.

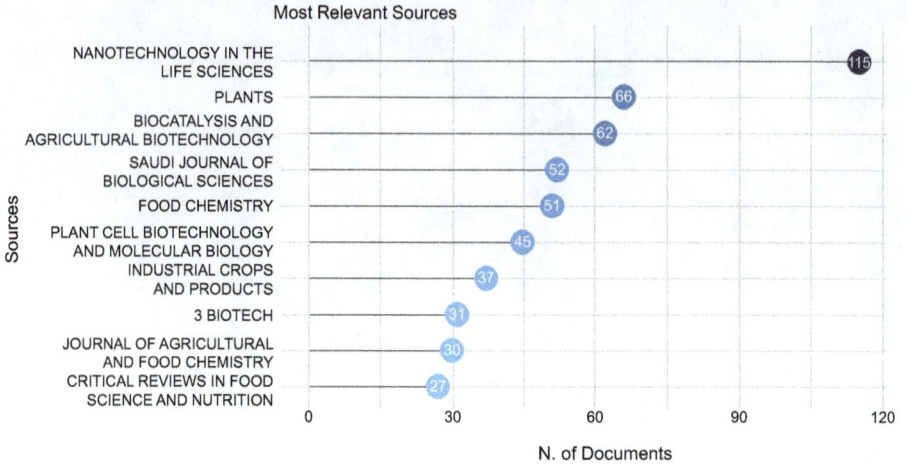

Figure 2.8: Most relevant sources.

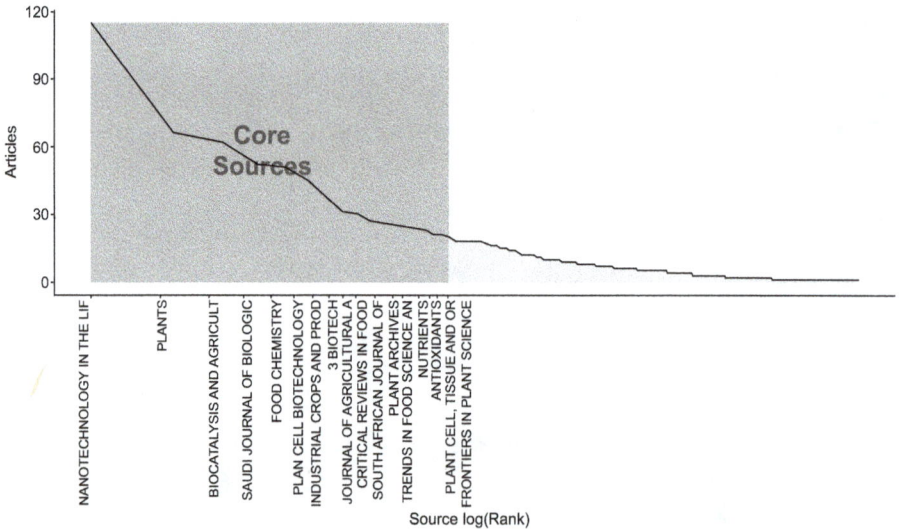

Figure 2.9: Source clustering through Bradford's Law.

authors to increase the visibility of their researches. As we noted above, Bradford's law also suggested the leading/core journals (Figure 2.9).

2.2.6 Word cloud analysis of the retrieved documents

We have further analyzed the data of the documents to extract the WordCloud. In this context, we have generated four WordClouds using Author Keywords, Keyword Plus,

Source Local Impact by H index

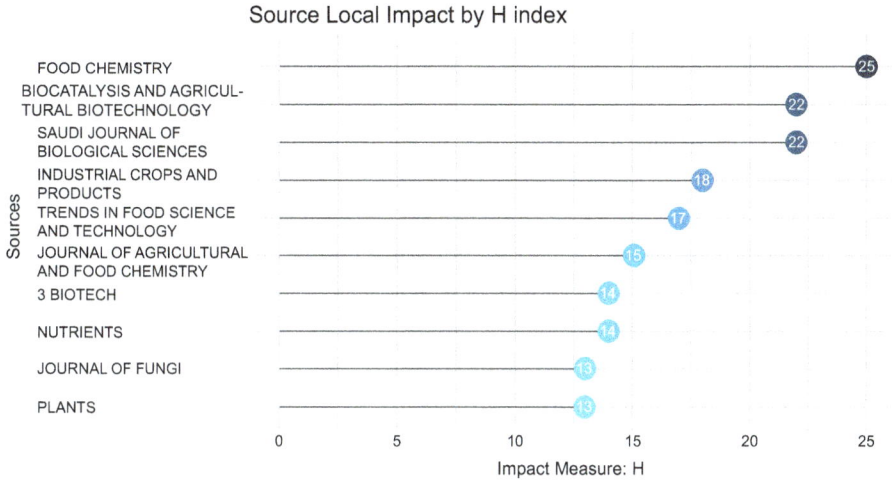

Figure 2.10: Source local impact.

Table 2.3: Source local impact.

Source	H-index	G-index	M-index	TC	NP
Food Chemistry	23	33	4.600	1,264	53
Biocatalysis and Agricultural Biotechnology	18	30	3.600	1,034	62
Saudi Journal of Biological Sciences	18	26	4.500	831	52
Trends in Food Science and Technology	17	25	3.400	1,281	25
Journal of Agricultural and Food Chemistry	16	26	3.200	691	35
Industrial Crops and Products	15	23	3.000	579	37
Food Hydrocolloids	13	18	2.600	559	18
3 Biotech	12	24	2.400	588	32
Nutrients	12	23	2.400	559	23
Critical Reviews in Food Science and Nutrition	11	25	2.200	668	29

H-index: "*The h-index is based on a list of publications ranked in descending order by the Times Cited,*" Clarivate Web of Science defines. **G-index**: Egghe [86] defines it as '*the largest number such that the top 'g' articles received together at least g^2 citations*' [87]. **M-index**: h-index per year since the first publications (https://pitt.libguides.com/bibliometricIndicators/AuthorMetrics); **TC**: Total citation; **NP**: Number of publications.

Titles, and Abstracts. For each analysis, the most frequently used 50 words were used for the construction of WordCloud. Nanoparticles (N = 331), silver nanoparticles (N = 186), green synthesis (N = 149) and antioxidant (N = 89) were of the leading words of Author Keywords, refer to Figure 2.11 Keyword Plus words were as nanoparticle (N = 339), nanoparticles (N = 279), non-human (N = 214), green synthesis-related words (plant extracts) and in vitro–related words, refer to Figure 2.12 as we also observed in thematic map of Keyword Plus. This WordCloud fits to the core theme of the current research.

Considering title WordCloud, as expected, general words such as nanoparticles (N = 3,759), silver (N = 367), activity (N = 232), synthesis (N = 226), extract (N = 196), and green (N = 196) were of the more pronounced words in the title of the documents considered for analysis; refer to Figure 2.13 on the other hand, Nanoparticles (N = 3,759), NPs (refer to Nanoparticles, N = 1,547), activity (N = 1,537), plant (N = 1,475),

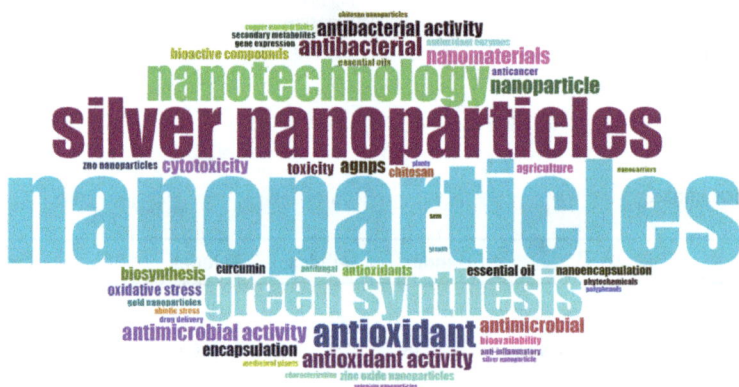

Figure 2.11: WordCloud of author's keyword (number of words: 50).

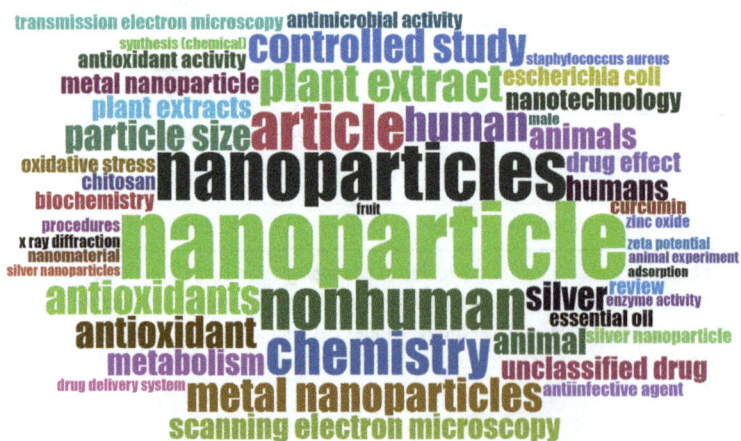

Figure 2.12: WordCloud of Keyword Plus (number of words: 50).

study (N = 1,298), AgNPs (N = 1,214), and other > 1,000 word (plants, extracts, silver) were of the frequently used words in Abstracts of the documents, refer to Figure 2.14.

Figure 2.13: WordCloud of titles (number of words: 50).

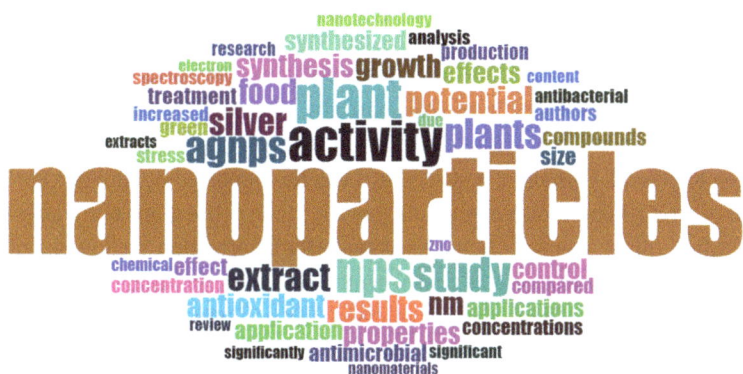

Figure 2.14: WordCloud of abstract (number of words: 50).

2.2.7 Co-occurrence network

The analysis of co-occurrences of the words are crucial in such bibliometric analysis [85, 88], suggesting that the high level of co-occurrence between the words reflects the higher level of relationship between the words. In addition to the common analysis of authors' keyword analysis, co-occurrence network of key plus and titles were constructed. As solidly requested by very many journals, the words used title and keyword should be different in order to disseminate the publication to more audiences. The analysis of the co-occurrence of keyword is shown in Figure 2.15, reporting the words with the highest co-occurrence are silver nanoparticles and green synthesis. Since the word "nanoparticle (s)" is quite common and corresponds to the wide fields, we excluded from the specific discussion. Instead, we further addressed our comments on silver nanoparticles and green synthesis or green synthesis of silver nanoparticles. The superior properties of silver nanoparticles endorsed them to have broad spectra of applications, especially as antimicrobial agents [89–91]. The words of Figure 2.15 clearly confirm the idea with various co-occurrences of "antimicrobial," "antibacterial activity," "antibacterial," "antifungal activity," and "antifungal." Not only being restricted to antimicrobial activities, the relevant nanoparticles are also strongly linked to the antioxidant activities [92], as also clearly seen from Figure 2.15 As of the highest co-occurrence, green synthesis of silver nanoparticles has been issued in an array of multiple original [93, 94] and review articles [74, 95, 96]. In the last decades, green synthesis has gained considerable interest due to the critical advantages such as low energy consumption and being noxious, being based on microorganisms (bacteria, yeast, mold, and algae) and plant extracts [89, 95]. In addition, even Figure 2.15 depicts two major cluster, both clusters do not include discriminative words for clear clustering. That can be explained with the number/frequency of items corresponding to the documents considered for the analysis. However, the co-occurrences of Keyword Plus (Figure 2.16) presented two major clusters with clear and sharp clustering. As well-reported by [97]; author keywords, the traditional ones, are provided by the authors of the relevant report, whereas Keywords Plus are retrieved/extracted from the titles of cited references by Thomson Reuters. Keywords Plus are the frequently used words or phrases which are automatically generated by a computer algorithm, providing greater depth and variety linked to the content of an article (by citing: Garfield, 1990). The Keywords Plus aided analysis (also will be supported by multiple correspondence analysis (MCA) and topic dendogram analysis in the subsections of the present chapter) identified the clusters as (I) synthesis and characterization techniques of nanoparticles and (II) the applications of the novel-synthesized particles. Interestingly; in comparison to the former two co-occurrences analysis, title analysis clearly fits to the thematic content of the current chapter (Figure 2.16) since we organized the chapter on application of nanomaterials on medicinal plants subjected to the either optimal or suboptimal growing conditions. The related Figure 2.17 presents words such as stress, bioactive metabolites, nanotechnology, and related characterization/synthesis methods.

2.2.8 Research hotspots

Conceptual structure of nanomaterials uses in medicinal plants is depicted in two thematic maps (I: Authors' keywords (Figure 2.18); II: Keyword Plus. The construction of the maps was based on keywords with minimum word frequency/occurrences of 250, providing depth insights in understanding the core content of nanomaterials and medicinal plants cases. The thematic map quadrants are identified by the co-occurrences networks. Di Cosmo et al. [98] clearly present the bases of the thematic maps, indicating that four themes are visualized through two dimensions (i.e., density and centrality). *Density* is shown on the Y-axis, being expressed as the "internal strength of cluster of a network" [98] or as "internal ties between the keywords used to define the research theme" [99, 100]. On the other hand, "*Centrality*" is symbolized at X-axis, being simply defined as the "interaction degree of a cluster in comparison to the other clusters" [98]. Xiao et al. [99] and Mumu et al. [100] specify the "*Centrality*" as "external tie strength to the other themes

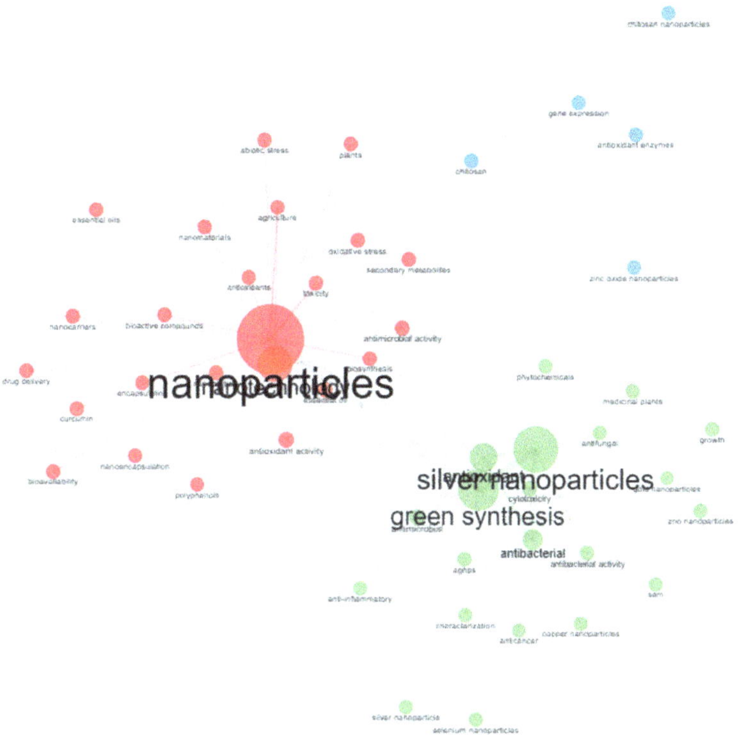

Figure 2.15: Co-occurrence network of author's keywords.

through keyword fields. To be clearer, *"Density"* indicates the "theme's development, whereas *"Centrality"* presents the information with respect to the importance of a theme [98]. Figure 2.18 show four quadrants (I: Motor Themes; II: Niche Themes; III: Emerging or Declining Themes; IV: Basic Themes). Briefly, "Motor Themes" identifies the well-developed and crucial themes in order to structure a research field. "Niche Themes" expresses the themes of the limited significance for the relevant field. "Emerging or Declining Themes" indicates the weakly developed or marginal themes for the research field. "Basic Themes" identifies the general topics which are transversal to the different research areas [98]. We herein constructed two thematic maps (I: Authors' Keyword; II: Keyword Plus). Considering thematic map of authors' keywords, motor themes were characterized with the "nanoparticles," "nanotechnology," and "antioxidant activity." As clearly documented, nanoparticles, in a concentration manner, improve/enhance antioxidant capacity of the plants, which are then translated to higher tolerance range of the plants suffering from stress conditions [101–103]. When we have a look at the general content of the studies, most of the documents are linked to the improvements in the oxidative status of the plants, fluorescence parameters, and agronomic traits of the plants, as can be supported from Keyword analysis constructed using VOSviewer software (Figure 2.19). As expected, the topics/themes with limited significances are located

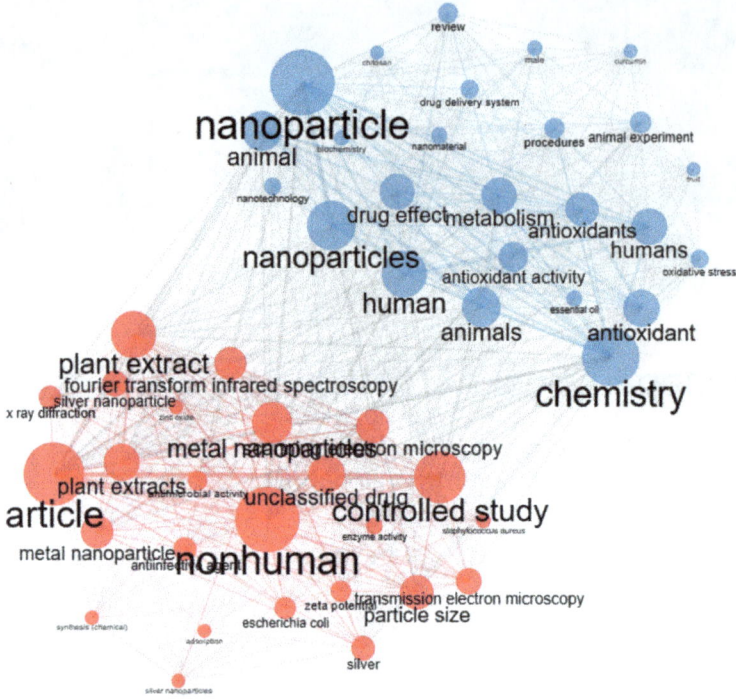

Figure 2.16: Co-occurrence network of Keyword Plus.

in "Niche Themes." The words are not directly linked to nanoparticle–agriculture interactions. An array of work concerned with the uses of nanoparticles and their fate on the
ecosystem and other living organisms have been documented [104, 105]. However, the
main issue of the nanoparticles is addressed on the *"eu-stress* effect," in a concentration
manner, which triggers/alerts the system of the plants, either as priming agent or post-
stress effects. In third quadrant (Emerging or Declining Themes), we observed the weakly
developed topics such as molecular analysis. We should note that weak-development
does not reflect the significance of the topic. As can be clearly deduced from the co-
occurrences analysis, most of the reports have been addressed on the physiological and
biochemical levels. For that reason, topics such as expression level analysis might be categorized as "isolated or marginal" topics. However, the categorization of those quadrants
might change by the time and research notes, as the case reported by Di Cosmo et al.
[98]. The last quadrant, basis themes, is as expected corresponding to the co-occurrences
networks. This quadrant is accompanied with two clusters such I: capsulation and II:
silver nanoparticles, green synthesis. As we also noted above, silver nanoparticles have
wide applications against pathogenic bodies but also have potential effects on plant systems. Those discriminations categorize the silver nanoparticles and green synthesis in
the basis and transversal themes.

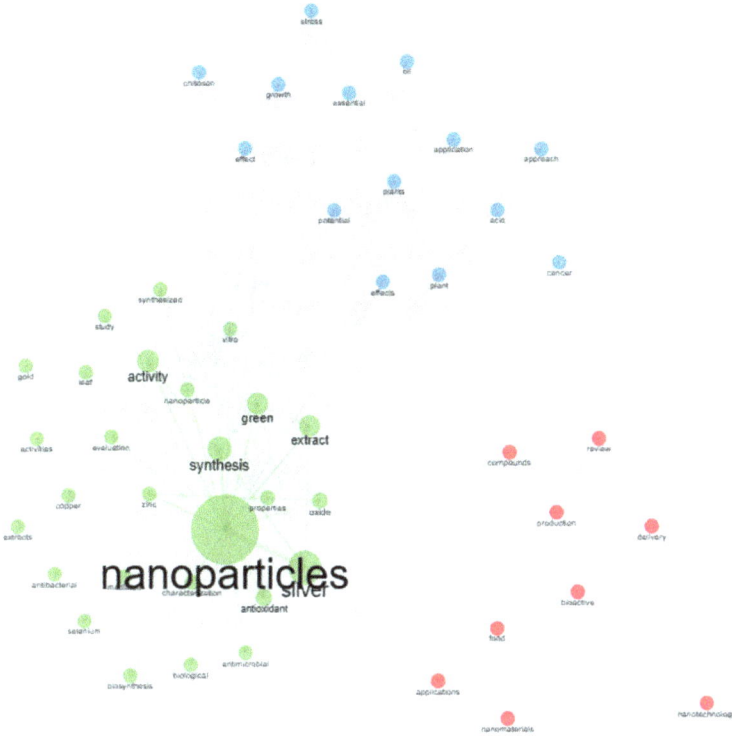

Figure 2.17: Co-occurrence network of titles.

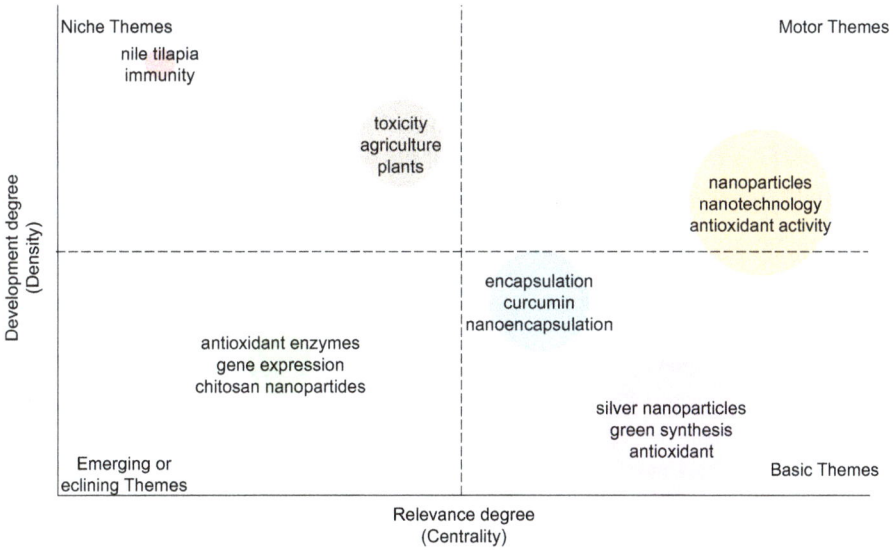

Figure 2.18: Thematic map of authors' keyword.

Figure 2.19: Author keyword analysis generated by VOSviewer software.

2.2.9 Thematic map of Keyword Plus

Figure 2.20 depicts the thematic map of Keyword Plus. We did not observe the "Niche Themes" or "Basis and transversal themes" according to the words extracted from Keyword Plus. Thematic map is based on the strongly related and weakly related topics. The Motor Theme comprised green synthesis of the particles and their application on nonhuman entities. However, the other quadrant is linked to metal nanoparticle synthesis and their applications in rat-model studies. The Keyword Plus analysis powerfully and successfully reveals the trend/direction of the studies. Recent reports have been already addressed on environmental-friendly approach, such as green synthesis.

2.2.10 Co-word analysis

A multiple correspondence analysis (MCA) was carried out to reveal the conceptual structure of the present study. Figure 2.21 depicts two clusters. The words for the *blue* cluster include *silver nanoparticle synthesis, metal nanoparticle, Fourier transform infrared spectroscopy, scanning electron microscopy, transmission electron microscopy, zeta potential,* and *X-Ray diffraction*. Such cumulative words can be simply described as synthesis and characterization techniques of the materials. This cluster also included in vitro/in vivo studies of antioxidant activity and enzyme activities and antimicrobial activities against

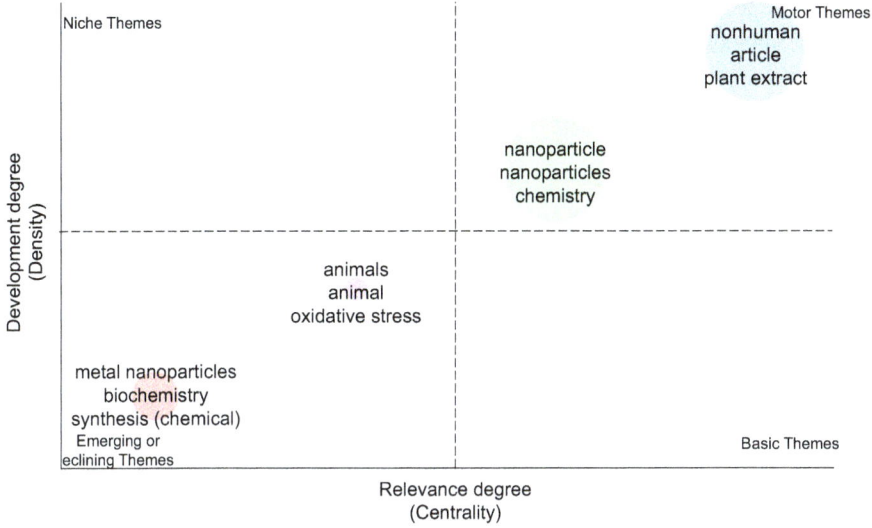

Figure 2.20: Thematic map of Keyword Plus.

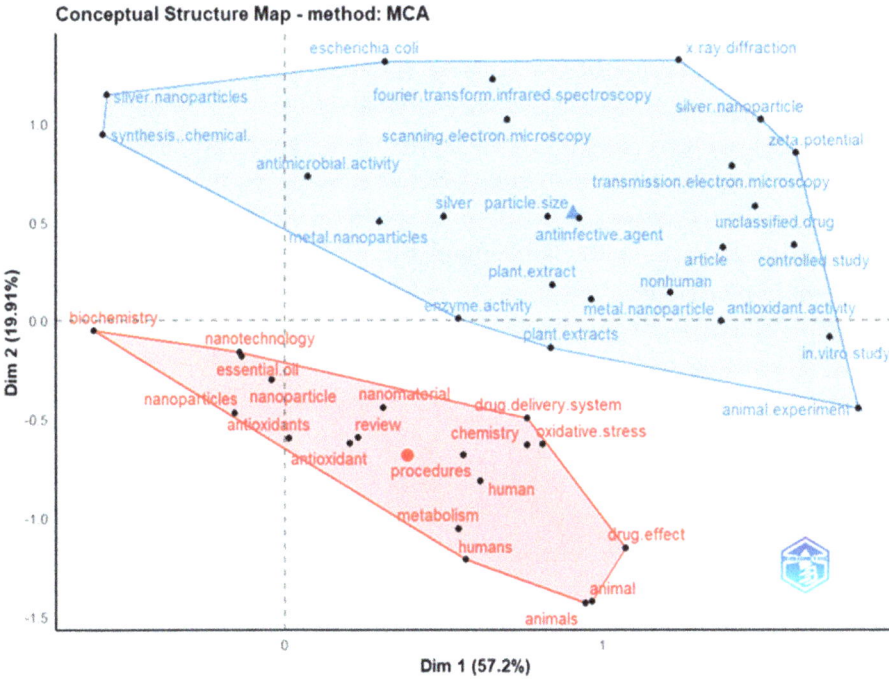

Figure 2.21: Conceptual structure of the documents.

microorganisms such as *Escherichia coli*. The words of red cluster were as the following: human (s), animal (s), nanomaterial (s), oxidative stress, drug delivery system, essential oil, metabolism, and drug effect. This cluster might be considered/deduced as biochemical responses of organisms. We also know that the nanomaterial behaves as "antioxidant" agents, which cause critical modifications in secondary metabolism of the plants via burst of reactive oxygen species [106]. Both clusters cumulatively explained the 77.11% of the total variation of the research topic considered for analysis. We furthermore constructed a topic dendogram to understand the linkages/closeness of the words (Figure 2.22). Topic dendogram analysis provided us how words are classified and how close to the other words. As being complementary analysis to MCA, two major clusters (also observed in MCA) with multiple subclusters were observed.

Topic Dendrogram

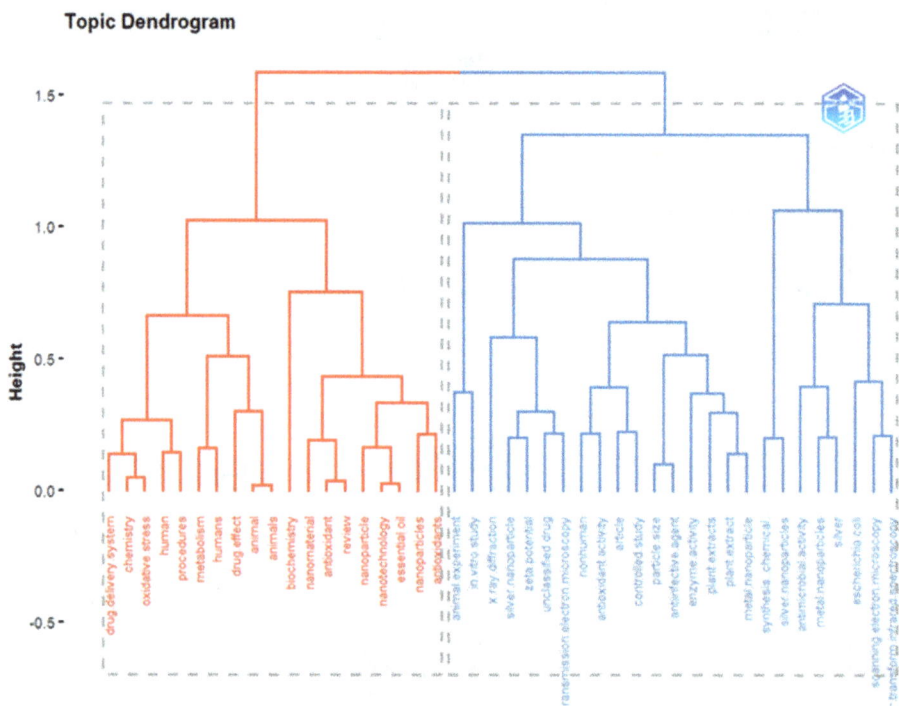

Figure 2.22: Topic dendogram of the documents.

2.3 Highlights and limitations of the present study

Within the scope of the nano-engineered materials, as we previously noted that there are about 892,938 documents in the SCOPUS-indexed journals. We, for the first time, analyzed 2,000 documents in the fields of agricultural and biological sciences, as the case of medicinal plants during the last five years. For the analysis, we applied Bibliometrix software for reducing the dimension and we critically categorized the main theme of the topics. However, limitation of the study is about not to uses of the documents in non-SCOPUS indexed journals. For that reason, we cannot strongly deduce or suggest postulations. Our comments are linked to the retrieved/extracted terms from the documents in SCOPUS indexed journals.

2.4 Conclusion

The present chapter is addressed on the nanomaterial uses in agricultural and biological sciences. The special comments are linked to the medicinal plants. The medicinal plants are of the special plants owing to their specialized plant metabolites, being approximately 20% of the plant kingdom in terms of species number. In relation to the ancient and iconic plant species, namely, rice, wheat, corn; medicinal plants have been less interest of scientific audience in plant stress physiology researches till recent years. The findings of the present bibliometric analysis revealed that either green synthesize nanomaterials or conventionally synthesized nanomaterials might be significant candidates in order to enhance the yield of plants or buffer the stress damage on the plants, but the effects are dependent on numerous factors such as concentration, mode of application, time of application, and plant species. In general, the improvements are directly linked to oxidative status of the plants and related to enzymatic and nonenzymatic components of the plants. The antioxidant activity or enhanced activities of the plant upon the nanomaterials were here clearly observed with the help co-occurrence, WordCloud, and thematic map of the documents. Considering the specialized metabolites, nanomaterials have been revealed to be elicitor of phenolic or flavonoids. Significantly, in many field of plant physiology, the molecular-based analysis has been the main concern of the researchers, supporting the well-known postulations of physiology and biochemistry of the plants. However, thematic map of the documents suggested us that the molecular analysis for nanomaterial and medicinal plant interaction is the weakly developed and marginal topics, whereas basic physiological and biochemical measurements are of the motor themes (strongly developed) in the relevant documents. Those findings clearly suggest that the reports are at their infancy, in case of medicinal plants. In order to step further, molecular analysis and metabolomics analysis are strongly required. However, the question how nanomaterials imprint the transgenerational memory and subsequently growth performance of the plants still remains unknown.

References

[1]   AbdElgawad, H., Zinta, G., Hegab, M. M., Pandey, R., Asard, H., Abuelsoud, W. High salinity induces different oxidative stress and antioxidant responses in maize seedlings organs. *Frontiers Plant Science* 2016, 7, 276.

[2]   Saini, S., Kaur, N., Pati, P. K. Reactive oxygen species dynamics in roots of salt sensitive and salt tolerant cultivars of rice. *Analytical Biochemistry* 2018, 550, 99–108.

[3]   Jhala, Y. K., Panpatte, D. G., Adetunji, C. O., Vyas, R. V., Shelat, H. N. Management of biotic and abiotic stress affecting agricultural productivity using beneficial microorganisms isolated from higher altitude agro-ecosystems: A remedy for sustainable agriculture. In Microbiological Advancements for Higher Altitude Agro-Ecosystems & Sustainability, Goel, R., Soni, R., Suyal, D., Eds. Singapore, Springer, 2020, 113–134.

[4]   Mariani, L., Ferrante, A. Agronomic management for enhancing plant tolerance to abiotic stresses-drought, salinity, hypoxia, and lodging. *Horticulturae* 2017, 3(4), 52.

[5]   Hafez, M., Abdallah, A. M., Mohamed, A. E., Rashad, M. Influence of environmental-friendly bio-organic ameliorants on abiotic stress to sustainable agriculture in arid regions: A long term greenhouse study in northwestern Egypt. *Journal of King Saud University* 2022, 34(6), 102212.

[6]   Kumari, R., Bhatnagar, S., Mehla, N., Vashistha, A. Potential of organic amendments (AM fungi, PGPR, vermicompost and seaweeds) in combating salt stress–a review. *Plant Stress* 2022, 100111.

[7]   Sekeroglu, N., Cimen, G., Kulak, M., Gezici, S. Plastic mulching or conventional cultivation of lavender flower: What influence on the yield, essential oil and their neuroprotective effects?. *Trakya University Journal of Natural Sciences* 2022, 23(1), 43–52.

[8]   Oladosu, Y., Rafii, M. Y., Arolu, F. Superabsorbent polymer hydrogels for sustainable agriculture: A review. *Horticulturae* 2022, 8(7), 605.

[9]   Musafiri, C. M., Kiboi, M., Macharia, J., et al. Does the adoption of minimum tillage improve sorghum yield among smallholders in Kenya? A counterfactual analysis. *Soil and Tillage Research* 2022, 223, 105473.

[10]   De Mastro, F., Brunetti, G., Traversa, A., Blagodatskaya, E. Fertilization promotes microbial growth and minimum tillage increases nutrient-acquiring enzyme activities in a semiarid agro-ecosystem. *Applied Soil Ecology* 2022, 177, 104529.

[11]   Rajora, N., Vats, S., Raturi, G. Seed priming with melatonin: A promising approach to combat abiotic stress in plants. *Plant Stress* 2022, 4, 100071.

[12]   Sen, A., Johnson, R., Puthur, J. T. Seed priming: A cost-effective strategy to impart abiotic stress tolerance. In Plant Performance Under Environmental Stress, Husen, A., Ed. Cham, Springer, 2021, 459–480.

[13]   Liu, X., Quan, W., Bartels, D. Stress memory responses and seed priming correlate with drought tolerance in plants: An overview. *Planta* 2022, 255(2), 1–14.

[14]   Pasala, R., Kulasekaran, R., Pandey, B. B. Recent advances in micronutrient foliar spray for enhancing crop productivity and managing abiotic stress tolerance. In Plant Nutrition and Food Security in the Era of Climate Change, Kumar, V., Kumar Srivastava, A., Suprasana, P., Eds. London, United Kingdom, Academic Press, 2022, 377–398.

[15]   Hamani, A. K. M., Li, S., Chen, J., et al. Linking exogenous foliar application of glycine betaine and stomatal characteristics with salinity stress tolerance in cotton (*Gossypium hirsutum* L.) seedlings. *BMC Plant Biology* 2021, 21(1), 1–12.

[16]   Chandra, H., Kumari, P., Bontempi, E., Yadav, S. Medicinal plants: Treasure trove for green synthesis of metallic nanoparticles and their biomedical applications. *Biocatalysis & Agricultural Biotechnology* 2020, 24, 101518.

[17]   Torrens, F., Haghi, A. K., Chakraborty, T. Chemical Nanoscience and Nanotechnology: New Materials and Modern Techniques, 1st ed. Palm Bay, Florida, USA, Apple Academic Press, 2019.

[18]   Khan, W. S., Hamadneh, N. N., Khan, W. A. Polymer nanocomposites–synthesis techniques, classification and properties. *Nanotechnology Science and Applications* 2016, 50.
[19]   Singh, A., Dubey, S., Dubey, H. K. Nanotechnology: The future engineering. *Nanotechnology* 2019, 6(2), 230–233.
[20]   Khan, F., Shariq, M., Asif, M., Siddiqui, M. A., Malan, P., Ahmad, F. Green nanotechnology: Plant-mediated nanoparticle synthesis and application. *Nanomaterials* 2022, 12(4), 673.
[21]   Sun, Y., Cheng, S., Lu, W., Wang, Y., Zhang, P., Yao, Q. Electrospun fibers and their application in drug controlled release, biological dressings, tissue repair, and enzyme immobilization. *RSC Advances* 2019, 9(44), 25712–25729.
[22]   Rizwan, M., Shoukat, A., Ayub, A., Razzaq, B., Tahir, M. B. Types and classification of nanomaterials. In Nanomaterials: Synthesis, Characterization, Hazards and Safety, Tahir, M., Sagir, M., Asiri, A., Eds. Alpharetta, USA, Elsevier, 2021, 31–54.
[23]   Bharathi, D., Diviya Josebin, M., Vasantharaj, S., Bhuvaneshwari, V. Biosynthesis of silver nanoparticles using stem bark extracts of *Diospyros montana* and their antioxidant and antibacterial activities. *Journal of Nanostructure in Chemistry* 2018, 8(1), 83–92.
[24]   Abid, N., Khan, A. M., Shujait, S., et al. Synthesis of nanomaterials using various top-down and bottom-up approaches, influencing factors, advantages, and disadvantages: A review. *Advances in Colloid Interface Science* 2021, 102597.
[25]   Alharbi, N. S., Alsubhi, N. S., Felimban, A. I. Green synthesis of silver nanoparticles using medicinal plants: Characterization and application. *Journal of Radiation Research and Applied Sciences* 2022, 15(3), 109–124.
[26]   Khan, M. S. A., Ahmad, I., Chattopadhyay, D. New Look to Phytomedicine: Advancements in Herbal Products as Novel Drug Leads. San Diego, CA, USA, Academic Press, 2018.
[27]   Shamaila, S., Sajjad, A. K. L., Farooqi, S. A., Jabeen, N., Majeed, S., Farooq, I. Advancements in nanoparticle fabrication by hazard free eco-friendly green routes. *Applied Materialstoday* 2016, 5, 150–199.
[28]   Gour, A., Jain, N. K. Advances in green synthesis of nanoparticles. *Artificial Cells, Nanomedicine, and Biotechnology* 2019, 47(1), 844–851.
[29]   Nikolaidis, P. Analysis of green methods to synthesize nanomaterials. In Green Synthesis of Nanomaterials for Bioenergy Applications, Srivastava, N., Srivastava, M., Mishra, P. K., Gupta, V. K., Eds. Wiley, 2020, 125–144.
[30]   Modena, M. M., Rühle, B., Burg, T. P., Wuttke, S. Nanoparticle characterization: What to measure?. *Advanced Materials Letters* 2019, 31(32), 1901556.
[31]   Begum, R., Farooqi, Z. H., Naseem, K., et al. Applications of UV/Vis spectroscopy in characterization and catalytic activity of noble metal nanoparticles fabricated in responsive polymer microgels: A review. *Critical Reviews in Analytical Chemistry* 2018, 48(6), 503–516.
[32]   Akbari, B., Tavandashti, M. P., Zandrahimi, M. Particle size characterization of nanoparticles–a practicalapproach. *Iranian Journal of Materials Science* 2011, 8(2), 48–56.
[33]   Bykkam, S., Ahmadipour, M., Narisngam, S., Kalagadda, V. R., Chidurala, S. C. Extensive studies on X-ray diffraction of green synthesized silver nanoparticles. *Advanced Nanoparticle* 2015, 4(1), 1–10.
[34]   Mohan, A. C., Renjanadevi, B. Preparation of zinc oxide nanoparticles and its characterization using scanning electron microscopy (SEM) and X-ray diffraction (XRD). *Proceeeding Technology* 2016, 24, 761–766.
[35]   Al-Radadi, N. S. Facile one-step green synthesis of gold nanoparticles (AuNp) using licorice root extract: Antimicrobial and anticancer study against HepG2 cell line. *Arabian Journal of Chemistry* 2021, 14(2), 102956.
[36]   Al-Radadi, N. S., Adam, S. I. Green biosynthesis of Pt-nanoparticles from Anbara fruits: Toxic and protective effects on CCl4 induced hepatotoxicity in Wister rats. *Arabian Journal of Chemistry* 2020, 13(2), 4386–4403.

[37]   Theophanides, T. Infrared Spectroscopy – Materials Science, Engineering and Technology. London, IntechOpen, 10.5772/2055, 2012.

[38]   Sharma, S. K., Verma, D. S., Khan, L. U., Kumar, S., Khan, S. B. Handbook of Materials Characterization. New York, NY, USA, Springer International Publishing, 2018.

[39]   Doane, T. L., Chuang, C. H., Hill, R. J., Burda, C. Nanoparticle ζ-potentials. *Accounts of Chemical Research* 2012, 45(3), 317–326.

[40]   Bhatia, N., Kumari, A., Chauhan, N., Thakur, N., Sharma, R. *Duchsnea indica* plant extract mediated synthesis of copper oxide nanomaterials for antimicrobial activity and free-radical scavenging assay. *Biocatalysis & Agricultural Biotechnology* 2023, 47, 102574.

[41]   Rajivgandhi, G. N., Chackaravarthi, G., Ramachandran, G. Synthesis of silver nanoparticle (Ag NPs) using phytochemical rich medicinal plant *Lonicera japonica* for improve the cytotoxicity effect in cancer cells. *Journal of King Saud University – Engineering Sciences* 2022, 34(2), 101798.

[42]   Alahmad, A., Al-Zereini, W. A., Hijazin, T. J., et al. Green synthesis of silver nanoparticles using hypericum perforatum L. aqueous extract with the evaluation of its antibacterial activity against clinical and food pathogens. *Pharmaceutics* 2022, 14(5), 1104.

[43]   Giri, A. K., Jena, B., Biswal, B., Pradhan, A. K., Arakha, M., Acharya, S., Acharya, L. Green synthesis and characterization of silver nanoparticles using *Eugenia roxburghii* DC. extract and activity against biofilm-producing bacteria. *Scientific Reports* 2022, 12(1), 1–9.

[44]   Davids, J. S., Ackah, M., Okoampah, E., Fometu, S. S., Guohua, W., Jianping, Z. Biocontrol of bacteria associated with Pine Wilt Nematode, *Bursaphelenchus xylophilus* by using plant mediated gold nanoparticles. *International Journal of Agriculture and Biology* 2021, 26, 517–526.

[45]   ElMitwalli, O. S., Barakat, O. A., Daoud, R. M., Akhtar, S., Henari, F. Z. Green synthesis of gold nanoparticles using cinnamon bark extract, characterization, and fluorescence activity in Au/eosin Y assemblies. *Journal of Nanoparticle Research* 2020, 22(10), 1–9.

[46]   Jayachandran, A., Aswathy, T. R., Nair, A. S. Green synthesis and characterization of zinc oxide nanoparticles using *Cayratia pedata* leaf extract. *Biochemistry and Biophysics Reports* 2021, 26, 100995.

[47]   Ahmad, H., Venugopal, K., Rajagopal, K., et al. Green synthesis and characterization of zinc oxide nanoparticles using Eucalyptus globules and their fungicidal ability against pathogenic fungi of apple orchards. *Biomolecules* 2020, 10(3), 425.

[48]   Narayanan, M., Devi, P. G., Natarajan, D., et al. Green synthesis and characterization of titanium dioxide nanoparticles using leaf extract of *Pouteria campechiana* and larvicidal and pupicidal activity on *Aedes aegypti*. *Environmental Research* 2021, 200, 111333.

[49]   Sethy, N. K., Arif, Z., Mishra, P. K., Kumar, P. Green synthesis of TiO$_2$ nanoparticles from *Syzygium cumini* extract for photo-catalytic removal of lead (Pb) in explosive industrial wastewater. *Green Processing and Synthesis* 2020, 9(1), 171–181.

[50]   Bathula, C., Subalakshmi, K., Kumar, A., et al. Ultrasonically driven green synthesis of palladium nanoparticles by Coleus amboinicus for catalytic reduction and Suzuki-Miyaura reaction. *Colloid Surfaces B: Biointerfaces* 2020, 192, 111026.

[51]   Fahmy, S. A., Fawzy, I. M., Saleh, B. M., Issa, M. Y., Bakowsky, U., Azzazy, H. M. E. S. Green synthesis of platinum and palladium nanoparticles using *Peganum harmala* L. Seed alkaloids: Biological and computational studies. *Nanomaterials* 2021, 11(4), 965.

[52]   Naguib, N. Y. M. Organic vs chemical fertilization of medicinal plants: A concise review of researches. *Advances in Environmental Biology* 2011, 5(2), 394–400.

[53]   Valussi, M., Scirè, A. S. Quantitative ethnobotany and traditional functional foods. *Forum Nutrition* 2012, 11(3), 85–93.

[54]   Gulmez, C., Kulak, M. New insights to enhance the desired anti-diabetic compounds in medicinal and aromatic plants exposed to abiotic stress factors. In Biotechnology of Anti-Diabetic Medicinal Plants, 1st ed, Gantait, S., Verma, S. K., Sharangi, A. B., Eds. Singapore, Springer, 2021, 285–306.

[55]   Sreenivasulu, N., Fernie, A. R. Diversity: Current and prospective secondary metabolites for nutrition and medicine. *Current Opinion in Biotechnology* 2022, 74, 164–170.

[56]   Ahammed, G. J., Li, X. Hormonal regulation of health-promoting compounds in tea (*Camellia sinensis L.*). *Plant Physiology and Biochemistry* 2022, 185, 390–400.

[57]   García-Pérez, P., Gallego, P. P. Plant phenolics as dietary antioxidants: Insights on their biosynthesis, sources, health-promoting effects, sustainable production, and effects on lipid oxidation. In Lipid Oxidation in Food and Biological Systems, 1st ed, Bravo-Diaz, C., Ed. Singapore, Springer, 2022, 405–426.

[58]   Fakhri, S., Moradi, S. Z., Farzaei, M. H., Bishayee, A. Modulation of dysregulated cancer metabolism by plant secondary metabolites: A mechanistic review. *Seminars in Cancer Biology* 2022, 80, 276–305.

[59]   Arimura, G. I., Maffei, M. Plant Specialized Metabolism: Genomics, Biochemistry, and Biological Functions. Boca Raton, USA, CRC press, 2016.

[60]   Harborne, J. B. Introduction to Ecological Biochemistry, 4th ed. New York, Academic Press, 1993.

[61]   Mammadov, R. Tohumlu Bitkilerde Sekonder Metabolitler. Ankara, Türkiye, Nobel Akademik Yayıncılık, 2014.

[62]   Caretto, S., Linsalata, V., Colella, G., Mita, G., Lattanzio, V. Carbon fluxes between primary metabolism and phenolic pathway in plant tissues under stress. *International Journal of Molecular Sciences* 2015, 16(11), 26378–26394.

[63]   Bartwal, A., Mall, R., Lohani, P., Guru, S. K., Arora, S. Role of secondary metabolites and brassinosteroids in plant defense against environmental stresses. *Journal of Plant Growth Regulation* 2013, 32(1), 216–232.

[64]   Jan, R., Asaf, S., Numan, M., Kim, K. M. Plant secondary metabolite biosynthesis and transcriptional regulation in response to biotic and abiotic stress conditions. *Agronomy* 2021, 11(5), 968.

[65]   Chung, I. M., Rajakumar, G., Thiruvengadam, M. Effect of silver nanoparticles on phenolic compounds production and biological activities in hairy root cultures of *Cucumis anguria*. *Acta Biologica Hungarica* 2018, 69(1), 97–109.

[66]   García-López, J. I., Zavala-García, F., Olivares-Sáenz, E., et al. Zinc oxide nanoparticles boosts phenolic compounds and antioxidant activity of *Capsicum annuum* L. during germination. *Agronomy* 2018, 8(10), 215.

[67]   Nourozi, E., Hosseini, B., Maleki, R., Mandoulakani, B. A. Pharmaceutical important phenolic compounds overproduction and gene expression analysis in *Dracocephalum kotschyi* hairy roots elicited by SiO_2 nanoparticles. *Industrial Crops and Products* 2019, 133, 435–446.

[68]   Singh, O. S., Pant, N. C., Laishram, L., et al. Effect of CuO nanoparticles on polyphenols content and antioxidant activity in Ashwagandha (Withania somnifera L. Dunal). *Journal of Pharmacognosy and Phytochemistry* 2018, 7(2), 3433–3439.

[69]   Yarizade, K., Hosseini, R. Expression analysis of ADS, DBR2, ALDH1 and SQS genes in *Artemisia vulgaris* hairy root culture under nano cobalt and nano zinc elicitation. *External Journal of Applied Sciences* 2015, 3(3), 69–76.

[70]   Hatami, M., Naghdi Badi, H., Ghorbanpour, M. Nano-elicitation of secondary pharmaceutical metabolites in plant cells: A review. *Journal of Medicinal Plants* 2019, 18, 6–36.

[71]   Rivero-Montejo, S. D. J., Vargas-Hernandez, M., Torres-Pacheco, I. Nanoparticles as novel elicitors to improve bioactive compounds in plants. *Agriculture* 2021, 11(2), 134.

[72]   Moradbeygi, H., Jamei, R., Heidari, R. Darvishzadeh, R. Investigating the enzymatic and non-enzymatic antioxidant defense by applying iron oxide nanoparticles in *Dracocephalum moldavica* L. plant under salinity stress. *Scientia Horticulturae* 2020, 272, 109537.

[73]   Kamalizadeh, M., Bihamta, M., Zarei, A. Drought stress and TiO_2 nanoparticles affect the composition of different active compounds in the *Moldavian dragonhead* plant. *Acta Physiologiae Plantarum* 2019, 41(2), 1–8.

[74] Sharma, V. K., Yngard, R. A., Lin, Y. Silver nanoparticles: Green synthesis and their antimicrobial activities. *Advances in Colloid Interface Science* 2009, 145(1–2), 83–96.

[75] Pestovsky, Y. S., Martínez-Antonio, A. The use of nanoparticles and nanoformulations in agriculture. *Journal for Nanoscience and Nanotechnology* 2017, 17(12), 8699–8730.

[76] Jiang, Y., Zhou, P., Zhang, P. Green synthesis of metal-based nanoparticles for sustainable agriculture. *Environmental Pollution* 2022, 119755.

[77] Santo Pereira, A. D. E., De Oliveira, J. L., Savassa, S. M., Rogério, C. B., De Medeiros, G. A., Fraceto, L. F. Lignin nanoparticles: New insights for a sustainable agriculture. *Journal of Cleaner Production* 2022, 131145.

[78] Aria, M., Cuccurullo, C. Bibliometrix: An R-tool for comprehensive science mapping analysis. *Journal of Informetrics* 2017, 11(4), 959–975.

[79] Campra, M., Esposito, P., Brescia, V. State of the art of COVID-19 and business, management, and accounting sector. A bibliometrix analysis. *International Journal of Business and Management* 2020, 16(1), 2021.

[80] Rodríguez-Soler, R., Uribe-Toril, J., Valenciano, J. D. P. Worldwide trends in the scientific production on rural depopulation, a bibliometric analysis using bibliometrix R-tool. *Land Use Policy* 2020, 97, 104787.

[81] Kulak, M. A bibliometric review of research trends in salicylic acid uses in agricultural and biological sciences: Where have been studies directed. *Agronomy* 2018, 61(1), 296–303.

[82] Kulak, M., Ozkan, A., Bindak, R. A bibliometric analysis of the essential oil-bearing plants exposed to the water stress: How long way we have come and how much further?. *Scientia Horticulturae* 2019, 246, 418–436.

[83] Kumlay, A. M., Kocak, M. Z., Gohari, G., et al. Agronomic traits, secondary metabolites and element concentrations of *Lavandula angustifolia* leaves as a response to single or reiterated drought stress: How effective is the previously experienced stress?. *Folia Horticulturae* 2022, 34(1), 1–16.

[84] Celik, E., Durmus, A., Adizel, O., Nergiz Uyar, H. A bibliometric analysis: What do we know about metals (loids) accumulation in wild birds?. *Environmental Science and Pollution Research* 2021, 28(8), 10302–10334.

[85] Virú-Vásquez, P., Pardavé, R. H., Coral, M. F. C., Bravo-Toledo, L., Curaqueo, G. Biochar and compost in the soil: A bibliometric analysis of scientific research. *Environmental Research Engineering and Management* 2022, 78(3), 73–95.

[86] Egghe, L. Theory and practise of the g-index. *Scientometrics* 2006, 69(1), 131–152.

[87] Ali, M. J. Understanding the 'g-index'and the 'e-index'. *Seminars in Ophthalmology* 2021, 36(4), 139.

[88] Ye, N., Kueh, T. B., Hou, L., Liu, Y., Yu, H. A bibliometric analysis of corporate social responsibility in sustainable development. *Journal of Cleaner Production* 2020, 272, 122679.

[89] Beyene, H. D., Werkneh, A. A., Bezabh, H. K., Ambaye, T. G. Synthesis paradigm and applications of silver nanoparticles (AgNPs), a review. *Sustainable Materials and Technologies* 2017, 13, 18–23.

[90] Marin, S., Mihail Vlasceanu, G., Elena Tiplea, R., et al. Applications and toxicity of silver nanoparticles: A recent review. *Current Topics in Medicinal Chemistry* 2015, 15(16), 1596–1604.

[91] Kim, J. S., Kuk, E., Yu, K. N. Antimicrobial effects of silver nanoparticles. *Nanomedicine: Nanotechnology, Biology and Medicine* 2007, 3(1), 95–101.

[92] Mittal, A. K., Kaler, A., Banerjee, U. C. Free radical scavenging and antioxidant activity of silver nanoparticles synthesized from flower extract of *Rhododendron dauricum*. *Nano Biomedicine and Engineering* 2012, 4(3).

[93] Awwad, A. M., Salem, N. M. Green synthesis of silver nanoparticles by Mulberry leaves extract. *Nanoscience and Nanotechnology Letters* 2012, 2(4), 125–128.

[94] Bar, H., Bhui, D. K., Sahoo, G. P., Sarkar, P., De, S. P., Misra, A. Green synthesis of silver nanoparticles using latex of *Jatropha curcas*. *Colloids and Surfaces A: Physicochemical and Engineering* 2009, 339(1–3), 134–139.

[95]   Mohammadlou, M., Maghsoudi, H., Jafarizadeh-Malmiri, H. J. I. F. R. J. A review on green silver nanoparticles based on plants: Synthesis, potential applications and eco-friendly approach. *International Food Research Journal* 2016, 23(2), 446.

[96]   Srikar, S. K., Giri, D. D., Pal, D. B., Mishra, P. K., Upadhyay, S. N. Green synthesis of silver nanoparticles: A review. *Green and Sustainable Chemistry* 2016, 6(1), 34–56.

[97]   Zhang, J., Yu, Q., Zheng, F., Long, C., Lu, Z., Duan, Z. Comparing keywords plus of WOS and author keywords: A case study of patient adherence research. *Journal of the Association for Information Science and Technology* 2016, 67(4), 967–972.

[98]   Di Cosmo, A., Pinelli, C., Scandurra, A., Aria, M., D'Aniello, B. Research trends in octopus biological studies. *Animals* 2021, 11(6), 1808.

[99]   Xiao, Z., Qin, Y., Xu, Z., Antucheviciene, J., Zavadskas, E. K. The journal buildings: A bibliometric analysis (2011–2021). *Buildings* 2022, 12(1), 37.

[100] Mumu, J. R., Tahmid, T., Azad, M. A. K. Job satisfaction and intention to quit: A bibliometric review of work-family conflict and research agenda. *Applied Nursing Research* 2021, 59, 151334.

[101] Fatemi, H., Pour, B. E., Rizwan, M. Isolation and characterization of lead (Pb) resistant microbes and their combined use with silicon nanoparticles improved the growth, photosynthesis and antioxidant capacity of coriander (*Coriandrum sativum* L.) under Pb stress. *Environmental Pollution* 2020, 266, 114982.

[102] Hernández-Hernández, H., González-Morales, S., Benavides-Mendoza, A., Ortega-Ortiz, H., Cadenas-Pliego, G., Juárez-Maldonado, A. Effects of chitosan–PVA and Cu nanoparticles on the growth and antioxidant capacity of tomato under saline stress. *Molecules* 2018, 23(1), 178.

[103] Hassan, F. A. S., Ali, E., Gaber, A., Fetouh, M. I., Mazrou, R. Chitosan nanoparticles effectively combat salinity stress by enhancing antioxidant activity and alkaloid biosynthesis in *Catharanthus roseus* (L.) G. Don. *Plant Physiology and Biochemistry* 2021, 162, 291–300.

[104] Shah, S. N. A., Shah, Z., Hussain, M., Khan, M. Hazardous effects of titanium dioxide nanoparticles in ecosystem. *Bioinorganic Chemistry and Applications* 2017, 4101735.

[105] Lei, C., Sun, Y., Tsang, D. C., Lin, D. Environmental transformations and ecological effects of iron-based nanoparticles. *Environmental Pollution* 2018, 232, 10–30.

[106] Marslin, G., Sheeba, C. J., Franklin, G. Nanoparticles alter secondary metabolism in plants via ROS burst. *Frontiers Plant Science* 2017, 8, 832.

Katarina Kráľová*, Josef Jampílek

Chapter 3
Nanofertilizers: recent approach in crop production

Abstract: Anthropogenic agricultural and industrial activities resulted in the global degradation of large areas, whereby the overuse of agrochemicals due to low nutrient use efficiency by crops since the mid-twentieth century contributed to a pronounced decline in agricultural soil fertility, resulting in deficiency of some essential nutrients needed for healthy growth and development of crops. With the boom of nanotechnologies over the last 20 years, the benefits of using nanosized fertilizers come to the fore over conventional ones, as they can slow the release of nutrients, ensure controlled release of nutrients, and lower doses of active ingredients that are sufficient to achieve comparable or better biological effects than the respective bulk conventional fertilizers. Lower environmental contamination during the application of nanosized fertilizers as well as biofortification of consumable plant organs and grains with some essential nutrients enables the production of safe and healthy food. This chapter presents an up-to-date overview of nanoscale formulations of essential macronutrients (N, P, K, Ca, Mg, and S) and micronutrients (Fe, Mn, Zn, Cu, B, Mo, and Ni) as well as nonessential micronutrients (Co, Si, and Se) needed for plants, which are used as fertilizers. Application of nanostructured materials serving as nutrient carriers such as hydroxyapatite and nanoclays, and polymeric materials used for coating of nanosized nutrients, enabling slow and controlled release of nutrients, is discussed as well. The beneficial impact of nanofertilizers on crop yield and nutritional quality, including the respective mode of action, is emphasized, together with the ability of nanofertilizers to alleviate the adverse effects of both abiotic and biotic stresses on crops.

Keywords: biofortification, coating, crops, environmental stresses, nanofertilizers, nanostructured nutrient carriers, nutrients

*Corresponding author: Katarina Kráľová, Institute of Chemistry, Faculty of Natural Sciences, Comenius University, Ilkovičova 6, 842 15 Bratislava, Slovakia, e-mail: kata.kralova@gmail.com
Josef Jampílek, Department of Analytical Chemistry, Faculty of Natural Sciences, Comenius University, Ilkovičova 6, 842 15 Bratislava, Slovakia, e-mail: josef.jampilek@gmail.com

https://doi.org/10.1515/9781501523229-003

3.1 Introduction

Fertilizers provide crops with nutrients essential for their growth and development, resulting in improved crop yields. Considering that nowadays majority of agricultural soils are depleted of important nutrients, and an expected increase in world population could achieve approximately 8.5 billion people in 2030, 9.7 billion in 2050, and 10.9 billion in 2100, compared to 7.7 billion people worldwide in 2019 [1], there should be a considerably increased agricultural productivity globally to ensure the increasing demand for food that can be implemented only using effective fertilizer management [2, 3]. Moreover, crops consumed by humans and animals provide them essential calories, proteins, vitamins, minerals, dietary fibers, antioxidants, and other bioactive compounds important for their health. Because fertilizers guarantee not only the quantity but also the quality of plant-based food, suppressing malnutrition and micronutrient deficiencies, the nutrition-sensitive agriculture should be focused on production of nutritionally rich foods, dietary diversity, and crops fortification besides conventional plant breeding or transgenic techniques and also agronomic biofortification, including nutripriming, fertilization with both mineral and organic fertilizers, and application of biofertilizers [4–10]. Such an approach can combat hidden hunger, which affects more than 2 billion people globally, particularly in low- and middle-income countries, and occurs when too low intake and absorption of vitamins and essential minerals (e.g., Zn, Fe, and I) cannot ensure good health and development [11–14].

Smart fertilizer management can utilize information/data, sensors, imaging technologies, and nutrient biomarkers able to quantitatively predict the status of a particular nutrient with high specificity allowing correct fertilization in precision agriculture, smart agriculture, and integrated nutrient management [3].

According to Sutton et al. [15], in the period 1960–2013, the human use of synthetic N fertilizers showed a 9-fold increase worldwide, and P use has tripled, and further increase of its usage by approx. 40–50% could be expected till 2050. It is obvious that not only pesticides but also overfertilization can contribute to the contamination of the environment due to leaching and low nutrient use efficiency (NUE) by plants. It was reported that at the use of nitrogen fertilizers, very low NUE (around 40% worldwide) is achieved; hence, approximately 60% of N remain in environmental matrices as a potential environmental pollutant [16]. Particularly, excess of N and P, which were not utilized by plants and were moved from soils to streams, lakes, and estuaries, causes eutrophication resulting in adverse impact on aqueous organisms. Erosion and runoff and subsurface drainage also contribute to nutrient loss [17, 18]. Moreover, considering the ongoing climate change characterized with enhanced frequency of extreme weather events, it is presumable that excessive fertilization will not be economically or environmentally sustainable [19].

Minerals and manure have already been used by Egyptians, Babylonians, Romans, and early Germans to increase crop yields [20]. Bogaard et al. [21] based on stable isotope determinations of charred cereals and pulses from 13 Neolithic sites across Europe (dating ca. 5900–2400 calibrated years BC) stated that livestock manure was used at that time to enhance crop yields. Nitrogen is the most important plant nutrient in commercial fertilizers also nowadays, and widespread application of nitrogen fertilizers in agriculture in the twentieth century was possible due to discovery of ammonia (the main constituent in N-based fertilizers) synthesis from its elements by Fritz Haber in 1909 and subsequent developing of commercial-scale production of ammonia using chemical high-pressure method by Carl Bosch; both researchers were honored with Nobel Prizes for their important scientific contributions, Fritz Haber in 1918 (awarded in 1919) and Carl Bosch in 1931 [22, 23]. History of chemical fertilizer development was presented by Russel and Williams [24].

Fertilizers can be applied to soils or directly to plants for increasing crop yield and/or quality by supplying essential nutrients and to improve soil fertility for future crop production [25–30]. This means that it is necessary to apply the **right source** of nutrients in the **right rate** required for the optimal yield of a particular crop in the **right place**, and at the **right time** [31–34]. In addition, thanks to emerging new technologies, including technological advances in nutrient and water management, agricultural production can become highly efficient, sustainable, energy efficient, and optimized for entry [33–35].

Besides carbon (C), hydrogen (H), and oxygen (O), which plants obtain from the atmosphere and water, they need six macronutrients (N, P, K, Ca, Mg, and S) and eight essential micronutrients (Fe, Mn, Zn, Cu, B, Mo, Cl$^-$, and Ni) for their growth and development. Although the other four micronutrients, namely Se, Si, I, and Co, which can be absorbed and accumulated by plants, are not essential for plants, they are essential for human health; therefore, their application as fertilizers is also highly desirable [11, 36, 37]. Macronutrient levels in plants are ≥2–50 g/kg of dry matter, and those of micronutrients are considerably lower, ranging from 0.1 to 100 µg/kg. In addition to the above-mentioned nutrients, the International Fertilizer Association (IFA) also lists Na and Si as minerals that can improve and protect plant health [11]. For example, soybean plants treated with Si fertilizer showed higher net photosynthesis, transpiration rate, and stomatal conductivity compared to control plants, resulting in higher yield [38] and Na may have a positive impact or may even be essential mainly for C_4/CAM, halophyte, and natrophilic species [39]. Essential macronutrients and micronutrients as well as nonessential micronutrients needed by plants are shown in Figure 3.1.

However, it could be mentioned that higher concentrations of above-mentioned mineral nutrients exhibit adverse impact on plants by inducing strong oxidative stress, which could not be overcome by plant's antioxidant defense system, resulting in impaired photosynthesis and plant growth [40–48]. Therefore, the used dose of nutrients applied to soil has to be considered besides plant species and also soil properties (e.g., pH, organic matter content, and presence of contaminants), which affect the bioavailability of nutrients [49–51]. On the other hand, at foliar fertilization, it is necessary to use nutrient formulations with superb adherence to leaves for avoiding discharges into the environment through rainwater washing or leaching [52–54].

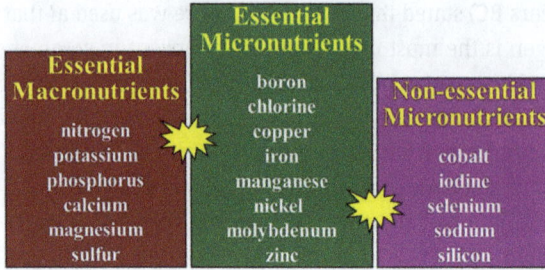

Figure 3.1: Essential macronutrients and micronutrients and nonessential micronutrients needed by plants.

Gorecki [55] summarized various criteria, according to which fertilizers can be classified as follows: (i) chemical composition of fertilizer (mineral or organic); (ii) source of obtaining (natural or commercial); (iii) number of ingredients (single, multicomponent, and complete fertilizers); (iv) physical form of fertilizer (solid, fluid, or gaseous); (v) type of fertilization (soil, fluid, or foliar application, hydroponics, and fertigation fertilizers); (vi) mode of plant availability (direct or indirect acting fertilizers); and (vii) concentration of fertilizer nutrients (major nutrients and micronutrients). Classification modes of fertilizer according to Gorecki [55] are shown in Figure 3.2.

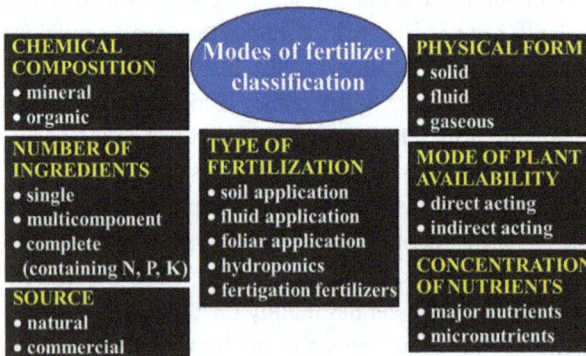

Figure 3.2: Classification modes of fertilizers according to Gorecki [55].

It is favorable to extend the release of nutrients from mineral fertilizers to grant crops' requirements optimally. For this purpose, slow-release fertilizers that break down gradually to release plant available nutrients or controlled-release fertilizers containing encapsulated nutrient coated with protective layer are convenient [56–59]. Classification of fertilizers showing controlled release of microelements into four basic groups, namely (i) low-solubility fertilizers, (ii) fertilizers with external coating, (iii) bio-based fertilizers, and (iv) nanofertilizers was presented by Mikula et al. [60].

In the twenty-first century, there is an unprecedented boom in the use of nano-technologies in various areas of economy, including medicine [61–66], food industry [65, 67–69], and agriculture [65, 67, 70–78], and recently nanoparticles (NPs) are increasingly used also as dietary supplements or nanonutraceuticals increasing the immunity [79, 80]. It is related to the fact that the physicochemical properties and biological activity of nanosized particles, which are characterized with high ratio of the surface to particle size, differ from those of their bulk counterpart [44, 70–72, 74, 81–83]. The NPs, nanocomposites, and nanosized formulations used in agricultural practice as plant growth stimulators or agents for plants protection (e.g., nanoherbi-cides, nanoinsecticides, nanobactericides, nanofungicides, and antivirucides) show mostly higher biological activity, and lower amount of active ingredient is needed to achieve the same of better activity, which is desirable also from the economical aspect [70, 75, 76, 78]. NPs of very small size (usually 1–100 nm) allow easy penetration into the plant cell and translocation between plant organs. Nanofertilizers containing NPs or nanocomposites of nutrients (essential macronutrients and micronutrients or nonessential micronutrients) showing beneficial impact on plant growth and development usually improve plant's morphological characteristics, levels of photosynthetic pigments, effectiveness of photosynthetic processes, as well as antioxidant defense system by enhancing activities of antioxidant enzymes and levels of nonenzymatic antioxidants along with altering the expression of enzymes responsible for plants growth, which is reflected in improved yield and increased production of valuable secondary metabolites [44, 71, 82–88]. Nanofertilizers showing slow and controlled release of nutrients can exhibit sustained delivery of required amounts of nutrients to plants. Slow-release fertilizers frequently use aluminosilicates with a mesoporous nanostructure as nutrient carriers and controlled release of nutrients can be achieved by their embedding in polymer matrix or using polymer coating on NPs. Nanomaterial coatings (e.g., nanomembrane) may reduce the release rate of nutrients, and a network of channels of porous nanofertilizers may act as a reservoir of nutrients releasing them on demand [71, 89, 90].

Nanofertilizers can be applied in the form of seed nanopriming, foliar spraying, or application to the soil [71, 74, 83, 91, 92]. Application of nanoscale fertilizers can also considerably contribute to alleviation of adverse impact of various abiotic and biotic environmental stresses on plants [74, 85, 92–94] as well as to pronounced reduction of environmental pollution, which is particularly important at changing environmental conditions [92]. Moreover, green synthesized nanofertilizers prepared without harmful chemicals and showing usually improved efficiency compared to synthetic nanofertilizers are increasingly preferred [44, 74, 77, 82, 83, 85, 92].

Benefits of the use of nanoscale fertilizers on growth and nutritional quality of plants were summarized by several researchers [71, 74, 95–109]. Most important benefits of nanoscale fertilizers to plants are shown in Figure 3.3.

Classification of nanoscale fertilizers was performed by several researchers. Whereas Kah et al. [110] classified nanofertilizers into three groups including nanomaterials made of micronutrients, nanomaterials made of micronutrients, or those used as

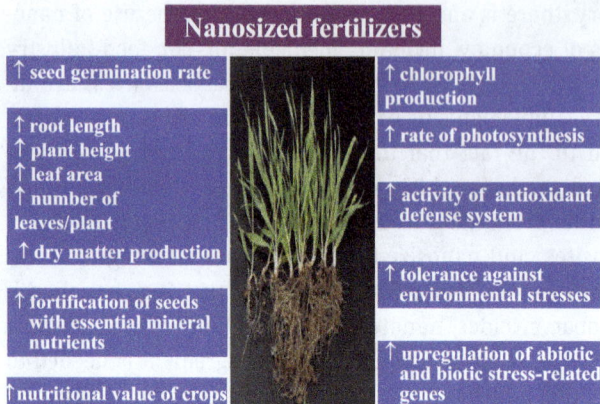

Figure 3.3: Most important benefits of nanoscale fertilizers to plants.

carriers for macronutrients, Liu and Lal [95] defined the following four groups of nano-fertilizers: (i) macronutrient nanofertilizers (e.g., apatite $Ca_5(PO_4)_3$ (F, Cl, OH) NPs, $CaCO_3$ NPs, and MgO NPs); (ii) micronutrient nanofertilizers (e.g., Fe_3O_4 NPs, MnO NPs, ZnO NPs, and CuO NPs); (iii) nutrient-loaded nanofertilizers (zeolites, SiO_2 NPs, and carbon nanotubes (CNTs)) and (iv) plant growth-stimulating nanomaterials (TiO_2 and CNTs). On the other hand, Mikkelsen [111] proposed three classes of nanofertilizers: (i) nanosized fertilizer (NPs containing nutrients); (ii) nanosized additives (traditional fertilizers supplemented with nanosized additives); and (iii) nanosized coating (obtained by coating or loading of traditional fertilizers with NPs).

This chapter presents an up-to-date overview of nanoscale formulations of essential macronutrients (N, P, K, Ca, Mg, and S) and micronutrients (Fe, Mn, Zn, Cu, B, Mo, and Ni) as well as nonessential micronutrients (Co, Si, and Se) needed for plants, which are used as fertilizers. Application of nanostructured materials serving as nutrient carriers such as hydroxyapatite (HA) and nanoclays, and polymeric materials used for coating of nanosized nutrients, enabling slow and controlled release of nutrients, is discussed as well. The beneficial impact of nanofertilizers on crop yield and nutritional quality, including the respective mode of action, is emphasized, together with the ability of nano-fertilizers to alleviate the adverse effects of both abiotic and biotic stresses on crops.

3.2 Macronutrient nanofertilizers

As macronutrients essential for plant growth and development, N, P, K, Ca, Mg, and S are considered [112]. Nitrogen is an essential component of nucleic acids, amino acids and is needed for protein synthesis and chlorophyll (Chl); P is indispensable for energy storage and transfer and membrane integrity; K is involved in enzyme activation, transpiration, and the transport of assimilates; Ca is necessary for biomembrane maintenance and contributes to cell wall stabilization as an enzyme activator; Mg is a constituent of Chl and sulfur is part of amino acids such as cysteine and methionine and helps to produce amino acids involved in Chl production, proteins, and vitamins [11, 37, 113]. Ca, Mg, and S are also called "secondary" nutrients because they are needed for plants in smaller amounts than N, P, and K but in larger quantities than the micronutrients.

3.2.1 Hydroxyapatite-based nitrogen and phosphorus nanofertilizers

Normal HA ($Ca_{10}(PO_4)_6(OH_2)$) and its micro- and nanoscale forms can increase soil pH and reduce soil exchangeable acid and exchangeable Al, and depending on the size of applied amendment, they can change the composition of soil microbial community as well [114]. HA NPs can serve as a source of phosphorus and as a carrier of nitrogen or micronutrients [90, 115–121].

HA NPs used as phosphate fertilizer were able to prevent phosphorus loss and ameliorated the growth and physiological properties of *Zea mays* plants more than the simple and triple superphosphate fertilizers [122]. On the other hand, needle-shaped HA NPs used as a PO_4^{3-} fertilizer did not affect considerably growth, biomass, total plant P levels, and yield of soybean compared to control; soil and rhizosphere community structures following HA NP treatment were like that observed in control [123].

Hybrid nanofertilizer fabricated by incorporation of Cu, Fe, and Zn NPs into urea-modified HA showed slow release during leaching studies and acted as slow-release fertilizer. It ensured the availability of macro- and micronutrients (Ca^{2+}, PO_4^{3-}, NO_2^-, NO_3^-, Cu^{2+}, Fe^{2+}, and Zn^{2+}) and pronouncedly higher uptake of Cu^{2+}, Fe^{2+}, and Zn^{2+} was observed in treated *Abelmoschus esculentus* plants. Improved swelling ratio and water absorption and retention capacities of this nanofertilizer resulted in considerably lower required dose compared to commercial fertilizer (50 mg/week vs 5 g/week) [90].

Soybean plants treated with HA NPs prepared from fish back bone and eggshell exhibited 1.88-fold higher growth compared to plants treated with superphosphate and showed higher levels of N, P, and K by 1.21%, 1.32%, and 1.17% compared to control plants; the effect of fish back bone- and eggshell-made HA NPs was comparable [124].

Nanofertilizer consisting of nanourea modified with HA NPs was able to serve as a source for N, P, and Ca nutrients under saline conditions and its use for pretreatment of seeds of bitter almond rootstock under saline conditions resulted in pronouncedly improved germination, stem length and diameter, elongation of secondary and primary roots, as well as number of secondary roots per plant compared to urea and $NH_4(SO_4)_2$, suggesting its beneficial impact on the moisture content of seeds [125].

HA NPs were applied as P fertilizer to *Glycine max* L. plants cultivated in greenhouse using foliar or soil treatment at various precipitation intensities. After 4 weeks at 100% precipitation intensity, plants fertilized with HA NPs contained higher P and Ca levels in shoots (by 32.6% and 33.2%), roots (by 40.6% and 45.4%), and pods (by 37.9% and 82.3%) compared to plants treated with PO_4^{3-} fertilizer, although fertilization did not affect plant biomass [126]. HA NPs (15.51–154.41 nm) were not phytotoxic to bean plants (up to 31 mg P/L), and treatment with 20 mmol/L HA NPs resulted in significant enhancement of plant height compared to plants watered with Hoagland solution [127].

The impact of HA NPs on plants was found to depend on the plant species and concentration of HA NPs, showing beneficial effects on *G. max*, *Sorghum bicolor*, *Pisum sativum*, and Pakchoi plants. This can be associated with longer persistence of HA NPs in the soil, resulting in longer P availability for plants, improved Ca nutrition, increased gibberellin hormone levels, and improved tolerance of tested plants to heavy metal stress [128]. On the other hand, adverse impact of HA NPs on mung bean sprouts, tomato, and Pakchoi plants can be explained by the structure of HA NPs and increased Ca^{2+} concentration in cells.

Raguraj et al. [129] applied conventional urea and urea–HA nanohybrid fertilizers to *Camellia sinensis* (L.) Kuntze plants at doses corresponding to half and full amounts of N recommendations at two and four splits per annum in three climatic zones of Sri Lanka, whereby the experiment lasted 3 years. When annual N requirement was ensured by urea–HA nanohybrid, in Low Country and UV A region, the observed yield increased by 10–17% and 14–16%, while in Mid-Country region, this increase was only 2–3%. In addition, the number of fertilizer applications could be reduced by treatment with urea–HA nanohybrid compared to conventional fertilizer. This slow-release nanofertilizer considerably enhanced soil P, leaf N levels, and P concentrations in tea plants cultivated in Low Country region.

Multifunctional P nanofertilizer fabricated by self-assembling natural or synthetic humic substances and HA NPs pronouncedly increased early plant growth and yield of maize plants cultivated in pots as well as rhizosphere bacteria, and tolerance of plants to saline-induced stress compared to commercial fused superphosphate and bare HA NPs. This beneficial impact was due to the synergistic co-release of PO_4^{3-} ions and humic substances, ensuring crop nutrition and plant growth stimulation. Moreover, by varying the amount of humic substances adsorbed on the HA NPs, the release patterns can be tuned [130]. Urea-doped HA NPs enhanced growth as well as fresh and dry weights of rice plants cultivated in soil and in sand columns; in agricultural soil, this nanofertilizer exhibited almost 2-fold retention capacity of plant nutrient efficiency compared to conventional P and N fertilizers [121].

HA NP solutions stabilized with carboxymethylcellulose (CMC) considerably stimulated elongation of *Solanum lycopersicum* L. roots (even by 97% using a dose of 500 mg/L), although increasing concentrations of HA NPs did not affect germination percentage of tomato seeds. At an application of 500 mg/L of CMC–HA NPs, the treated plants showed 4-fold lower Ca concentration than control plants suggesting inhibitory effect of CMC–HA NPs on the uptake of Ca by roots. On the other hand, observed P concentration estimated in shoots of plants treated with 500 mg/L of CMC–HA NPs was by 25% higher than in control plants [118].

HA-based nanostructured P fertilizer (NPF) translated from refractory calcium phosphate (i.e., $CaHPO_4$ and $Ca_2P_2O_7$) via controlling the associated dissolution–precipitation processes for HA nanocrystal self-assembly was prepared by Tang and Fei [131], whereby the researchers used an alkali-enhanced hydrothermal process with added biomass. Phosphorus use efficiency for the $CaHPO_4$-derived NPF with rod-like morphology and $Ca_2P_2O_7$-derived NPF showing hexagonal morphology was 45.87% and 46.21%, respectively, compared to that of the chemical phosphorus fertilizer (23.44%), and NPFs were able to effectively deliver HA NPs to root zones of the plants. Urea-modified HA nanohybrid with urea molecules incorporated into a matrix of HA NPs and nitrogen weight of 40% was reported as fertilizer enabling slow release of nitrogen, which was able to maintain yield along with reduced amount of applied urea [116].

Treatment with citrate-stabilized HA NPs of hydroponically cultivated P-deficient barley plants resulted in considerable enhancement of leaf P concentration (from 6,000 ppm), and plant's P functionality was subsequently restored in less than 48 h. At the beginning of the treatment, HA NPs adhered to the root epidermis without dissolving, for 24 h and following 24 h, HA NPs penetrated the roots via the apoplast of mature epidermal and cortical cells and dissolved here due to the acidic environment of the cell wall matrix. The mucilage layer on the root cap prevented the entrance of HA NPs, which dissolved stepwise without penetrating the deeper cell layers. The similar plant bioavailability as synthetic HA NPs showed citrate-stabilized nanorock phosphate [132].

Treatments with a slow-release NPK fertilizer fabricated by Rop et al. [133] by incorporating HA NPs, urea, $(NH_4)_2HPO_4$, and K_2SO_4 into water hyacinth cellulose-graft-poly(acrylamide) polymer hydrogel showed pronouncedly enhanced content of mineral nitrogen between the 8th and 12th weeks than conventional fertilizer, followed by a decline in the 16th week. However, in the first 4 weeks, considerably higher mineral N content was achieved using conventional fertilizer compared to slow-release NPK fertilizer or control. Increasing content of soluble P and decreasing content of HA NPs in the fabricated fertilizer improved availability of P, which reached maximum in the 8th week; at 12th and 16th weeks, no changes in available P were observed. On the other hand, minor variations in exchangeable K were observed, suggesting short release time. Similarly, also a water hyacinth cellulose-g-poly(ammonium acrylate-co-acrylic acid)/HA NP polymer hydrogel composite was described as a suitable source of P, which was available for plants after microbial degradation of cellulose-grafted copolymer and solubilization of HA NPs, whereby it contributed to moisture retention as well [134].

Comparison of the effect of fertilization on *Lactuca sativa* plants grown on low and high calcareous soil using HA NPs or soluble P source (H_3PO_4) at a dose of 200 mg/kg showed that not only the fertilized first lettuce plants but also the second lettuce plants cultivated after first ones were characterized with pronouncedly higher dry biomass.

Treatment with HA NPs more effectively improved growth and P content of plants than the use of soluble P source [135].

Evaluation of availability of P from three fertilizers, HA NPs, bulk HA, and triple superphosphate (TSP), to *Triticum aestivum* plants cultivated in two andisols from Chile and New Zealand and two oxisols from Australia showed that 5% of HA NPs were leached in the andisol and the P uptake, and the percentage of P in the plants decreased in the following order: triple superphosphate > HA NPs > bulk HAP. Higher effectiveness of HA NPs than of bulk HA was explained by their faster dissolution [115].

HA NP nanorods synthesized by sol-gel technique showed a beneficial impact on seed germination, and growth of chickpea plants and growth rate of plants treated with a dose of 1 mg/mL were >2.5 higher than that of control [136].

In a field experiment, the broccoli (*Brassica oleracea* var. *italica*) plants cultivated on a heavy clay soil for two seasons (2016/2017 and 2017/2018) were fertilized with HA NPs and nanoscale boron oxide (NBO) or with calcium superphosphate (CSP) and boric acid (BA). It was observed that the leaf area and total head yield of plants fertilized with HA NPs exceeded that of CSP-fertilized plants by 14.2–17.8% and 13.6–15.8%, respectively, and similarly, foliar spraying with NBO resulted in higher leaf area, head yield, and vitamin C content in heads than spraying with BA. Moreover, treatment with nanofertilizers led to higher P and B levels in broccoli shoots compared to application of conventional fertilizers, and combined treatment with both nanofertilizers provided by 16% better effectiveness than combined application of both conventional fertilizers [137].

Synthesized raw HA NPs markedly enhanced shoot elongation, root elongation, and dry biomass of *Raphanus sativus* plants grown in sand and coir media, but they had any pronounced impact on the soluble protein and indole acetic acid content in the roots of *R. sativus* up to a concentration of 10,000 mg/L. The HA NPs were internalized and transformed to $Ca_3(PO_4)_3$ and $Ca_4(PO_4)_2O$ [138].

Zn-doped HA–urea nanoseed coating increased the germination rate and growth rate of maize seeds by 19% and 69%, respectively, suggesting that it can serve as an efficient macronutrient–micronutrient delivery agent for plants [139]. Urea–HA nanohybrid showed 11.5-fold slower urea release compared to bare urea. On treatment with bare urea, >85% of the applied dose was converted to NH_4^+ within 12 days, while on application of urea–HA nanohybrid, only 58.1% of urea was converted to NH_4^+ 20 days after the application, suggesting that slow urea release from this formulation can decrease N loss and improve its use efficiency in calcareous paddy soils [128]. A nanohybrid prepared by encapsulation of urea-modified HA NPs into the nanolayers of montmorillonite (MMT) exhibited slow and sustained release of N at different pH with considerably slower N release rate compared to nanohybrid composite fabricated using solvent-assisted grinding techniques. Application of this nanohybrid resulted in pronounced improvement of rice yield [140].

3.2.2 Macronutrient nanofertilizers based on nanoscale zeolites and nanoclays

Zeolites (ZEO), that is, crystalline aluminosilicates with 3D microporous structures, have general empirical formula of $M_{2n} \cdot Al_2O_3 \cdot xSiO_2 \cdot yH_2O$, where M is alkaline or alkaline earth metal; n is the degree of its oxidation; x is the number of SiO_2 molecules (from 2 to 10); y is the number of H_2O molecules (from 2 to 7) [141]. SiO_4^{4-} rings in ZEO contain open cavities creating nanoscale channels in which compounds showing smaller diameter than that of the channel can be adsorbed, whereby some of SiO_4^{4-} rings can be replaced by AlO_4^{5-} ones [142]. Because these channels are characterized with a huge inner surface area corresponding to hundreds of square meters per gram of ZEO, these aluminosilicates show superior ion exchanging properties and due to their powerful affinity for NH_4^+ and K^+ they are utilized in agriculture as slow-release fertilizer to improve NUE [143]. Their beneficial impact on soil is also reflected in the reduction of soil acidity and amelioration of the soil water retention ability [144, 145]. Moreover, ZEO can also be loaded with K^+ along with Ca^{2+} and some other micronutrients [95], although due to dimensions of interior channels as well as density of negative charge, they can selectively adsorb NH_4^+ and K^+ ions with high efficiency compared to Na^+ or Ca^{2+} and Mg^{2+}. Hence, ZEO acts as soil conditioners, which ameliorate soil's physical and chemical properties such as infiltration rate, saturated hydraulic conductivity, water holding capacity, and cation exchange capacity and can efficaciously hold water and nutrients (e.g., NH_4^+, NO_3^{3-}, PO_4^{3-}, K^+, and SO_4^{2-}) in their porous structure [146].

Several researchers focused attention on fabrication of nanosized ZEO [147–150]. Kuznetsov et al. [150] overviewed methods used for fabrication of nanoscale zeolites (50–200 nm) using mechanical milling on bead mills. Due to milling degradation of ZEO lattice, a decrease in the ZEO crystallinity is observed; however, ZEO structure can be restored by recrystallization and dealumination. Another review paper discussed the synthesis, crystallization mechanism, and application of colloidal ZEO, and the use of nanozeolites for the preparation of functionalized materials [149]. ZEO-based nanocomposites suitable as environment-friendly slow-release fertilizer stimulating germination, growth, flowering, and fruiting of crops were synthesized and characterized by Lateef et al. [151].

In a field study, combined treatment of nanozeolite and bioinoculant (indigenous *Bacillus* spp.) considerably stimulated morphological characteristics as well as contents of photosynthetic pigments, total sugar, and phenols in *Z. mays* plants compared to control, and enhanced levels of antioxidant enzymes such as catalase (CAT), peroxidase (POD), and superoxide dismutase (SOD) were observed as well, whereby maize productivity increased even by 29.80% [152]. Macronutrients incorporated in slow-release nanozeolite composite fertilizer prepared by stirring of the mixture of ZEO NPs with $NaH_2PO_4 \cdot 2H_2O$, $MgSO_4 \cdot 7H_2O$, $Ca_3(PO_4)_2$, KCl, and $NaNO_3$ showed pronounced enhancement of water retention compared to commercial fertilizer and exhibited long-term release pattern of the macronutrients, suggesting its potential to be used for plant growth stimulation [153]. Application of nanozeolites with sizes >200–400 nm at doses up to 2,000 ppm was found to be safe for soil microbes *Azotobacter chroococcum*, *Rhizobium leguminosarum*, *Bacillus megaterium*, and *Pseudomonas fluorescens* [154]. A composite of ZEO NPs–CS/sago starch-based biopolymer with mean particle size of 12.80 nm releasing 64.00% P and 41.93% of urea, which was able to maintain water level, was reported to be suitable for the use as slow-release fertilizer [155].

Clinoptilolite is a natural ZEO. For example, in natural and modified mesoporous clinoptilolites $((Na,K,Ca)_{2-3}Al_3(Al,Si)_2Si_{13}O_6 \cdot 12H_2O)$, pores of fracture-type ranging from 25 50 nm to 100 nm between clinoptilolite grains were estimated, while pores between crystal aggregates achieved up to 500 nm [156]. Combined application of nano-clinoptilolite with HA NPs can be used as an effective P fertilizer in calcareous soil [157]. Nanoscale clinoptilolite (90–110 nm) fabricated by ball milling loaded with $ZnSO_4$ exhibited Zn release for 5.4-fold longer period compared to free $ZnSO_4$ (1,176 h vs 216 h), ensuring improved NUE by crops [158].

Addition of macro- and nano-clinoptilolite particles (20, 2.0, and 0.2 μm) at a dose of 1% to loamy sand soil modified its pore-size distribution, resulting in enhanced amount of microporosity, and thus improved water retention. The increases of available water (by 3.6–14.7%) and soil water storage (by 6.8–10.5%) were greater with smaller NP sizes; the best results were observed with nanosized particles, application of which can ameliorate efficiency of water use in dry areas [146].

Micro- and nanosized clinoptilolite or surfactant-modified ZEO applied at a dose of 60 g/kg to soil lysimeters exhibited reduced leaching of NO_3–N compared to control by 22% (clinoptilolite) and 26% (surfactant-modified ZEO), respectively, whereby the effect of particle size was not pronounced. On the other hand, grain yield, grain nitrogen content, stover dry matter, and N uptake by maize plants were considerably higher at clinoptilolite application [159].

Clay minerals are hydrous aluminum phyllosilicate minerals, in which Al and Si ions are bonded into tiny, thin plates by interconnecting O and OH ions. To clay minerals belong MMT $(R_{0.33}Al_2Si_4O_{10}(OH)_2 + nH_2O$, where R represents one or more of the cations Na^+, Ka^+, Mg^{2+}, Ca^{2+}, and possibly others), bentonite $(Si_4(Al_{(2-x)} R_x)O_{10}(OH)_2(CE)_x \cdot nH_2O$, where CE represents exchangeable cations Ca^{2+}, Na^+, Mg^{2+}, and R_x represents Mg, Fe, Mn, Zn, Ni), kaolinite $(Al_2Si_2O_5(OH)_4)$, and illite $(K,H_3O)(Al,Mg,Fe)_2(Si,Al)_4O_{10})$ [160].

Hydrogel fertilizer was prepared by Azarian et al. [161], in which MMT–urea nanocomposite was encapsulated in nanofibers with poly(vinyl alcohol) (PVA) used as the wall matrix and intercalated MMT–urea contained 30.4% nitrogen. K-containing microspheres based on chitosan (CS) and MMT clay were characterized with two specific periods of fertilizer release: at the first 3 days, the external fertilizer on the microspheres was released followed by the release of internal fertilizer, suggesting that CS–MMT microspheres can be considered as a sustainable potassium fertilizer delivery system [162]. Nanobentonite particles applied to sandy soil at a dose of 500 kg/ha increased available water and water-holding capacity, ensured considerable increase of wheat yield, and enhanced N, P, and K content in the soil as well as in the grains and straw of wheat crops grown both in pots and under field conditions [163]. Nanoclay/superabsorbent polymer composites (NCPCs) fabricated using nanoclays kaolinite, illite, and smectite, which were loaded with $(NH_4)_2HPO_4$ and urea solution and tested in Alfisol, Inceptisol, and Vertisol soils, showed higher effectiveness related to cumulative P and total mineral N recovery than conventional fertilizer, achieving even by 88% and 27% higher values for nanocomposites doped with smectite, which were characterized with good slow-release property as well [164]. Noh et al. [165] in their review paper discussed slow-release fertilizers based on MMT, ZEO, and layered double hydroxide phases serving as a host for nutrients, especially nitrogen.

3.2.3 Other potassium and phosphorus nanofertilizers

Potassium NPs (KNPs) (21–30 nm) biosynthesized using leaf extract of *Morus alba*, which were sprayed on wheat plants in a field experiment, stimulated the plant growth, which was reflected in the increased number of spikelets per spike, number of spikes, crop yield, total protein content, and photosynthetic pigment levels compared to treatment with K_2SO_4. KNPs increased the activities of acid phosphatase and alkaline phosphatase, suggesting better P use efficiency by the treated plant. Moreover, wheat plants, which were sprayed with KNPs, were able to utilize the maximum amount of exchangeable K^+ from soil, ensuring decline of K loss via leaching [166]. K-humate nanofertilizer applied as foliar spray at a dose corresponding to 60%, 30%, and 10% of a dose recommended for traditional K fertilizer (control treatment) to tomato plants, increased height, fresh and dry biomass of plants, number of fruit/plant, fruit yield, macronutrient (N, K, P) and micronutrient (Fe, Mn, Zn) content of fruits, and levels of assimilation pigments in leaves; in addition, it increased K^+/Na^+ ratio, total soluble solids, total soluble sugar, and vitamin C in fruits compared to control plants. Moreover, foliar spraying at a rate of 10% was able to reduce losses of K through soil depths [167]. Similarly, foliar treatment of broad bean grown in sandy soil with P (250 and 500 ppm) and K (150 and 300 ppm) nanofertilizers increased the plant yield, and best results were observed using spraying with 500 and 300 ppm of P and K, respectively. Increasing rates of applied nanofertilizer resulted in increased concentrations of nutrients in plants compared to control, and increased the amount of N, P, and K in soil after harvesting [168].

By coating of conventional fertilizers with nanomaterial ameliorated NUE can be achieved. Diammonium phosphate (DAP) fertilizer coated with $KFeO_2$ NPs of 7–18 nm (10%) exhibited controlled release of P, which increased from 14.5 µg/g (day 1) to 178.6 µg/g at day 60 in loam soil, and after 30 days of incubation, the maximum release of 50.4 µg/g NH_4^+–N was observed; also the release of K and Fe ions from this nanofertilizer in 60 days was higher compared to traditional DAP in clay loam soil [169]. The beneficial impact of 10% $KFeO_2$ NP-coated DAP fertilizer on the growth attributes, yield parameters, and uptake of nutrients by wheat crop was observed by Saleem et al. [170]. Nanoscale K fertilizer applied either as soil dressing or foliar spray to Zaghloul date palm resulted in considerably increased fruit retention, bunch weight, and yield per palm compared to application of K_2SO_4. Moreover, at treatment with KNPs increased fruit weight, total soluble sugar, and anthocyanin levels, while decreased total acidity and total tannin levels were observed compared to K_2SO_4 application [171].

3.2.4 Calcium nanofertilizers

Investigation of the impact of nitrogen (urea, consisting of 46% N) and calcium (Ca chelate containing 10% Ca) fertilizers and nanoscale Lithovit fertilizer (consisting of 80% $CaCO_3$, 4.6% $MgCO_3$, and 0.75% Fe) on yield and quality of strawberries showed that both N and Ca fertilization affected sugars, organic acids, volatile, and phenolic contents in strawberry fruits, and decreased the strength of ketone and terpenoid fruity aroma; however, higher doses of fertilizers increased the levels of hexanal (up to 3.8-fold) and (E)-2-hexen-1-al (up to 3.7-fold) belonging to unpleasant aroma aldehydes. On the other hand, fertilization with nanoscale Lithovit ensured the highest content of total phenols and individual hydroxycinnamic and hydroxybenzoic acid derivatives, suggesting its beneficial impact on the levels of phenolic secondary metabolites and aroma of strawberry [172]. CaO NPs prepared using shells of marine mollusks *Limalima* (29 nm), *Lottioidea* (32 nm), and *Oliva reticularis* (25 nm) and applied at a dose of 250 ppm were found to be suitable to be used as priming agents for improving green gram seed germination, root and shoot growth, as well as vigor index, and they were more effective than $CaCO_3$ and normal control. The highest stimulating effect on root growth and seed quality characteristics showed CaO NPs fabricated from *Lima lima*. The utilization of waste molluscan shells as fertilizers can also contribute to reducing environmental pollution [173].

Soils amended with CaNPs (100, 75, and 50 mg CaNPs/kg) showed improved fertility, reduced salinity, and caused more effective immobilization of heavy metals, and CaNPs improved morphological and physiological characteristics and photosynthetic efficiency of cultivated *Moringa oleifera* plants. Moreover, CaNPs considerably improved both macro- and micronutrient levels of plants and contributed to higher translocation rates of essential nutrients [174]. Vidak et al. [175] assessed the impact of calcite NPs on sweet pepper (*Capsicum annuum* L.) fruit quality and found that they reduced the yield but increased fruit firmness. However, the beneficial impact of foliar-applied calcite NPs on the content of macronutrients and micronutrients depended on the cultivar and technological/physiological maturity, and fruit's morphological properties were affected by NP treatment only at the second harvest.

Foliar treatment of *Oryza sativa* with calcite NPs (40, 80, 120, and 160 ppm) had beneficial impact on growth, yield, seed quality, and insect resistance of plants; best results were observed with a dose of 120 ppm, resulting in reduced period needed for 50%

flowering by 6 days and yield increase by 25% as well as improved seed quality attributes compared to control plants. The best insecticidal effects against leaf mites, stem borer damage, and paddy bugs were observed at an application of 160 ppm calcite NPs [176]. Combined application of foliar spraying with calcite NPs with recommended soil-added fertilizer improved the rice yield by ca. 1 ton/ha, and even reduction of soil-amended fertilizer by 20% did not affect considerably the yield [177].

Improvement of the growth of tomato seedlings using a 1.5% amendment of nano-composite hydrogel with calcium MMT combined with peat moss-based substrate was reported by Melo et al. [178]. The beneficial impact on the yield of *Arachis hypogaea* L. plants by both foliar and in soil applications of Ca nanochelates was observed by Nobahar et al. [179]. Ca content in plant organs of groundnut considerably increased after foliar application of CaO NPs (69.9 nm), suggesting that in contrast to bulk Ca, the NPs moved through phloem tissue, and consequently, they can be used as an appropriate Ca source for plants via foliar spraying [180]. N-nanofertilizer prepared by postsynthetic modification of NO_3^--doped amorphous $Ca_3(PO_4)_2$ NPs with urea, which was applied at a dose corresponding to half absolute N-content used in conventional urea treatment, showed comparable stimulation of root and shoot biomass with urea application and exhibited up to 69% NUE [181]. At soil and foliar treatment of Pinot gris grapevines with urea-doped $Ca_3(PO_4)_2$ NPs using a dose of 36 kg N/ha, qualitative and quantitative characteristics of vine plants were comparable with those estimated by using conventional fertilizer at a dose of 45 kg N/ha, suggesting that such nanofertilization can be used as an alternative nitrogen fertilization strategy in modern viticulture [182].

3.2.5 Magnesium nanofertilizers

Root dip application of 100 ppm MgO NPs increased plant growth, Chl, carotenoid, seed protein, and root and shoot N contents in *Vigna unguiculata* L. infected with *Meloidogyne incognita*. Treatment with MgO NPs also reduced nematode fecundity, decreased the number of galls, and smaller size of galls were observed [183]. Foliar application of 0.5 g/L bulk Fe + 0.5% Mg NPs on black-eyed pea plants considerably improved yield, leaf Fe content, stem Mg content, plasma membrane stability, and Chl content, likely due to more efficient photosynthesis [184]. Green MgO NPs of 12 nm synthesized using marine brown alga, *Turbinaria ornate*, which were applied as a nanopriming agent, improved germination and seedling vigor of *Vigna radiata* to a greater extent than the application of hydropriming [185].

MgNPs foliar applied at a dose of 50 ppm to bean plants effectively enhanced the plant biomass and content of photosynthetic pigments, and increased the activity of nitrate reductase (NR); improved yield of pods was observed when MgNPs were doubled. However, increasing concentration of MgNPs was reflected in a declined content of carotenes [186]. Foliar treatment of cotton plants with 60 ppm MgO NPs showing a size of 50 nm, which was performed at vegetative and boll formation stages, resulted in markedly increased number of opened bolls per plant and single plant yield; seed cotton yield increased even by 42.2% compared to control, while with application of bulk MgO and MgSO$_4$, this increase achieved only 39.9% and 24.8%, respectively. Further, the beneficial impact on MgO NP treatment was reflected in increased N, P, and K levels as well as Mg content in cotton plants [187].

Treatment with 100 mg/L of green-synthesized MgO NPs (5–25 nm) ameliorated germination and growth of Z. *mays* plants, and enhanced Chl content in plants [188]. The same dose of Mg(OH)$_2$ NPs was able to increase biomass and reduce Cd concentration in Chinese cabbage grown under Cd stress to a greater extent than treatment with 100 mg/L of bulk Mg(OH)$_2$ [189]. Mg(OH)$_2$ NPs green synthesized by Shinde et al. [190] and applied at a dose of 500 ppm promoted seed germination and ameliorated growth characteristics of Z. *mays* plants. Improved growth characteristics of *Hordeum vulgare* plants were observed at an application of MnFe$_2$O$_4$ NPs, which after uptake from the roots translocated to the leaves, where they reached 4- to 7-fold higher levels than in the control plants [191].

3.2.6 Nanosized sulfur fertilizers

Synthesis, modifications, and applications of zero-dimensional sulfur nanomaterials, including their beneficial impact on plant growth, were overviewed by Jin et al. [192]. Kurmanbayeva et al. [193] fabricated a long-acting fertilizer by encapsulating molten elemental sulfur and impregnating with a solution of calcium polysulfide. Among three types of S-containing NPs, that is, powdered, pasty sulfur-containing composition, and a solution of calcium polysulfide, the powdered and dissolved S-containing fertilizers showed a beneficial impact on the early ripeness and increased productivity of wheat. Moreover, these NPs increased the resistance of wheat plant against phytopathogens. Burkitbayev et al. [194] evaluated the impact of powdery, solute, and pasty sulfur-containing agrochemicals, in which elemental S was in a nanostructured state and found that powdered and solute sulfur-containing agrochemicals contributed to increased yield of soy and improved supply of proteins in the grains.

In saline soil treatment with nanosulfur at a dose of 15 kg/fed pronouncedly increased the content of available nutrients (N, P, K) by 17%, 29%, and 20% and micronutrients Fe, Mn, and Zn by 24%, 27%, and 29%, respectively, in maize plants compared to control; a considerable increase of these nutrients was also estimated in seeds of treated plants. Moreover, a strong increase in Z. *mays* productivity was reflected in 38% and 34% increase of grain and straw yields. This may be associated with oxidizing of amended S to SO$_4^{2-}$ and reduced soil pH; the formed H$_2$SO$_4$ increased the solubility of nutrients, resulting in their higher availability for plant uptake [195].

Priming of wheat seeds with 100 μM SNPs showed a beneficial impact on photosynthetic pigments, nitrogen metabolism, antioxidant status, and ionic relations of T. *aestivum* plants, and contributed to the enhancement of growth attributes under salinity stress (100 and 200 mM NaCl). Although treatment with SNPs at saline conditions did not affect photosynthesis, it induced glutathione (GSH) levels and increased activities of antioxidant enzymes (CAT, POD, SOD, ascorbate peroxidase, and polyphenol oxidase) and proline content. SNPs reduced the uptake of N and P, the P contents were rebalanced, and Na uptake diminished, resulting in improved growth attributes in wheat plants [196]. Priming of *Helianthus annuus* L. seeds with 12.5–200 µM of green-synthesized SNPs and subsequent irrigation of 14-day-old plants with 100 mM $MnSO_4$ causing metal stress considerably stimulated activities of antioxidant enzymes (CAT and SOD) and increased the levels of nonenzymatic antioxidants such as flavonoids and phenolics, while pronouncedly decreased the reactive oxygen species (ROS; $O_2^{\cdot-}$, H_2O_2) levels and lipid peroxidation, resulting in reduced oxidative damage of sunflower plants. In contrast, pronounced reduction in GSH content observed in the presence of SNPs can be explained with the consumption and incorporation of GSH into biosynthesis of other chelating ligands [197]. Kaya et al. [198] tested the effects of SNPs used as an amendment to soil, seed coating, seed coating + application of booting stage, and seed coating + application of heading stage on growth characteristics and protein ratio of bread wheat plants. The best results were recorded with the application of SNPs in seed coating + booting period, increasing grain yield by 14% compared to control.

3.3 Micronutrient nanofertilizers

As essential micronutrients of vascular plants Zn, Cu, Fe, B, Mn, Mo, Ni, and Cl are considered, nonessential nutrients for plants are Co, I, Na, and Se. Application of iodine derivatives supports biomass production and contributes to increased antioxidant levels in plants, enabling to mitigate abiotic stresses [11, 199]; Na shows positive impact on plants, especially when K^+ is deficient [11, 200]; and Se besides ameliorating plant growth increases plant tolerance to environmental stresses via improving activities of antioxidant enzymes and nonenzymatic antioxidants [37, 201, 202]. Co is not classified as an essential nutrient for all vascular plants; however, it is considered as a beneficial element for plants because it is required for bacterial nitrogen (N) fixation within the nodules of leguminous plants. It is a component of some enzymes and coenzymes affecting plant growth. It inhibits the production of ethylene, which regulates germination, stimulates growth, and reduces an adverse impact of abiotic stresses on plants [11, 36, 203]. The essential micronutrient chlorine is involved in several physiological metabolism processes related to plant growth and development, in osmotic and stomatal regulation, and in photosynthetic O_2 evolution, and can contribute to increased resistance and tolerance of plants [36].

3.3.1 Nanofertilizers with essential micronutrients

3.3.1.1 Iron nanofertilizers

Iron is an essential element for vascular plants, which has a crucial role in metabolic processes, including DNA synthesis, respiration, and photosynthesis. It serves as a component of cytochromes and ferredoxins, which function as electron carriers in the photosynthetic electron transport chain, respiratory complexes, and it is a component of many vital enzymes involved in redox reactions. Moreover, it is involved in Chl biosynthesis and is indispensable for the maintenance of chloroplast structure and function [36, 113, 204].

Green-synthesized FeNPs used as fertilizer of *S. bicolor* L. plants grown in calcareous soils increased the growth, levels of photosynthetic pigments, as well as Fe uptake of the sorghum plants [205]. Treatment of *Morus alba* L. plants with ethylenediaminetetraacetic acid (EDTA) functionalized iron oxide NPs, applied at a dose of 10 mg/kg soil, enhanced the number of leaves, root biomass, and shoot biomass by 52.73, 37.20%, and 90.24%, respectively, compared to control plants, and the period needed for the first leaf appearance was shortened as well. Moreover, in treated plants, the Chl and sugar contents increased by 42% and 15%, respectively, compared to control plants along with increasing the activities of antioxidant enzymes [206]. Fe^{2+} encapsulated in egg-derived phosphatidylcholine liposomes of 200 nm, used for foliar treatment of *Ocimum basilicum* L. plants grown in pots with calcareous loamy soil, increased plant biomass, total leaf area, Chl and Fe levels, as well as essential oil (EO) content compared to $FeSO_4$ fertilizer, suggesting that such nanoformulation can be used as an excellent fertilizer for delivery of Fe^{2+} to plants [207].

Priming of rice seeds with 40–80 mg/L of Fe^0NPs fabricated using fruit peel waste of *Punica granatum* L. effectively stimulated early plant growth due to increased ROS generation, increased activities of hydrolytic enzymes, and improved Fe uptake. Using seeds nanoprimed with Fe^0NPs in field experiments, it was found that treated plans had higher crop yield and grain nutrient concentrations compared to hydroprimed seeds, and on application of a dose of 80 mg/L Fe^0NPs, even 1.53-fold higher crop yield was observed [208]. In another field study, it was found that nanopriming with Fe^0NPs resulted not only in higher growth and biomass but also in broader leaves and tiller number of rice plants compared to control plants but also in improved photosynthetic efficiency and higher accumulations of starch, soluble sugars, proteins, lipids, phenol, ascorbic acid, thiamine, and riboflavin as well as macro- and micronutrients in grains; a dose of 20 mg Fe^0NPs/L used for nanopriming ensured even 3.8-fold higher yield compared to control [209]. The beneficial impact of priming of rice seeds with 20 mg/L iron oxide NP green synthesized using *Cassia occidentalis* L. flower extract on germination rate and seedling growth, sugar and amylase content, and activity of antioxidant enzymes was also reported by Afzal et al. [210]. The positive effect of nanopriming with Fe_3O_4 NPs on germination and early growth of maize plants was described by Neto et al. [211]. In contrast to Fe_2O_3 NPs, a dose of 50 mg/L Fe^0NPs and Fe_3O_4 NPs stimulated growth of *O. sativa* plants cultivated in hydroponium, higher Fe concentrations accumulated in treated plants compared to those exposed to Fe_2O_3 NPs, and reduced oxidative stress was observed [212].

Maize and cucumber plants fertilized with $FePO_4$ NPs used nutrients with much higher efficiency compared to bulk $FePO_4$. However, while cucumber plants used $FePO_4$ NPs mainly as a source of P, maize plants favored their use as a source of Fe [213]. Green-fabricated Fe nanorods applied at a dose of 50 mg/L to *Z. mays* plants ameliorated leaf area, number of leaves per plant, total Chl content, and nitrate content by 13, 26, 80%, and 6%, respectively, compared to control plants, and positively affected plant's antioxidative activity due to their ability to form complexes with metal ions [214]. Exposure of wheat plants cultivated in hydroponium to Fe_3O_4 NPs increased root length, plant height, biomass, and Chl content of plants, the NPs were translocated to the leaves, and a strong increase of Fe_3O_4 NPs in treated plants was assumed to be responsible for improved plant growth. Fe_3O_4 NPs showing sizes of 20–40 nm were found to be more effective than NPs with sizes of 8–10 nm and 30–50 nm, respectively [215]. A 100% germination of rice and maize seeds was observed on treatment with 500 ppm γ-Fe_2O_3, whereby values of seedling vigor index achieved 27,500 and 36,500, respectively, and foliar application of the same dose was also effective at field conditions, which was reflected in improved grain characteristics, including yield (kg/ha) [216].

Uncoated Fe_2O_3 NPs as well as fulvic acid-coated Fe_2O_3 NPs applied to soil or as foliar treatment to *G. max* plants strongly increased Chl content, plant biomass, and root developmental indices, being more effective as Fe–EDTA (conventional fertilizer); by foliar application of NPs, 2–4-fold higher shoot Fe content was achieved compared to soil amendment. Moreover, fulvic acid-coated Fe_2O_3 NPs explicitly promoted biological N fixation, which was particularly manifested in the formation of root nodules [217].

High efficiency of EDTA-grafted Fe_3O_4 NPs on improvement of morphological characteristics and Chl levels of *H. annuus* plants along with an increase in Fe content by 37% relative to the control was reported by Shahrekizad et al. [218]; generally, soil amendment was found to be more effective than foliar application. Positive impact of citric acid-coated Fe_3O_4 NPs (50 and 100 mg Fe/L) on growth of *G. max* and *Medicago sativa* plants reflected in pronouncedly higher root surface and shoot weight was observed by Iannone et al. [219].

Presoaking of seeds of legumes with α-Fe_2O_3 NPs containing 5.54×10^{-3} mg Fe/L resulted in improved root growth by 88–366% [220]. Although foliar application of γ-Fe_2O_3 NPs on *Citrus maxima* plants had only negligible impact, strong adsorption ability of these NPs was able to diminish nutrient loss, suggesting their suitability to be used as foliar fertilizer [221]. γ-Fe_2O_3 NPs applied at a dose of 50 mg/L enhanced soluble sugar, soluble protein, and Chl levels in the watermelon plants. They induced oxidative stress in watermelon leaves, but this stress was eliminated with the growth of watermelon. Moreover, application of γ-Fe_2O_3 NPs at optimal concentration was able to improve chlorosis due to Fe deficiency and stimulated the growth of *C. maxima* plants

[222]. γ-Fe_2O_3NPs and Fe_3O_4NPs stimulated growth of muskmelon plants, increased Fe levels in plant organs, and a dose of 200 mg/L γ-Fe_2O_3NPs and 100 and 200 mg/L Fe_3O_4 NPs enhanced fruit weight by 9.1%, 9.4%, and 11.5%, respectively [223].

Fe_2O_3 NPs increased root length, plant height, and biomass of *A. hypogaea* plants by affecting phytohormone levels and activities of antioxidant enzymes; adsorption of Fe_2O_3 NPs on sandy soil ameliorated Fe bioavailability resulting in increased Fe levels in peanut plants [224]. Foliar application of magnetic Fe_3O_4 NPs green fabricated using *O. basilicum* extract was reported to improve both morphological characteristics and chemical composition of *H. annuus* and *Vicia faba* plants more than Fe_3O_4 NPs prepared using extracts of *Mentha varidis* L and *Vinca rosea* plants [225]. Chitosan (CS)-coated Fe_3O_4 NPs (3–22 nm) applied at doses of 200 and 400 mg/kg considerably improved seed germination and seedling growth of *Capsicum annuum* L., being more effective than bare NPs [226]. Also, polyethylene glycol (PEG)-coated Fe_3O_4 NPs applied for soaking of *Phaseolus vulgaris* L. seeds at a dose of 1,000 mg Fe/L increased root growth of plants, likely due to enhanced water content mediated by PEG coating [227].

3.3.1.2 Copper nanofertilizers

Copper is an essential metal for normal plant growth and development. It is a constituent of plastocyanin, an electron carrier in photosynthetic electron transport chain. It is cofactor for numerous metalloproteins, and component of oxidative enzymes involved in many physiological processes [77, 228, 229]. Impact of copper-based NPs on economically important plants was overviewed by Jampílek and Kráľová [77].

Pronounced improvement of morphological characteristics along with amelioration of the levels of chemical constituents (e.g., macro- and micro-nutrients, total soluble solids, amino acids, and vitamins) of onion plants was achieved by spraying of plants with 10 ppm CuO NPs, whereby this treatment was more effective than the application of 20 ppm $CuSO_4$ or 20 ppm of Cu chelates [230].

Biogenic CuNPs sprayed on leaves of *L. sativa* plants (20 mg/plant) increased the dry weight, number of leaves, improved CO_2 assimilation, and enhanced concentrations of nutrients in leaves; no translocation of CuNPs was observed. Similarly, when plants were exposed to CuNPs via soil irrigation, CuNPs did not move from roots to shoots [231].

Spherical CuO NPs with a biodegradable shell consisting of CS–alginate with a size of 300 nm showed controlled release of nutrient and exhibited favorable impact on germination and seedling growth, and a synergic effect on developing of both above- and underground parts of *Fortunella margarita* Swingle plants [232].

Plants of *Allium fistulosum* cultivated in greenhouse and exposed to 150 mg/kg of the CuNPs accumulated higher Cu levels in roots compared to treatment with bulk CuO and $CuSO_4$, and pronouncedly higher Cu and Fe levels in roots and Ca and Mg levels in bulb of treated plants than in control plants were observed; at application of 75–600 mg/kg CuNPs leaf allicin content increased by 56–187% over control plants

[233]. In field experiments, foliar spraying of winter wheat and maize plants in spring vegetation period with CuNPs at a dose of 1% applied twice increased the grain yield of plants, while improved yield of sugar beet roots was achieved using twice a dose of 2 L/ha [234]. Treatment with 100 ppm of biogenic CuNPs with mean size of 41 ± 21 nm and zeta potential of -18.2 mV, prepared using leaf extract of *Azadirachta indica*, stimulated germination and growth of *Vigna radiata* L. plants and increased Cu levels in treated plants over control [235].

On the other hand, treatment of *C. annuum* plants at the vegetative stage with 500 mg CuNPs/kg resulted in pronounced increase or root length and root dry biomass by 58.46% and 187.18%, respectively, compared to control. When such treatment was realized at plant's reproductive stage, photosynthesis and stomatal conductance increased by 42% and 51%, respectively, and Cu leaf concentration was higher by 1,510% compared to treatment with bulk Cu. In contrast, plants fertilized with CuNPs in vegetative stage accumulated in their tissue much less Cu compared to treatment with bulk CuO [236].

Treatment of soil-cultivated *L. sativa* plants with CuO NPs increased shoot's biomass by 16.3–19.1%, likely due to increased transpiration rate and higher stomatal conductance, while the application of bulk CuO improved biomass of lettuce plants. In treated plants, there is higher Cu accumulation in roots, but no changes in shoot Cu content were observed [237].

Biogenic CuNPs with mean size of 11.60 ± 4.65 nm containing Cu_2O and $Cu(OH)_2$ species, which were fabricated using crude aqueous extract of *Avicennia marina* leaves and were applied at a dose of 0.06 mg/mL, stimulated root and shoot growth of wheat seedlings by $172.78 \pm 23.11\%$ and $215.94 \pm 37.76\%$, respectively. Three weeks after foliar treatment of seedlings with this dose of CuNPs, enhanced Chl contents (Chl*a*, Chl*b*, total Chl) were observed [238]. Treatment of *Allium sativum* L. seedlings cultivated in pots with 1 and 2 g/L CuNPs resulted in total yield of 26.91 and 27.83 t/ha, respectively, compared to 22.78 t/ha observed with control plants [239].

3.3.1.3 Zinc nanofertilizers

Zinc is an essential micronutrient for plants; it is a key structural and catalytic component of numerous proteins. Zn activates enzymes responsible for the synthesis of certain proteins and DNA replication, and is involved in Chl biosynthesis, synthesis of cytochrome, and in the formation of auxins affecting growth regulation and stem elongation [36, 113, 240].

Spraying of wheat plants at harvesting stage with 80 ppm ZnO NPs resulted in improved biomass accumulation, seeds per spike, 100 seed weight and biomass accumulation, and the yield was 2.86-fold higher than with application of $Zn(NO_3)_2$, and even 4.55-fold higher compared to control [241]. Investigation of the effect of three ZnO NPs with different morphology, namely, spherical (38 nm), floral-like (59 nm), and rod-like (500 nm) having negative surface charge as well as Zn^{2+} ions on soybean plants cultivated in soil inoculated with *Rhizobium japonicum* for 120 days showed that up to concentration 160 mg Zn/kg the spherical NPs exhibited the best protection against oxidative stress, resulting in higher yield [242]. ZnO NPs (16–20 nm) biosynthesized using *Bacillus* sp. applied to *Z. mays* plants at doses 8 mg/L increased shoot and root length by 61.7% and 56.9%, respectively, and a considerable increase of protein contents and leaf area was observed as well, suggesting that ZnO NPs have great potential to be used as a nutrient in Zn-deficient soils [243].

Priming of *Z. mays* seeds with 80 mg/L ZnO NPs enhanced germination, root length, and dry biomass of maize seedlings by 17%, 25%, and 12%, respectively, compared to control, while at priming with bulk ZnO, the evaluated characteristics were comparable with those of control [244]. Foliar treatment of habanero pepper plants with 1,000 mg/L ZnO NPs improved plant height, stem diameter, and Chl content, and enhanced fruit yield and biomass accumulation compared to control and treatment with $ZnSO_4$; using a dose of 2,000 ZnO NPs mg/L, total phenolics and total flavonoid contents (TFCs) in fruits increased by 14.50% and 26.9%, respectively [245]. ZnO NPs sprayed on leaves of *O. sativa* plants grown in Zn-deficient soil at 15-day interval using a dose of 5.0 g/L exhibited the greatest improvement of plant vegetative characteristics, yield, and grain Zn contents along with the highest amelioration of the chemical and microbial characteristics of Zn-deficient soil such as viable cell counts and dehydrogenase activity. After spraying, the ZnO NPs were observed on the leaf lamina near stomatal openings [246]. Comparison of the effects of three Zn-based fertilizers on biomass production and nitrogen assimilation of *Phaseolus vulgaris* L. cv. Strike plants grown in acid soil (pH 6.8) showed that best results were observed with 25 ppm ZnO NPs, 50 ppm $ZnSO_4$, and 100 ppm of diethylenetriamine pentaacetate (DTPA)-Zn chelate suggesting the potential of ZnO NPs to be used as a nanofertilizer [247]. Foliar treatment of coffee plants with ZnO NPs at a dose corresponding to 10 mg Zn/L increased the dry biomass of roots, stems, and leaves by 28%, 85%, and 20%, respectively, and the photosynthetic rate by 55% when compared to control, and treated leaves contained considerably higher Zn levels (1267.1 ± 367.2 mg/kg dry weight) than plants exposed to $ZnSO_4$ and control plants. This beneficial impact of ZnO NPs on coffee plants can be associated with their increased penetration in leaf tissue [248].

Priming of rice seeds with 20 mg/L ZnO NPs promoted relative water uptake of seeds and root length of seedlings by >50% compared to hydroprimed seeds and caused an enhancement of total soluble sugar levels by 23%, and amylase activity by 45% along with pronounced increase in antioxidant enzyme activities [249]. Foliar spraying of *Mentha piperita* plants in the flowering stage with ZnO NPs using a dose of 1 g/L improved dry biomass and EO content of this medicinal plants, while at an application of 1.5 g/L ZnO NPs, the contents of secondary metabolites such as menthol, menthone, and menthofuran in EO were enhanced up to 28%, 61%, and 237%, respectively [250].

Pullagurala et al. [251] overviewed findings related to the effects of ZnO NPs on terrestrial plants and found that lower concentrations (approx. 50 mg/kg) of ZnO NPs have beneficial impact, while concentrations >500 mg/kg show deleterious effects, with the exception that the growing medium suffers from Zn deficiency.

3.3.1.4 Boron nanofertilizers

Boron (B) is indispensable for the physiological functioning of vascular plants because it is involved, for example, in the structural and functional integrity of the cell wall and membranes, ion fluxes across the membranes, cell division, and N and carbohydrate metabolism, and its deficiency adversely affects the metabolism and growth of plants [113, 252]. Foliar spraying of *L. sativa* and *Cucurbita pepo* plants cultivated in hydroponium for 60 days on a modified Hoagland solution with 30 mg/L of nanosized B fertilizer at 10 days increased the root and shoot growth of lettuce 1.9- and 2.7-fold and production of biomass by 58%, while growth of zucchini increased by 18% and 66% compared to control plants. Moreover, in the presence of B nanofertilizers, the 2,2-diphenyl-1-picrylhydrazyl (DPPH) activity in lettuce was reduced by 32%, while at the application of bulk B only 21% reduction was observed [253]. The use of membrane vesicles derived from plant material and applied to *Prunus dulcis* L. trees in foliar spray as Fe and B nanofertilizers showed that due to invaginations in the plasma membrane of the leaf cells caused by vesicles B and Fe levels in leaves increased, particularly when they were applied in an encapsulated form [254]. Foliar spraying with B nanofertilizer considerably enhanced the root and shoot growth and biological yield of *Beta vulgaris* L. plants, and combined application with fulvic acid and NPK NPs, supplemented also in form of spray, resulted in further improvement of these growth characteristics along with ameliorated parameters of plants [255]. Foliar spraying of olive cv. "Zard" with 300 mg/L of nanochelated B, corresponding to 270 mg of pure B per liter, effectively increased fruit yield, although lower amount of oil accumulation in fruits was observed due to heavy fruit weight. This treatment with nanochelated B generally ameliorated oil yield, free fatty acids, antioxidant activity, and total phenol content of olives but the beneficial impact of nanofertilizer was reflected rather in the growth, and fruit and oil yield than in oil percentage [256]. Zn/B nanofertilizer fabricated by loading $ZnSO_4$ and H_3BO_3 on a CS NP emulsion prepared by ionic gelation with tripolyphosphate and applied as foliar spray on coffee plants using doses of 10–40 ppm enhanced their leaf area, and height and stem diameter, and increased Chl content and photosynthesis of plants as well as uptake of nutrients such as Zn, N, and P [257].

ZnAl-double hydroxides (LDH) associated with borate exhibited controlled-release of Zn and B micronutrients initiating from the anionic exchange of BO_3^{3-} and the transformation of Zn^{2+} from LDH basal plane, and ameliorated plant growth, suggesting their suitability to be used as micronutrient fertilizer [258]. Foliar spraying with nano-B chelate fertilizer applied as a single spray before full bloom of *Punica granatum* cv. Ardestani tree showed that already a dose of 34 mg B per tree increased the pomegranate fruit yield due to increased number of fruits per tree. At the application of nanofertilizer containing 6.5 mg B/L pronounced amelioration of fruit quality reflected in the increase of total soluble solids, a decrease of titratable acidity, strong increase in maturity index, as well as pH unit increase in the juice was observed; total sugars and total phenolic compounds were only slightly affected, and no effect on the antioxidant activity and total anthocyanins was estimated [259]. Foliar fertilization of globe artichoke plants with a mixture of 3,000 ppm of nanosized K and 50 ppm of nanosized B ensured superb vegetative growth parameters, the highest nutrient contents in leaves of globe artichoke [260], and plants irrigated every 20 days achieved very good yield parameters and the highest quality [261]. Three times application of boron NPs at a dose of 0.025% to flame seedless grapevines resulted in pronouncedly ameliorated yield and berries quality [262]. Excellent effectiveness of nanosized boron fertilizer applied at a dose of 4 g/L on the yield of garlic plants was reported by Yassir and Yassen [239].

3.3.1.5 Molybdenum nanofertilizers

Arnon and Stout [263] declared molybdenum as an essential element already in 1939. Many enzymes, including NR, xanthine dehydrogenase, aldehyde oxidase, and sulfite oxidase require Mo for the activity; it is also needed for purine degradation, hormone synthesis, and sulfite detoxification [113]. Foliar spraying with Mo can solve Mo deficiency in plants, enhance activities of molybdoenzymes, and ensure required Mo nutrition needed to optimize the yield [264, 265]. Foliar treatment of common bean plants with 40 ppm MoO_3 NPs showed a beneficial impact on the number of leaves and branches per plant as well as fresh and dry weights of plants and caused a strong increase in the seed yield of up to 84.1% [266]. *O. sativa* plants cv. HUR 3022 cultivated in hydroponics, which were exposed to 100 ppm of α-MoO_3 and MoS_2 NPs showing nanosheet and nanoflower-like structures with crystallite size of 21.34 nm and 4.32 nm, respectively, showed improved growth and enhanced protein levels. Application of MoS_2 NPs resulted in increased Chl*a* level, but in contrast, treatment with 100 ppm α-MoO_3 led to a decrease of Chl content. Lower translocation factor estimated for MoS_2 NPs (0.42–0.65) compared to that of α-MoO_3 NPs (0.6–2.0) indicated higher accumulation of MoS_2 NPs in roots, suggesting that they can be considered as more environmentally safe than α-MoO_3 NPs [267]. Foliar application of MoS_2 sheets loaded with copper NPs (MoS_2-CuNPs) using a dose of 4–32 µg/mL ameliorated the growth of *O. sativa* seedlings and increased Mo and Chl content (up 30.85%) in rice plants [268]. Green-synthesized MoNPs strongly affected fresh and dry biomass, NO_3^- concentration, NR activity, Chl*a*, Chl*b*, and Chl levels, as well as the plant height of spinach plants treated with 3 mol/L NH_4NO_3, whereby the effect of MoNPs on the increase of NR activity and pronounced reduction of NO_3^- accumulation exceeded that of elemental Mo [269]. On the other hand, octahedral hexamolybdenum clusters did not affect germination of *Brassica napus* L. seeds but greatly inhibited plant growth; higher inhibition was observed at the treatment of rapeseed plants with nanosized entities compared to microsized cluster aggregates [270]. However, Mo-based nanomaterials applied at high concentration showed epigenetic toxicity to soybean plants and affected the agronomical and physiological characteristics in *G. max* plants, including the microstructure of the roots and the activity of rhizobium in the symbiotic system, which resulted in a decline of the N fixation capacity of the soybean–rhizobia symbiotic system [271].

3.3.1.6 Manganese nanofertilizers

Manganese (Mn) is classified as an essential metal for vascular plants. It is indispensable for photosynthetic processes in plants. It activates some enzymes involved in tricarboxylic acid cycle and electron transport systems, triggers some enzymes involved in Chl synthesis, and is a constituent of oxygen-evolving complex situated on the donor side of photosystem II. Mn also plays a role in ATP synthesis, biosynthesis of fatty acids, synthesis of amino acids and hormone activation, and it is a cofactor of several enzymes (e.g., Mn–SOD and Mn–CAT) [36, 113, 272]. Mn-based NPs were found to mitigate an adverse impact of abiotic stresses in plants more than their bulk or ionic counterparts [273].

Spraying of squash plants with 20 ppm Mn_3O_4 NPs (20–60 nm) resulted in considerably improved vegetative growth characteristics, levels of photosynthetic pigments, and fruit yield (kg/plant and t/ha), whereby the yield could be further enhanced via co-application of Mn_3O_4 NPs with Fe_2O_3 NPs [274]. Foliar treatment of *C. pepo* L. during planting process with $Mn_{0.5}Zn_{0.5}Fe_2O_4$ NPs synthesized at 160 °C using a dose of 10 ppm increased squash yield by 49.3% and 52.9%, respectively, compared to control plants for two consecutive seasons, while the application of 20 ppm ensured the highest organic matter content and total energy in fruits [275]. MnNPs were found to affect the assimilation process in mung bean plants by enhancing the net flux of nitrogen assimilation via nitrate reductase–nitrite reductase and glutamine synthetase–glutamate synthase pathways and can be considered as a suitable alternative to $MnSO_4$ fertilizer [276]. Foliar treatment of wheat plants with MnNPs considerably increased shoot and grain Mn contents by 37% and 12%, respectively. In addition, an increase of P concentration in soil and shoots by 17% and 43%, respectively, along with 40% reduction of soil nitrate–N was observed [277]. MnO_x NPs applied at a concentration of <50 ppm pronouncedly promoted the growth of *L. sativa* seedlings by 12–54%, showing the potential to be used as effective nanofertilizers [278]. Priming of watermelon seeds with 20 mg/L of Mn_2O_3 NPs (22–39 nm) modified Chl and antioxidant profiles of seedlings, and using a dose of <40 mg/L superb impact on the phenolic acid and phytohormone profiles of the *Citrullus lanatus* seedlings was observed [279]. Nanopriming of *Capsicum annuum* seeds with 0.1–1 mg/L of Mn_3O_4 markedly increased the root growth of seedlings under normal as well as salt-stressed (100 μM) conditions, and Mn_3O_4 was found to penetrate through the seed coat and form NP corona complex [280]. A dose of 25 mg/L Mn_2O_3 NPs was reported to promote growth and metabolic processes, including production of alkaloids in *Atropa belladonna* shoot tip culture grown in Murashige and Skoog (MS) medium [281].

3.3.1.7 Nickel nanofertilizers

Although Ni was declared by Brown et al. [282] as a micronutrient essential for vascular plants already in 1987, it was recognized as the 17th essential element for plant growth and development in 2001 [283]. It is a functional constituent of eight enzymes, including urease that is needful for N metabolism in plants [113, 284]. NiO NPs fabricated using leaf extract of *Berberis balochistanica* as a reducing and stabilizing agent applied at low concentration considerably improved germination compared to control [285]. Barley plants cultured in hydroponium, which were treated with spherical $Ni_{0.4}Cu_{0.2}Zn_{0.4}Nd_{0.05}Y_{0.05}Fe_{1.9}O_4$ NPs of 18 nm at a dose of 500 mg/L for 21 days, increased the levels of K, Ca, Mg, and P in roots but with exception of Ca, their concentrations in the leaves were considerably reduced; incorporation of Zn, Ni, Cu, and Fe into the plant body via the inclusion of NPs was observed as well. Consequently, these NPs containing micronutrients have the potential to be applied as fertilizer to plants showing single or multiple nutrient deficiencies [286]. Treatment with 0.01 and 0.1 mg/L NiNPs of 5 nm had no effect or stimulated young wheat seedlings, and considerably enhanced the intensity of photosynthesis [287].

3.3.2 Nonessential micronutrient fertilizers

3.3.2.1 Cobalt nanofertilizers

An electrospun cowpea seed coating prepared by incorporation of 2,2,4,4,6,6-hexa-aminocyclotriphosphatriene into polyvinylpyrrolidone along with CoNPs showed controlled release of nutrient from the fiber scaffolds and improved nutrient usage by the seedling, whereby the coated seeds were characterized with increased imbibition [288]. CoNPs were also reported to act as an elicitor, and they increased diosgenin production in roots and shoots of fenugreek seedlings cultivated in vitro in MS medium by inducing increased expression of the genes involved in diosgenin biosynthesis pathway, whereby the highest levels of this secondary metabolite were observed using a dose of 300 μmol/L [289]. Co_3O_4 NPs sprayed on 2-week-old *Brassica napus* L. seedlings using doses of 50 and 100 mg/L increased fresh and dry weights of leaves, leaf area, relative water content (RWC), as well as relative Chl content in *B. napus* leaves compared to control, while application of higher doses resulted in strong oxidative damage due to release of Co ions, ROS generation, and increased peroxidation of lipids, resulting in ion leakage [290, 291]. CoNPs applied at a concentration of 1 g/ha functioned as a regulator of the growth and development of cucumber plantlets, and increased length and mass of the plantlets aerial part over control by 7.9% and 17.1%, respectively, while length and mass of the plantlets underground part increased by 9.3% and 41.9%, respectively [292].

Exposure of barley plants grown in hydroponium to 125–1,000 mg/L of Tb-substituted $CoFe_2O_4$ NPs on the growth, physiological indices, and magnetic character of barley (*Hordeum vulgare* L.) for 3 weeks increased the growth (approx. by 38–65%) and biomass (approx. by 2–133%) of seedlings. Based on 26- and 75-fold higher leaf photoluminescence observed in treated plants compared to control, it can be stated that the NPs translocated to leaves improved seedling growth, likely due to incorporation of Fe into tissues and changes in photoluminescence [293]. Treatment of soybean seeds with 0.08 g/ha of nanocrystalline Co powder in a laboratory experiment increased germination percentage to 80% compared to 55.4% observed in control, and root length, shoot length, and seeding vigor index showed a considerable increase over control by 90.9%, 57.3%, and 36.2%, respectively. The beneficial impact of nanocrystalline Co powder reflected in increased Chl index and pronouncedly increased number of nodules compared to control was also observed in the field experiment [294].

3.3.2.2 Silicon nanofertilizers

Si belongs to elements required by plants due to its ability to activate growth and development and regulate overall physiological and metabolic characteristics of plants, and it can attenuate adverse effects of abiotic and biotic stresses on plants [295, 296]. Findings related to favorable impact of SiNPs on growth and development as well as on plant productivity were summarized by Rastogi et al. [297]. Si bioavailability was reported to modify NUE and C and P status, C:N:P stoichiometry, as well as productivity of *T. aestivum* L. plants [298]. A dose of 150 mg/L SiNPs was found to be the most convenient for treatment of *Cymbopogon flexuosus* aromatic grass to stimulate photosynthesis and activities of antioxidant enzymes such as CAT, SOD, and POD as well as geraniol dehydrogenase involved in EO metabolism, and NR catalyzing the first step in reduction of nitrate nitrogen to organic forms, resulting in improved plant growth and yield [299]. Bhat et al. [300] overviewed the uptake and beneficial impact of SiNPs on the growth and development of plants, and application of SiNPs in agriculture, and discussed the superiority of the use of nanosized Si compared to bulk Si fertilizers.

SiO_2 NPs applied before (50 or 100 mg/L) or after (50 mg/L) flowering of strawberry plants, which were exposed to salt stress (25 and 50 mM NaCl), diminished an adverse impact of salinity, which was reflected in improved cell wall thickness, higher content of water, and increased Chl content. In addition, treated plants effectively transferred the absorbed K, P, and Ca from the nutrient solution to leaves. However, in the presence of SiO_2 NPs, negative effects on phyllochron, flowering, and fruit set were observed, and fruit production parameters were reduced as well [301].

Foliar application of SiNPs on *Coriandrum sativum* plants using irrigation regime causing a moderate water stress pronouncedly ameliorated growth and yield of plants as well as yield of EO via improved RWC. They enhanced the content of total soluble solids, total phenolics content (TPC), and TFC and induced effective antioxidant defense

[302]. Foliar treatment of *Theobroma cacao* L. with SiO_2 NPs increased the photosynthetic rate and electron transport and decreased stomatal conductance due to increased levels of N, P, and total soluble protein in leaves [303].

SiO_2 NP uptake suppressed ROS accumulation in roots and leaves, and reduced lipid peroxidation, and SiO_2 NPs were able to restrict the entry of Na^+ ions and other toxic metals in treated plants [304]. Foliar co-treatment with SiNPs and Zn of on-time planted *Z. mays* plants increased the grain yield by 37% compared to control, while at application of Zn alone, the increase was 33%. On the other hand, treatment at delayed planting date resulted in reduction of leaf area index, grain weight, oil percentage, biological yield, and oleic acid content and increased linoleic acid content [305].

The adhesion capacity of foliar nitrogen fertilizer containing as a carrier sea urchin-like micro-nanostructured hollow SiO_2 spheres of ca. 500 nm characterized with high surface roughness on leaves of *A. hypogaea* and *Z. mays* plants increased 5.9- and 2.2-fold, resulting in 2.29-fold higher utilization rate of the nitrogen fertilizer [53]. Mesoporous SiO_2 NPs (50 nm) with a pore size of about 14.7 nm functionalized with NH_2, which were applied at doses of 20 and 50 µg/mL, pronouncedly improved the biomass, photosynthetic pigments, photosynthetic capacity, and seed yield in *Arabidopsis thaliana* plants [306]. Foliar treatment of *O. sativa* and *G. max* plants with SiO_2 NPs using a dose of 1.5 g/L showed a beneficial impact on their shoot dry weight and grain production [307]. Foliar spraying of *Mentha piperita* plants with 50 and 100 mg/L SiNPs exhibited considerable increase of peltate glandular trichome density and diameter, and increased Chl content and net photosynthetic rate, TPC and EO content at 150 days after plantation by 11.5%, 21.1%, 11.9%, and 26.5%, respectively, and an increase of menthol content by 8.85% was observed as well [308]. Both foliar and soil applications of SiNPs and ZnNPs showed favorable impact on growth, yield, and nutrient accumulation in *O. sativa* plant tissue [309]. The treatment of L. sativa plants with 30 mg/L of fluorescent, NH2-functionalized water-soluble Si quantum dots measuring 2.4 nm had a significant impact. As a result, the treated plants exhibited increased root length and plant height. Furthermore, there was a substantial increase in soluble sugar and water content, as well as Chl a and Chl b levels, which increased by 49.8%, 40.9%, 41.0%, and 114.8% respectively. Additionally, the treatment led to an improved activity of photosystem II, resulting in improvement of photosynthesis. [310].

3.3.2.3 Selenium nanofertilizers

Although Se is not essential for growth and development of vascular plants, when applied at low concentrations, it shows a beneficial impact on plant growth, and can function as a quasiessential micronutrient via altering some physiological and biochemical traits [202]. Se acts as an antioxidant either directly or via enhancing the activities of antioxidant enzymes, and in such a way ameliorates the antioxidative capacity of plants and increases the tolerance of plants exposed to stresses [201]. El-

Ramady et al. [311] in their review paper focused on the production, biological effects, and use of SeNPs in agriculture.

Foliar spraying of 20 mg/L SeNPs on pepper leaves resulted in enhanced Chl and soluble sugar contents in plants, enabling activation of phenylpropane and branched-chain fatty acid pathways, as well as *AT3*-related enzymes and gene expressions, resulting in increased production of capsaicinoids, flavonoids, and total phenols along with enhancement of jasmonic, abscisic, and salicylic acid levels due to considerably stimulated expression of phytohormone synthesis genes [312].

Co-exposure of *Melissa officinalis* plants to 10 mg/L SeNPs and ZnO NPs (100 or 300 mg/L), or treatment with 10 mg/L SeNPs obtained in irrigation water, strongly increased the biomass, activation of lateral buds, and promoted development of lateral roots. It enhanced K, Fe, and Zn concentrations in plant organs, and pronouncedly improved the levels of nonprotein thiols, ascorbate levels, and soluble phenols in roots. The treatments also markedly induced the expression of rosmarinic acid synthase and hydroxyphenylpyruvate reductase genes [313]. Foliar spraying of pomegranate tree with 2 µmol/L SeNPs provided over 2 consecutive years resulted in an increase of some morphological characteristics in the first/second year by 22%/25% for leaf area, 34%/24% for Chl content, 35%/28% for number of fruits, and 17%/15% for yield compared to control; levels of total sugars, phenolic compounds, antioxidants, and anthocyanins were enhanced by 17%/19%, 5%/8%,23/25%, and 23%/27%, respectively; a pronounced increase in leaf nutrient (N, P, K, Ca, Fe, Se, and Zn) levels were observed as well [314]. Foliar application of 20 and 40 ppm SeNPs provided during vegetative stage to three groundnut cultivars (NC, Gregory, and Giza) ameliorated their growth, but the treatment had negative impact on the growth of NC cultivar. SeNPs showed a beneficial impact on antioxidant defense systems in tested cultivars, which was reflected in ameliorated plant tolerance under sandy soil conditions [315]. Treatment of radish seeds with nanocomposites of Se embedded in arabinogalactan, starch, and carrageenan matrices promoted root growth during germination, and in root tissues lower levels of diene conjugates were observed. Priming of radish seeds with a solution containing Se nanocomposite embedded in arabinogalactan increased glutathione peroxidase activity in root tissues by 40%. Amelioration of soybean root growth due to treatment with Se-carrageen nanocomposite could also be associated with the activation of other antioxidant enzymes, suggesting that Se acts as a plant growth stimulant [316]. The beneficial impacts of foliar treatment of *Pisum sativum* plants with 10 and 20 ppm SeNPs fabricated using probiotic lactic acid bacteria and mineral Se prepared using Na_2SeO_3 on vegetative growth, yield, and quality as well as mineral contents in pea leaves and seeds were reported by Shedeed et al. [317]. The researchers highlighted the advantage of the use of SeNPs as an environment-friendly product, showing higher safety compared to chemical fertilizer.

3.4 Mitigation of adverse impact of environmental stresses by nanofertilizers

Supplementation of nanoscale nutrients to crops can pronouncedly attenuate the negative impact of abiotic [85, 92, 109, 318–322] as well as biotic stresses [75, 78, 322–325] and improve adaptive mechanisms in plants, resulting in higher yields. For example, alleviation of drought stress by application of SiO_2 NPs [326, 327], SiNPs [302], ZnNPs [328], ZnO NPs [329], nanochelated Si, B, and Zn fertilizers [330], and nanochelated nitrogen and urea fertilizers [331] was reported. Nanoscale nutrients such as HA NPs [130], SNPs [196], SiO_2 NPs [301], SiNPs [332], SeNPs [332, 333], CuNPs [332], Fe_2O_3 NPs [334], and ZnO NPs [335, 336] mitigated negative impact of saline stress on crops, while SeNPs [337] and SiNPs suppressed adverse effects caused by high-temperature stress. SiNP-attenuated UV-B stress-induced damage [338] and radioprotective effects of ZnO NPs were described as well [339]. Nanosized nutrients are also able to increase crop resistance to stress induced by heavy metals [340]; this was reported for HA NPs [128], SNPs [197], FeNPs [341, 342], ZnO NPs [341], CuO NPs, [343], SiNPs [344, 345], and SiO_2 NPs [346]. Favorable impact of nanosized nutrients applied as fertilizers on crops exposed to environmental stresses is mainly due to their ability to induce antioxidative defense system along with upregulation of the expression of abiotic- and biotic stress-related genes.

3.5 Conclusions

Due to increasingly growing human population, the main challenge for the agricultural sector and politicians worldwide is to ensure sufficient food for all, considering the balance between the sustainability of ecosystems and higher crop productivity. Given the global reduction in arable land and water resources, the introduction of emerging environmental technologies, such as nanofertilization, which increase soil fertility, yields, and crop quality, while reducing environmental pollution, is inevitable. The rapid development of nanotechnologies over the last 20 years attracted increased attention to the development and application of nanofertilizers enabling improved NUE, reduction of nutrient doses required by crops, and attenuation of adverse impact of fertilization on the environment. The encapsulation of nutrients in biodegradable matrices of natural origin or polymer coatings contributes significantly to improved stability of nanoformulations and slower and controlled release of active ingredients. As above-mentioned, nanostructured fertilizer formulations are characterized with increased nutrients utilization efficiency and can provide targeted delivery of nutrient, release of nutrients for longer period compared to conventional bulk fertilizers, while in smart nanoformulations, the precise release of nutrients can be induced by environmental triggers, and the timing of nutrient release is adjusted to the plant nutrient demand, which results in improved agronomic yields and reduced environmental contamination. In addition to increasing nutrient utilization efficiency and soil fertility, nanofertilizers are able to significantly suppress soil toxicity, and lower doses of nanofertilizers along with reduced treatment frequencies can greatly diminish their potential adverse effects on nontarget organisms. Moreover, application of nanofertilizers can suppress overfertilization and related adverse environmental effects, such as eutrophication. As higher doses of bulk and nanosized nutrients are generally toxic to living organisms and the range of concentrations between their stimulatory and inhibitory effects may be tight, it is necessary to estimate optimal doses of nanofertilizers for individual crop species with respect to soil properties and methods of application. Moreover, nanosized formulations should also be tested for potential nanospecific toxicity, and it would be desirable to introduce regulations and an evaluation system concerning nanofertilizers that would be useful for introduction of their cost-effective mass production. The ability of nanoscale nutrients used as nanofertilizers to alleviate both abiotic and biotic stresses is particularly important nowadays when crops are increasingly threatened by long periods of drought, elevated salinity levels, floods, or plant diseases caused by phytopathogens. According to the World Wildlife Federation's forecast, by 2025 two-thirds of the world's population will be affected by water scarcity, also due to ongoing climate changes, which will also have a serious impact on ecosystems. By addition of nanofertilizers to the irrigation water, crop nutrient requirements can be met throughout their growth cycle, thereby reducing overall water consumption, even when using micro-irrigation systems that allow precise placement and timing of nanofertilizers, which can be particularly favorable in arid and semi-arid agricultural areas. Due to micronutrient deficiencies (mainly Zn, Fe, or Se) in crop-cultivated soils, approximately 2 billion people worldwide suffer from micronutrient deficiencies, which have a detrimental impact on their health. Nanofertilizers can be effectively applied for agronomic biofortification of crops with essential micronutrients, thus ensuring increased levels of these health and immunity-enhancing minerals in the consumable plant organs and grains. This is particularly important in developing countries, where the population does not have access to appropriate commercial dietary supplements for completing the missing nutrients. Consequently, nanofertilizers can be considered as a valuable tool in the fight against hidden hunger.

References

[1]   World Population Prospects 2019 – Highlights. United Nations, New York, 2019. (Assessed August 23, 2021 at https://population.un.org/wpp/Publications/Files/WPP2019_Highlights.pdf).
[2]   Nkebiwe, P. M., Weinmann, M., Bar-Tal, A., Müller, T. Fertilizer placement to improve crop nutrient acquisition and yield: A review and meta-analysis. *Field Crops Research* 2016, 196, 389–401.
[3]   Agrahari, R. K., Kobayashi, Y., Tanaka, T. S. T., Panda, S. K., Koyama, H. Smart fertilizer management: The progress of imaging technologies and possible implementation of plant biomarkers in agriculture. *Soil Science and Plant Nutrition* 2021, 67, 248–258.

[4]   Bouis, H. E., Hotz, C., McClafferty, B., Meenakshi, J. V., Pfeiffer, W. H. Biofortification: A new tool to reduce micronutrient malnutrition. *Food Nutrition Board* 2011, 32, S31–S40.

[5]   Saha, S., Roy, A. Whole grain rice fortification as a solution to micronutrient deficiency: Technologies and need for more viable alternatives. *Food Chemistry* 2020, 326, 127049.

[6]   Izydorczyk, G., Ligas, B., Mikula, K., Witek-Krowiak, A., Moustakas, K., Chojnacka, K. Biofortification of edible plants with selenium and iodine – A systematic literature review. *Science of the Total Environment* 2021, 754, 141983.

[7]   Pandey, N. Role of plant nutrients in plant growth and physiology. In Plant Nutrients and Abiotic Stress Tolerance. Springer, Singapore, 2018, 51–93.

[8]   Marques, E., Darby, H. M., Kraft, J. Benefits and limitations of non-transgenic micronutrient biofortification approaches. *Agronomy-Basel* 2021, 11, 464.

[9]   Praharaj, S., Skalicky, M., Maitra, S., Bhadra, P., Shankar, T., Brestic, M., Hejnak, V., Vachova, P., Hossain, A. Zinc biofortification in food crops could alleviate the zinc malnutrition in human health. *Molecules* 2021, 26, 3509.

[10]   Veena, M., Puthur, J. T. Seed nutripriming with zinc is an apt tool to alleviate malnutrition. *Environmental Geochemistry and Health* 2022, 44, 2355–2373.

[11]   Micronutrients: How Fertilizers help to address the problems of hunger and malnutrition. IFA, 2020. (Assessed August 23, 2021 at https://www.fertilizer.org/key-priorities/human-nutrition/micronutrients/).

[12]   Siwela, M., Pillay, K., Govender, L., Lottering, S., Mudau, F. N., Modi, A. T., Mabhaudhi, T. Biofortified crops for combating hidden hunger in South Africa: Availability, acceptability, micronutrient retention and bioavailability. *Foods* 2020, 9, 815.

[13]   Giller, K. E., Zingore, S. Mapping micronutrients in grain and soil unearths hidden hunger in Africa. *Nature* 2021, 594, 31–32.

[14]   Lowe, N. M. The global challenge of hidden hunger: Perspectives from the field. *The Proceedings of the Nutrition Society* 2021, 80, 283–289.

[15]   Sutton, M., Bleeker, A., Howard, C. M., Bekunda, M., Grizzetti, W., De Vries, W., Van Grinsven, H. J. M., Adhya, Y. P., Billen, G., Davidson, E. A., Datta, A., Diaz, R., Erisman, J. W., Liu, X. J., Oenema, O., Palm, C., Raghuram, N., Reis, S., Scholz, R. W., Sima, T., Westhoek, H., Zhang, F. S. Our nutrient world: The challenge to produce more food and energy with less pollution. Global overview of nutrient management. Centre for Ecology and Hydrology, Edinburgh on Behalf of the Global Partnership on Nutrient Management and the International Nitrogen Initiative, 2013. (Assessed August 23, 2021 at http://nora.nerc.ac.uk/id/eprint/500700/1/N500700BK.pdf).

[16]   Martínez-Dalmau, J., Berbel, J., Ordóñez-Fernández, R. Nitrogen fertilization. A review of the risks associated with the inefficiency of its use and policy responses. *Sustainability* 2021, 13, 5625.

[17]   Karthik, R., Robin, R. S., Anandavelu, I., Purvaja, R., Singh, G., Mugilarasan, M., Jayalakshmi, T., Samuel, V. D., Ramesh, R. Diatom bloom in the Amba River, west coast of India: A nutrient-enriched tropical river-fed estuary. *Regional Studies in Marine Science* 2020, 35, 101244.

[18]   Grenon, G., Singh, B., De Sena, A., Madramootoo, C. A., Von Sperber, C., Goyal, M. K., Zhang, T. Q. Phosphorus fate, transport and management on subsurface drained agricultural organic soils: A review. *Environmental Reserach Letters* 2021, 16, 013004.

[19]   Hendricks, G. S., Shukla, S., Roka, F. M., Sishodia, R. P., Obreza, T. A., Hochmuth, G. J., Colee, J. Economic and environmental consequences of overfertilization under extreme weather conditions. *Journal of Soil and Water Conservation* 2019, 74, 160–171.

[20]   Kiiski, H., Scherer, H. W., Mengel, K., Kluge, G., Secerin, K. Fertilizers, 1. General. In Ullmann's Encyclopedia of Industrial Chemistry. Weinheim, Wiley-VCH, 2016, (Assessed August 23, 2021 at https://doi.org/10.1002/14356007.a10_323.pub4).

[21] Bogaard, A., Fraser, R., Heaton, T. H. E., Vaiglova, M. W. P., Charles, M., Jones, G., Evershed, R. P., Styring, A. K., Andersen, N. H., Arbogast, R. M., Bartosiewicz, L., Gardeisen, A., Kanstrup, M., Maier, U., Marinova, E., Ninov, L., Schäfer, M., Stephan, E. Crop manuring and intensive land management by Europe's first farmers. *Proceedings of the National Academy of Sciences of the United States of America* 2013, 110, 12589–12594.

[22] Louchheim, J. Fertilizer history: The Haber-Bosch process. 2014. (Assessed August 23, 2021 at https://www.tfi.org/the-feed/fertilizer-history-haber-bosch-process).

[23] All Nobel Prizes in Chemistry. 2021. (Assessed August 23, 2021 at https://www.nobelprize.org/prizes/lists/all-nobel-prizes-in-chemistry/).

[24] Russel, D. A., Williams, G. G. History of chemical fertilizer development. *Soil Science Society of America Journal* 1977, 41, 260–265.

[25] Agwe, J., Morris, M., Fernabdes, E. Africa's growing soil fertility crisis: What role for fertilizer? Agricultural and Rural Development Notes, No. 21. World Bank. 2007. (Assessed August 23, 2021 at https://openknowledge.worldbank.org/handle/10986/9573).

[26] Rajasekar, M., Udhaya Nandhini, D., Suganthi, S. Supplementation of mineral nutrients through foliar spray-A review. *International Journal of Current Microbiology and Applied Sciences* 2017, 6, 2504–2513.

[27] Dossa, E. L., Arthur, A., Dogbe, W., Mando, A., Snoeck, D., Afrifa, A., Acquaye, S. Improving fertilizer recommendations for cocoa in Ghana based on inherent soil fertility characteristics. In Improving the Profitability, Sustainability and Efficiency of Nutrients through site Specific Fertilizer Recommendations in West Africa Agro-Ecosystems. Vol. 1. Cham, Switzerland, Springer, 2018, 287–299.

[28] Ejigu, W., Selassiea, Y. G., Eliasb, E., Damte, M. Integrated fertilizer application improves soil properties and maize (*Zea mays* L.) yield on Nitisols in Northwestern Ethiopia. *Heliyon* 2021, 7, e06074.

[29] Niu, J. H., Liu, C., Huang, M. G., Liu, K. Z., Yan, D. Y. Effects of foliar fertilization: A review of current status and future perspectives. *Journal of Soil Science and Plant Nutrition* 2021, 21, 104–118.

[30] Krasilnikov, P., Taboada, M. A., Amanullah. Fertilizer use, soil health and agricultural sustainability. *Agriculture* 2022, 12, 462.

[31] Mikkelsen, R. L. The "4R" nutrient stewardship framework for horticulture. *Horttechnology* 2011, 21, 658–662.

[32] Snyder, C. S. Enhanced nitrogen fertiliser technologies support the '4R' concept to optimise crop production and minimise environmental losses. *Soil Research* 2017, 55, 463–472.

[33] Integrated Plant Nutrient Management. IFA, 2018 (Assessed August 23, 2021 at https://www.fertilizer.org/resource/integrated-plant-nutrient-management/).

[34] The crucial role of plant nutrition in the 2030 sustainable development agenda. IFA, 2018 (Assessed August 23, 2021 at https://www.fertilizer.org/resource/the-crucial-role-of-plant-nutrition-in-the-2030-sustainable-development-agenda/).

[35] Singh, J. P. Soil Fertility, Nutrient and Water Management. Write and Print Publications, 2017.

[36] Marschner, H. Mineral Nutrition of Higher Plants, 2nd ed. Amsterdam Academic Press, 1995.

[37] 19 essential nutrients for improving and protecting plant health. IFA, 2020 (Assessed August 23, 2021 at https://www.fertilizer.org/resource/19-plant-nutrients-for-improving-and-protecting-plant-health/).

[38] Tripathi, P., Na, C. I., Kim, Y. Effect of silicon fertilizer treatment on nodule formation and yield in soybean (*Glycine max* L.). *European Journal of Agronomy* 2021, 122, 126172.

[39] Piccolo, E., Ceccanti, C., Guidi, L., Landi, M. Role of beneficial elements in plants: Implications for the photosynthetic process. *Photosynthetica* 2021, 59, 349–360.

[40] Masarovičová, E., Kráľová, K., Šeršeň, F. Plant responses to toxic metal stress. In Handbook of Plant and Crop Stress, 3rd ed. Boca Raton, CRC Press, 2010, 595–634.

[41] Masarovičová, E., Kráľová, K. Essential elements and toxic metals in some crops, medicinal plants, and trees. In Phytoremediation. Cham, Switzerland, Springer International Publishing AG, 2017, 183–255.

[42] Fernando, D. R., Lynch, J. P. Manganese phytotoxicity: New light on an old problem. *Annals of Botany* 2015, 116, 313–319.

[43] Ruttkay-Nedecky, B., Krystofova, O., Nejdl, L., Adam, V. Nanoparticles based on essential metals and their phytotoxicity. *Journal of Nanobiotechnology* 2017, 15, 33.

[44] Kráľová, K., Masarovičová, E., Jampílek, J. Plant responses to stress induced by toxic metals and their nanoforms. In Handbook of Plant and Crop Stress, 4th ed. Boca Raton, CRC Press, 2019, 479–522.

[45] Zaid, A., Ahmad, B., Jaleel, H., Wani, S. H., Hasanuzzaman, M. A critical review on iron toxicity and tolerance in plants: Role of exogenous phytoprotectants. In Plant Micronutrients. Cham, Switzerland, Springer, 2020, 83–99.

[46] Kaur, H., Garg, N. Zinc toxicity in plants: A review. *Planta* 2021, 253, 129.

[47] Hassan, M. U., Chattha, M. U., Khan, I., Chattha, M. B., Aamer, M., Nawaz, M., Ali, A., Khan, M. A. U., Khan, T. A. Nickel toxicity in plants: Reasons, toxic effects, tolerance mechanisms, and remediation possibilities – A review. *Environmental Science and Pollution Research* 2019, 26, 12673–12688.

[48] Kumar, V., Pandita, S., Sidhu, G. P. S., Sharma, A., Khanna, K., Kaur, P., Bali, A. S., Raj, S. Copper bioavailability, uptake, toxicity and tolerance in plants: A comprehensive review. *Chemosphere* 2021, 262, 127810.

[49] Schroth, G., Lehman, J., Barrios, E. Soil nutrient availability and acidity. In Trees, Crops and Soil Fertility: Concepts and Research Methods. CABI Publishig 2002, 93–130.

[50] Gao, X. Y., Rodrigues, S. M., Spielman-Sun, E., Lopes, S., Rodrigues, S., Zhang, Y., Avellan, A., Duarte, R. M. B. O., Duarte, A., Casman, E. A., Lowry, G. V. Effect of soil organic matter, soil pH, and moisture content on solubility and dissolution rate of CuO NPs in soil. *Environmental Science Technology* 2019, 7, 4959–4967.

[51] Li, J. Q., Nie, M., Powell, J. R., Bissett, A., Pendall, E. Soil physico-chemical properties are critical for predicting carbon storage and nutrient availability across Australia. *Environmental Reserach Letters* 2020, 15, 094088.

[52] Wang, M., Sun, X., Zhang, N. Q., Cai, D. Q., Wu, Z. Y. Promising approach for improving adhesion capacity of foliar nitrogen fertilizer. *ACS Sustainable Chemistry & Engineering* 2015, 3, 499–506.

[53] Li, W. C., Fan, R. Y., Zhou, H. J., Zhu, Y. F., Zheng, X., Tang, M. G., Wu, X. S., Yu, C. Z., Wang, G. Z. Improving the utilization rate of foliar nitrogen fertilizers by surface roughness engineering of silica spheres. *Environmental Science Nano* 2020, 7, 3526–3535.

[54] Macedo, L. O., Mattos-Jr, D., Jacobassi, R. C., Petená, G., Quaggio, J. A., Boaretto, R. M. Characterization and use efficiency of sparingly soluble fertilizer of boron and zinc for foliar application in coffee plants. *Bragantia* 2021, 80, e3421.

[55] Górecki, H. Ammonia and fertilizers. In Chemical Engineering and Chemical Process Technology, Vol. V, Encyclopedia of Life Support Systems, EOLSS Publications, 2010, 132–164

[56] Fu, J. J., Wang, C., Chen, X. X., Huang, Z. W., Chen, D. M. Classification research and types of slow controlled release fertilizers (SRFs) used – A review. *Communications in Soil Science and Plant Analysis* 2018, 49, 2219–2230.

[57] Ransom, C. J., Jolley, V. D., Blair, T. A., Sutton, L. E., Hopkins, B. G. Nitrogen release rates from slow- and controlled-release fertilizers influenced by placement and temperature. *PLoS ONE* 2020, 15, e0234544.

[58] Al-Rawajfeh, A. E., Alrbaihat, M. R., AlShamaileh, E. M. Characteristics and types of slow- and controlled-release fertilizers. In Controlled Release Fertilizers for Sustainable Agriculture. Academic Press, 2021, 57–78.

[59]   Lawrencia, D., Wong, S. K., Low, D. Y. S., Goh, B. H., Goh, J. K., Ruktanonchai, U. R., Soottitantawat, A., Lee, K. H., Tang, S. Y. Controlled release fertilizers: A review on coating materials and mechanism of release. *Plants* 2021, 10, 238.

[60]   Mikula, K., Izydorczyk, G., Skrzypczak, D., Mironiuk, M., Moustakas, K., Witek-Krowiak, A., Chojnacka, K. Controlled release micronutrient fertilizers for precision agriculture – A review. *Science of the Total Environment* 2020, 712, 136365.

[61]   Jampílek, J., Kráľová, K. Nano-antimicrobials: Activity, benefits and weaknesses. In Nanostructures in Therapeutic Medicine, Vol. 2. Nanostructures for Antimicrobial Therapy. Amsterdam, Elsevier, 2017, 23–54.

[62]   Jampílek, J., Kráľová, K. Nanoformulations – Valuable tool in therapy of viral diseases attacking humans and animals. In Nanotheranostics – Applications and Limitations. Cham, Switzerland, Springer Nature, 2019, 137–178.

[63]   Jampílek, J., Kráľová, K., Novák, P., Novák, M. Nanobiotechnology in neurodegenerative diseases. In Nanobiotechnology in Neurodegenerative Diseases. Cham, Switzerland, Springer Nature Switzerland AG, 2019, 65–138.

[64]   Jampílek, J., Kráľová, K. Natural biopolymeric nanoformulations for brain drug delivery. In Nanocarriers for Brain Targeting, Principles and Applications. Apple Academic Press, 2019, 131–203.

[65]   Jampílek, J., Kráľová, K., Campos, E. V. R., Fraceto, L. F. Bio-based nanoemulsion formulations applicable in agriculture, medicine and food industry. In Nanobiotechnology in Bioformulations. Cham, Switzerland, Springer, 2019, 33–84.

[66]   Jampílek, J., Kráľová, K. Nanoweapons against tuberculosis. In Nanoformulations in Human Health – Challenges and Approaches. Cham, Switzerland, Springer Nature, 2020, 469–502.

[67]   Jampílek, J., Kráľová, K. Application of nanotechnology in agriculture and food industry, its prospects and risks. *Ecological Chemistry and Engineering S* 2015, 22, 321–361.

[68]   Jampílek, J., Kráľová, K. Nanomaterials applicable in food protection. In Nanotechnology Applications in Food Industry. Boca Raton, CRC Press, 2018, 75–96.

[69]   Jampílek, J., Kráľová, K. Benefits of chitosan-based and cellulose-based nanocomposites in food protection and food packaging. In Biobased Nanotechnology for Green Applications, Nanotechnology in the Life Sciences. Cham, Switzerland, Springer Nature Switzerland AG, 2021, 121–160.

[70]   Jampílek, J., Kráľová, K. Nanopesticides: Preparation, targeting and controlled release. In Nanotechnology in the Agri-Food Industry. New Pesticides and Soil Sensors, Vol. 10. London, UK, Elsevier, 2017, 81–127.

[71]   Jampílek, J., Kráľová, K. Nanomaterials for delivery of nutrients and growth-promoting compounds to plants. In Nanotechnology: An Agricultural Paradigm. Singapore, Springer Nature Singapore Pte Ltd, 2017, 177–226.

[72]   Jampílek, J., Kráľová, K. Benefits and potential risks of nanotechnology applications in crop protection. In Nanobiotechnology Applications in Plant Protection. Cham, Switzerland, Springer Nature Switzerland AG, 2018, 189–246.

[73]   Jampílek, J., Kráľová, K. Nano-biopesticides in agriculture: State of art and future opportunities. In Nano-Biopesticides Today and Future Perspectives. Amsterdam, Academic Press & Elsevier, 2019, 397–447.

[74]   Jampílek, J., Kráľová, K. Beneficial effects of metal- and metalloid-based nanoparticles on crop production. In Nanotechnology for Agriculture. Singapore, Springer Nature, 2019, 161–219.

[75]   Jampílek, J., Kráľová, K. Nanocomposites: Synergistic nanotools for management mycotoxigenic fungi. In Nanomycotoxicology – Treating Mycotoxins in the Nano Way. London, UK, Academic Press & Elsevier, 2020, 349–383.

[76]   Jampílek, J., Kráľová, K., Fedor, P. Bioactivity of nanoformulated synthetic and natural insecticides and their impact on the environment. In Nanopesticides – From Research and Development to

Mechanisms of Action and Sustainable Use in Agriculture. Cham, Switzerland, Springer Nature Switzerland AG, 2020, 165–225.

[77] Jampílek, J., Kráľová, K. Impact of copper-based nanoparticles on economically important plants. In Copper Nanostructures: Next-Generation of Agrochemicals for Sustainable Agroecosystems. Amsterdam, Elsevier, 2022, 293–339.

[78] Kráľová, K., Jampílek, J. Applications of nanomaterials in plant disease management and protection. In Nanotechnology for Agro-Ecosystem. Amsterdam, Elsevier, 2023, 239–296.

[79] Jampílek, J., Kráľová, K. Potential of nanonutraceuticals in increasing immunity. *Nanomaterials* 2020, 10, 2224.

[80] Jampílek, J., Kos, J., Kráľová, K. Potential of nanomaterial applications in dietary supplements and foods for special medical purposes. *Nanomaterials* 2019, 9, 296.

[81] Masarovičová, E., Kráľová, K. Metal nanoparticles and plants. *Ecological Chemistry and Engineering S* 2013, 20, 9–22.

[82] Masarovičová, E., Kráľová, K., Zinjarde, S. S. Metal nanoparticles in plants. Formation and action. In Handbook of Plant and Crop Physiology, 3rd edn. Boca Raton, CRC Press, 2014, 683–731.

[83] Kráľová, K., Masarovičová, E., Jampílek, J. Risks and benefits of metal-based nanoparticles for vascular plants. In Handbook of Plant and Crop Physiology, 4th ed. Boca Raton, CRC Press, 2021, 923–963.

[84] Kráľová, K., Jampílek, J. Responses of medicinal and aromatic plants to engineered nanoparticles. *Applied Sciences* 2021, 11, 1813.

[85] Jampílek, J., Kráľová, K. Nanoparticles for improving and augmenting plant functions. In Advances in Nano-Fertilizers and Nano-pesticides Application for Crop Improvement. Kidlington, UK, Woodhead Publishing & Elsevier Inc., 2020, 171–227.

[86] Tiwari, P. K., Shweta, S. A. K., Singh, V. P., Prasad, S. M., Ramawat, N., Tripathi, D. K., Chauhan, D. K., Rai, A. K. Liquid assisted pulsed laser ablation synthesized copper oxide nanoparticles (CuO-NPs) and their differential impact on rice seedlings. *Ecotoxicology & Environmental Safety* 2019, 176, 321–329.

[87] Tombuloglu, H., Slimani, Y., Tombuloglu, G., Almessiere, M., Baykal, A. Uptake and translocation of magnetite (Fe_3O_4) nanoparticles and its impact on photosynthetic genes in barley (*Hordeum vulgare* L.). *Chemosphere* 2019, 226, 110–122.

[88] Hoang, S. A., Nguyen, K. Q., Nguyen, N. H., Tran, C. Q., Nguyen, D. V., Le, N. T., Ha, C. V., Vu, Q. N., Phan, C. M. Metal nanoparticles as effective promotors for maize production. *Scientific Reports* 2019, 9, 13925.

[89] Mikula, K., Izydorczyk, G., Skrzypczak, D., Mironiuk, M., Moustakas, K., Witek-Krowiak, A., Chojnacka, K. Controlled release micronutrient fertilizers for precision agriculture – A review. *Science of the Total Environment* 2020, 712, 136365.

[90] Tarafder, C., Daizy, M., Alam, M. M., Ali, M. R., Islam, M. J., Islam, R., Ahommed, S., Saad Aly, M. A., Hossain Khan, M. Z. Formulation of a hybrid nanofertilizer for slow and sustainable release of micronutrients. *ACS Omega* 2020, 5, 23960–23966.

[91] Paparella, S., Araújo, S. S., Rossi, G., Wijayasinghe, M., Carbonera, D., Balestrazzi, A. Seed priming: State of art and new perspectives. *Plant Cell Reports* 2015, 34, 1281–1293.

[92] Kráľová, K., Jampílek, J. Nanotechnology as effective tool for improved crop production under changing climatic conditions. In Biobased Nanotechnology for Green Application. Cham, Switzerland, Springer Nature, 2021, 463–512.

[93] Khan, Z., Upadhyaya, H. Impact of nanoparticles on abiotic stress responses in plants: An overview. In Nanomaterials in Plants, Algae and Microorganisms. Concepts and Controversies, Vol. 2. Amsterdam, Elsevier, 2019, 305–322.

[94] Singh, S., Husen, A. Role of nanomaterials in the mitigation of abiotic stress in plants. In Nanomaterials and Plant Potential. Cham, Switzerland, Springer Nature Switzerland AG, 2019, 441–471.

[95] Liu, R. Q., Lal, R. Potentials of engineered nanoparticles as fertilizers for increasing agronomic productions. *Science of the Total Environment* 2015, 514, 131–139.

[96] Dimkpa, C. O., Bindraban, P. S. Fortification of micronutrients for efficient agronomic production: A review. *Agronomy for Sustainable Development* 2016, 36, 7.

[97] Morales-Díaz, A. B., Ortega-Ortíz, H., Juárez-Maldonado, A., Cadenas-Pliego, G., González-Morales, S., Benavides-Mendoza, A. Application of nanoelements in plant nutrition and its impact in ecosystems. *Advances in Natural Sciences: Nanoscience and Nanotechnology* 2017, 8, 013001.

[98] Preetha, P. S., Balakrishnan, N. A review of nano fertilizers and their use and functions in soil. *International Journal of Current Microbiology and Applied Sciences* 2017, 6, 3117–3133.

[99] Guo, H. Y., White, J. C., Wang, Z., Xing, B. S. Nano-enabled fertilizers to control the release and use efficiency of nutrients. *Current Opinion in Environmental Science & Health* 2018, 6, 77–83.

[100] López-Valdez, F., Miranda-Arámbula, M., Ríos-Cortés, A. M., Fernández-Luqueño, F., De-la-luz, V. Nanofertilizers and their controlled delivery of nutrients. In Agricultural Nanobiotechnology. Cham, Switzerland, Springer, 2018, 35–48.

[101] Pacheco, I., Buzea, C. Nanoparticle uptake by plants: Beneficial or detrimental? In Phytotoxicity of Nanoparticles. Cham, Switzerland, Springer, 2018, 1–61.

[102] Qureshi, A., Singh, D. K., Dwivedi, S. Nano-fertilizers: A novel way for enhancing nutrient use efficiency and crop productivity. *International Journal of Current Microbiology and Applied Sciences* 2018, 7, 3325–3335.

[103] Vázquez-Núñez, E., López-Moreno, M. L., De la Rosa Álvarez, G., Fernández-Luqueño, F. Incorporation of nanoparticles into plant nutrients: The real benefits. In Agricultural Nanobiotechnology. Cham, Switzerland, Springer, 2018, 49–76.

[104] Verma, S. K., Das, A. K., Patel, M. K., Shah, A., Kumar, V., Gantait, S. Engineered nanomaterials for plant growth and development: A perspective analysis. *Science of the Total Environment* 2018, 630, 1413–1435.

[105] Juárez-Maldonado, A., Ortega-Ortíz, H., Morales-Díaz, A. B., González-Morales, S., Morelos-Moreno, A., Cabrera-de la Fuente, M., Rangel, A. S., Cadenas-Pliego, G., Benavides-Mendoza, A. Nanoparticles and nanomaterials as plant biostimulants. *International Journal of Molecular Sciences* 2019, 20, 162.

[106] Ghosh, S. K., Bera, T. Molecular mechanism of nanofertilizer in plant growth and development: A recent account. In Advances in Nano-Fertilizers and Nano-Pesticides in Agriculture. A Smart Delivery System for Crop Improvement. Kidlington, UK, Woodhead Publishing & Elsevier Inc., 2021, 535–560.

[107] Gomes, M. H. F., Duran, N. M., Pereira de Carvalho, H. W. Challenges and perspective for the application of nanomaterials as fertilizers. In Advances in Nano-Fertilizers and Nano-Pesticides in Agriculture. A Smart Delivery System for Crop Improvement. Kidlington, UK, Woodhead Publishing & Elsevier Inc., 2021, 331–359.

[108] Nandini, B., Geetha, N. Smart delivery mechanisms of nanofertilizers and nanocides in crop biotechology. In Advances in Nano-Fertilizers and Nano-Pesticides in Agriculture. A Smart Delivery System for Crop Improvement. Kidlington, UK, Woodhead Publishing & Elsevier Inc., 2021, 385–414.

[109] Kráľová, K., Jampílek, J. Metal- and metalloid-based nanofertilizers and nanopesticides for advanced agriculture. In Inorganic Nanopesticides and Nanofertilizers: A View from the Mechanisms of Action to Field Applications. Cham, Springer Nature Switzerland AG, 2022, 295–361.

[110] Kah, M. Nanopesticides and nanofertilizers: Emerging contaminants or opportunities for risk mitigation? *Frontiers in Chemistry* 2015, 3, 64.

[111] Mikkelsen, R. Nanofertilizer and nanotechnology: A quick look. *Better Crops with Plant Food* 2018, 102, 18–19.

[112] Nutritional requirements of plants. Boundless Biology, 2013. (Assessed August 23, 2021 at https://courses.lumenlearning.com/boundless-biology/chapter/nutritional-requirements-of-plants/.)

[113] Shrivastav, P., Prasad, M., Singh, T. B., Yadav, A., Goyal, D., Ali, A., Dantu, P. K. Role of nutrients in plant growth and development. In Contaminants in Agriculture. Switzerland, Springer Nature Switzerland AG, 2020, 43–59.

[114] Cui, H. B., Shi, Y., Zhou, J., Chu, H., Cang, L., Zhou, D. M. Effect of different grain sizes of hydroxyapatite on soil heavy metal bioavailability and microbial community composition. *Agriculture, Ecosystems & Environment* 2019, 267, 165–173.

[115] Montalvo, D., McLaughlin, M. J., Degryse, F. Efficacy of hydroxyapatite nanoparticles as phosphorus fertilizer in andisols and oxisols. *Soil Science Society of America Journal* 2015, 79, 551–558.

[116] Kottegoda, N., Sandaruwan, C., Priyadarshana, G., Siriwardhana, A., Rathnayake, U. A., Arachchige, D. M. B., Kumarasinghe, A. R., Dahanayake, D., Karunaratne, V., Amaratunga, G. A. J. Urea-hydroxyapatite nanohybrids for slow release of nitrogen. *ACS Nano* 2017, 11, 1214–1221.

[117] Xiong, L., Wang, P., Hunter, M. N., Kopittke, P. Bioavailability and movement of phosphorus applied as hydroxyapatite nanoparticles (HA-NPs) in soils. *Environmental Science Nano* 2018, 5, 2888–2898.

[118] Marchiol, L., Filippi, A., Adamiano, A., Esposti, L. D., Iafisco, M., Mattiello, A., Petrussa, E., Braidot, E. Influence of hydroxyapatite nanoparticles on germination and plant metabolism of tomato (*Solanum lycopersicum* L.): Preliminary evidence. *Agronomy-Basel* 2019, 9, 161.

[119] Phan, K. S., Nguyen, H. T., Huong Le, T. T., Thuy Vu, T. T., Do, H. D., Oanh Vuong, T. K., Nguyen, H. N., Tran, C. H., Hang Ngo, T. T., Ha, P. T. Fabrication and activity evaluation on *Asparagus officinalis* of hydroxyapatite based multimicronutrient nano systems. *Advances in Natural Sciences: Nanoscience and Nanotechnology* 2019, 10, 025011.

[120] Fellet, G., Pilotto, L., Marchiol, L., Braidot, E. Tools for nano-enabled agriculture: Fertilizers based on calcium phosphate, silicon, and chitosan nanostructures. *Agronomy* 2021, 11, 1239.

[121] Pradhan, S., Durgam, M., Mailapalli, D. R. Urea loaded hydroxyapatite nanocarrier for efficient delivery of plant nutrients in rice. *Archives of Agronomy and Soil Science* 2021, 67, 371–382.

[122] Sajadinia, H., Ghazanfari, D., Naqhavii, K., Naghavi, H., Tahamipur, B. A comparison of microwave and ultrasound routes to prepare nano-hydroxyapatite fertilizer improving morphological and physiological properties of maize (*Zea mays* L.). *Heliyon* 2021, 7, e06094.

[123] McKnight, M. M., Qu, Z., Copeland, J. K., Guttman, D. S., Walker, V. K. A practical assessment of nano-phosphate on soybean (*Glycine max*) growth and microbiome establishment. *Scientific Reports* 2020, 10, 9151.

[124] Mohamed, S. E., Mohamed, H. I., Mubarak, M., Sallam, A. M. Efficacy of natural hydroxyapatite nano-particles as a phosphorus fertilizer for soybean. *Journal of Biosciences* 2019, 16, 1094–1103.

[125] Badran, A., Savin, I. Effect of nano-fertilizer on seed germination and first stages of bitter almond seedlings' growth under saline conditions. *Bionanoscience* 2018, 8, 742–751.

[126] Li, Q. Q., Ma, C. X., White, J. C., Xing, B. S. Effects of phosphorus ensembled nanomaterials on nutrient uptake and distribution in Glycine max L. under simulated precipitation. *Agronomy-Basel* 2021, 11, 1086.

[127] De la Vega-Garcia, N. L., Pena-Valdivia, C. B., Gonzalez-Chavez, D. A., Padilla-Chacon, D., Carrillo-Gonzalez, R. Synthesis and effect of hydroxyapatite nanoparticles on the germination and growth of common bean. *Agrociencia* 2020, 54, 1009–1029.

[128] Maghsoodi, M. R., Ghodszad, L., Lajayer, B. A. Dilemma of hydroxyapatite nanoparticles as phosphorus fertilizer: Potentials, challenges and effects on plants. *Environmental Technology & Innovation* 2020, 19, 100869.

[129] Raguraj, S., Wijayathunga, W. M. S., Gunaratne, G. P., Amali, R. K. A., Priyadarshana, G., Sandaruwan, C., Karunaratne, V., Hettiarachchi, L. S. K., Kottegoda, N. Urea-hydroxyapatite nanohybrid as an efficient nutrient source in *Camellia sinensis* (L.) Kuntze (tea). *Journal of Plant Nutrition* 2020, 43, 2383–2394.

[130] Yoon, H. Y., Lee, J. G., Degli Esposti, L., Iafisco, M., Kim, P. J., Shin, S. G., Jeon, J. R., Adamiano, A. Synergistic release of crop nutrients and stimulants from hydroxyapatite nanoparticles functionalized with humic substances: Toward a multifunctional nanofertilizer. *ACS Omega* 2020, 5, 6598–6610.

[131] Tang, S. Q., Fei, X. C. Refractory calcium phosphate-derived phosphorus fertilizer based on hydroxyapatite nanoparticles for nutrient delivery. *ACS Applied Nano Materials* 2021, 4, 1364–1376.

[132] Szameitat, A. E., Sharma, A., Minutello, F., Pinna, A., Er-Rafik, M., Hansen, T. H., Persson, D. P., Andersen, B., Husted, S. Unravelling the interactions between nano-hydroxyapatite and the roots of phosphorus deficient barley plants. *Environmental Science Nano* 2021, 8, 444–459.

[133] Rop, K., Karuku, G. N., Mbui, D., Michira, I., Njomo, N. Formulation of slow release NPK fertilizer (cellulose-graft-poly(acrylamide)/nano-hydroxyapatite/soluble fertilizer) composite and evaluating its N mineralization potential. *Annals of Agricultural Science* 2021, 66, 93–94.

[134] Rop, K., Mbui, D., Karuku, G. N., Michira, I., Njomo, N. Characterization of water hyacinth cellulose-g-poly(ammonium acrylate-co-acrylic acid)/nano-hydroxyapatite polymer hydrogel composite for potential agricultural application. *Results in Chemistry* 2020, 2, 100020.

[135] Taskin, M. B., Sahin, Ö., Taskin, H., Atakol, O., Inal, A., Gunes, A. Effect of synthetic nano-hydroxyapatite as an alternative phosphorus source on growth and phosphorus nutrition of lettuce (*Lactuca sativa* L.) plant. *Journal of Plant Nutrition* 2018, 41, 1148–1154.

[136] Bala, N., Dey, A., Das, S., Basu, R., Nandy, P. Effect of hydroxyapatite nanorod on chickpea (*Cicer arietinum*) plant growth and its possible use as nano-fertilizer. *Iranian Journal of Plant Physiology* 2014, 4, 1061–1069.

[137] Shams, A. S., Abbas, M. H. H. Calcium superphosphate and boric acid for broccoli (*Brassica oleracea* var. *italica*) grown on a heavy clay soil? *Egyptian Journal of Horticulture* 2019, 46, 215–234.

[138] Madanayake, N. H., Adassooriya, N. M., Salim, N. The effect of hydroxyapatite nanoparticles on *Raphanus sativus* with respect to seedling growth and two plant metabolites. *Environmental Nanotechnology, Monitoring and Management* 2021, 15, 100404.

[139] Abeywardana, L., De Silva, M., Sandaruwan, C., Dahanayake, D., Priyadarshana, G., Chathurika, S., Karunaratne, V., Kottegoda, N. Zinc-doped hydroxyapatite–urea nanoseed coating as an efficient macro–micro plant nutrient delivery agent. *ACS Agricultural Science & Technology* 2021, 1, 230–239.

[140] Madusanka, N., Sandaruwan, C., Kottegoda, N., Sirisena, D., Munaweera, I., Alwis, A. D., Karunaratne, V., Amaratunga, G. A. J. Urea–hydroxyapatite-montmorillonite nanohybrid composites as slow release nitrogen compositions. *Applied Clay Science* 2017, 150, 303–308.

[141] Pavelic, S. K., Medica, J. S., Gumbarevic, D., Filosevic, A., Przulj, N., Pavelic, K. Critical review on zeolite clinoptilolite safety and medical applications in vivo. *Frontiers in Pharmacology* 2018, 9, 1350.

[142] McCusker, L. B., Baerlocher, C. Zeolite structures. In Introduction to Zeolite Science and Practice, Studies in Surface Science and Catalysis, vol. 137. London, Elsevier, 2001, 37–67.

[143] Eslami, M., Khorassani, R., Fotovat, A., Halajnia, A. NH_4^+-K^+ co-loaded clinoptilolite as a binary fertilizer. *Archives of Agronomy and Soil Science* 2020, 66, 33–45.

[144] Hasbullah, N. A., Ahmed, O. H., Ab Majid, N. M. Effects of amending phosphatic fertilizers with clinoptilolite zeolite on phosphorus availability and its fractionation in an acid soil. *Applied Sciences* 2020, 10, 3162.

[145] Ibrahim, H. M., Alghamdi, A. G. Effect of the particle size of clinoptilolite zeolite on water content and soil water storage in a loamy sand soil. *Water* 2021, 13, 607.

[146] Nakhli, S. A. A., Delkash, M., Bakhshayesh, B. E., Kazemian, H. Application of zeolites for sustainable agriculture: A review on water and nutrient retention. *Water, Air, & Soil Pollution* 2017, 228, 464.

[147] Wakihara, T., Ichikawa, R., Tatami, J., Endo, A., Yoshida, K., Sasaki, Y., Komeya, K., Meguro, T. Bead-milling and postmilling recrystallization: An organic template-free methodology for the production of nano-zeolites. *Crystal Growth & Design* 2011, 11, 955–958.

[148] Wakihara, T., Tatami, J. Top-down tuning of nanosized zeolites by bead-milling and recrystallization. *Journal of the Japan Petroleum Institute* 2013, 56, 206–213.

[149] Kianfar, E. Nanozeolites: Synthesized, properties, applications. *Journal of Sol-Gel Science and Technology* 2019, 91, 415–429.

[150] Kuznetsov, P. S., Dementiev, K. I., Palankoev, T. A., Kalmykova, D. S., Malyavin, V. V., Sagaradze, A. D., Maximov, A. L. Synthesis of highly active nanozeolites using methods of mechanical milling, recrystallization, and dealumination (A review). *Petroleum Chemistry* 2021, 61, 649–662.

[151] Lateef, A., Nazir, R., Jamil, N., Alam, S., Shah, R., Khan, M. N., Saleem, M. Synthesis and characterization of zeolite based nano-composite: An environment friendly slow release fertilizer. *Microporous and Mesoporous Materials* 2016, 232, 174–183.

[152] Chaudhary, P., Khati, P., Chaudhary, A., Gangola, S., Kumar, R., Sharma, A. Bioinoculation using indigenous *Bacillus* spp. improves growth and yield of *Zea mays* under the influence of nanozeolite. *3 Biotech* 2021, 11, 11.

[153] Khan, M. Z. H., Islam, M. R., Nahar, N., Al-Mamun, M. R., Khan, M. A. S., Matin, M. A. Synthesis and characterization of nanozeolite based composite fertilizer for sustainable release and use efficiency of nutrients. *Heliyon* 2021, 7, e06091.

[154] Sivashankari, L., Rajkishore, S. K., Lakshmanan, A., Subramanian, K. S., Praghadeesh, M. Bio-safety assessment of nanozeolites of varying size and dose on soil beneficial microorganisms. *Journal of Environmental Biology* 2021, 42, 1181–1190.

[155] Pimsen, R., Porrawatkul, P., Nuengmatcha, P., Ramasoot, S., Chanthai, S. Efficiency enhancement of slow release of fertilizer using nanozeolite-chitosan/sago starch-based biopolymer composite. *Journal of Coatings Technology and Research* 2021, 18, 1321–1332.

[156] Kowalczyk, P., Sprynskyy, M., Terzyk, A. O., Lebedynets, M., Namiesnik, J., Buszewski, B. Porous structure of natural and modified clinoptilolites. *Journal of Colloid and Interface Science* 2006, 297, 77–85.

[157] Mikhak, A., Sohrabi, A., Kassaee, M. Z., Feizian, M. Effect of nanoclinoptilolite/ nanohydroxyapatite mixtures on phosphorus solubility in soil. *Journal of Plant Nutrition* 2018, 41, 1227–1239.

[158] Yuvaraj, M., Subramanian, K. S. Development of slow release Zn fertilizer using nano-zeolite as carrier. *Journal of Plant Nutrition* 2018, 41, 311–320.

[159] Malekian, R., Abedi-Koupai, J., Eslamian, S. S. Influences of clinoptilolite and surfactant-modified clinoptilolite zeolite on nitrate leaching and plant growth. *Journal of Hazardous Materials* 2011, 185, 970–976.

[160] Barton, C. D., Karathanasis, A. D. Clay minerals. In Encyclopedia of Soil Science. New York, Marcel Dekker, Inc., 2002, 187–192.

[161] Azarian, M. H., Mahmood, W. A. K., Kwok, E., Fathilah, W. F. B. W., Ibrahim, N. F. B. Nanoencapsulation of intercalated montmorillonite–urea within PVA nanofibers: Hydrogel fertilizer nanocomposite. *Journal of Applied Polymer Science* 2018, 135, 45957.

[162] Dos Santos, B. R., Bacalhau, F. B., Pereira, T. D., Souza, C. F., Faez, R. Chitosan-montmorillonite microspheres: A sustainable fertilizer delivery system. *Carbohydrate Polymers* 2015, 127, 340–346.

[163] El-Nagar, D. A., Sary, D. H. Synthesis and characterization of nano bentonite and its effect on some properties of sandy soils. *Soil and Tillage Research* 2021, 208, 104872.

[164] Sarkar, S., Datta, S. C., Biswas, D. R. Effect of fertilizer loaded nanoclay/superabsorbent polymer composites on nitrogen and phosphorus release in soil. *Proceedings of the National Academy of Sciences, India, Section B: Biological Sciences* 2015, 85, 415–421.

[165] Noh, Y. D., Komarneni, S., Park, M. Mineral-based slow release fertilizers: A Review. *Korean Journal of Soil Science and Fertilizer* 2015, 48, 1–7.

[166] Sheoran, P., Goel, S., Boora, R., Kumari, S., Yashveer, S., Grewal, S. Biogenic synthesis of potassium nanoparticles and their evaluation as a growth promoter in wheat. *Plant Genetic* 2021, 27, 100310.

[167] Abd El-Razik, E. M., Abd El-Aziz, Z. H., Akl, A. A. Using of nanotechnology to evaluate some potassium sources and methods application on soil and plant. *Menoufia Journal of Soil Science* 2021, 111–132.

[168] El-Azizy, F. A., Habib, A. A. M., Abd-el Baset, A. M. Effect of nano phosphorus and potassium fertilizers on productivity and mineral content of broad bean in North Sinai. *Journal of Soil Sciences and Agricultural Engineering* 2021, 12, 239–246.

[169] Saleem, I., Maqsood, M. A., Rehman, M. Z. U., Aziz, T., Bhatti, I. A., Ali, S. Potassium ferrite nanoparticles on DAP to formulate slow release fertilizer with auxiliary nutrients. *Ecotoxicology & Environmental Safety* 2021, 2015, 112148.

[170] Saleem, I., Maqsood, M. A., Aziz, T., Bhatti, I. A., Jabbar, A. Potassium ferrite nano coated DAP fertilizer improves wheat (*Triticum aestivum* L.) growth and yield under alkaline calcareous soil conditions. *Pakistan Journal of Agricultural Sciences* 2021, 58, 485–492.

[171] Alsalhy, A. M., Al-Wasfy, M. M., Badawy, I. F. M., Gouda, F. M., Shamroukh, A. A. Effect of nano-potassium fertilization on fruiting of Zaghloul date palm. *SVU-International Journal of Agricultural Sciences* 2021, 3, 1–9.

[172] Weber, N. C., Koron, D., Jakopič, J., Veberič, R., Hudina, M., Česnik, H. B. Influence of nitrogen, calcium and nano-fertilizer on strawberry (*Fragaria × ananassa* Duch.) fruit inner and outer quality. *Agronomy* 2021, 11, 997.

[173] Anand, K. V., Reshma, M., Kannan, M., Selvan, S. M., Chaturvedi, S., Shalan, A. E., Govindaraju, K. Preparation and characterization of calcium oxide nanoparticles from marine molluscan shell waste as nutrient source for plant growth. *Journal of Nanostructure in Chemistry* 2021, 11, 409–422.

[174] Azeez, L., Lateef, A., Adetoro, R. O., Adeleke, A. E. Responses of *Moringa oleifera* to alteration in soil properties induced by calcium nanoparticles (CaNPs) on mineral absorption, physiological indices and photosynthetic indicators. *Beni-Suef University Journal of Basic and Applied Sciences* 2021, 10, 39.

[175] Vidak, M., Lazarevic, B., Petek, M., Gunjaca, J., Satovic, Z., Budor, I., Carovic-Stanko, K. Multispectral assessment of sweet pepper (*Capsicum annuum* L.) fruit quality affected by calcite nanoparticles. *Biomolecules* 2021, 11, 832.

[176] Kumara, K. H. C. H., Hafeel, R. F., Wathugala, D. L., Kumarasinghe, H. K. M. S. Effect of nano-calcite foliar fertilizer on growth and yield of *Oryza sativa* variety at 362. *Tropical Agricultural Research and Extension* 2017, 20, 59–65.

[177] Kumara, K. H. C. H., Wathugala, D. L., Hafeel, R. F., Kumarasinghe, H. K. M. S. Effect of nano calcite foliar fertilizer on the growth and yield of rice (*Oryza sativa*). *Journal of Agricultural Science* 2019, 14, 154–164.

[178] Melo, R. A. C., Jorge, M. H. A., Bortolin, A., Boiteux, L. S., Ribeiro, C., Marconcini, J. M. Growth of tomato seedlings in substrates containing a nanocomposite hydrogel with calcium montmorillonite (NC-MMt). *Horticultura Brasileira* 2019, 37, 199–203.

[179] Nobahar, A., Zakerin, H. R., Rad, M. M., Sayfzadeh, S., Valadabady, A. R. Response of yield and some physiological traits of groundnut (*Arachis hypogaea* L.) to topping height and application methods of Zn and Ca nano-chelates. *Communications in Soil Science and Plant Analysis* 2019, 50, 749–762.

[180] Deepa, M., Sudhakar, P., Nagamadhuri, K. V., Reddy, K. B., Krishna, T. G., Prasad, T. N. V. K. V. First evidence on phloem transport of nanoscale calcium oxide in groundnut using solution culture technique. *Applied Nanoscience* 2015, 5, 545–551.

[181] Carmona, F. J., Dal Sasso, G., Ramirez-Rodriguez, G. B., Pii, Y., Delgado-López, J. M., Guagliardi, A., Masciocchi, N. Urea-functionalized amorphous calcium phosphate nanofertilizers: Optimizing the synthetic strategy towards environmental sustainability and manufacturing costs. *Scientific Reports* 2021, 11, 3419.

[182] Gaiotti, F., Lucchetta, M., Rodegher, G., Lorenzoni, D., Longo, E., Boselli, E., Cesco, S., Belfiore, N., Lovat, L., Delgado-López, J. M., Carmona, F. J., Guagliardi, A., Masciocchi, N., Pii, Y. Urea-doped calcium phosphate nanoparticles as sustainable nitrogen nanofertilizers for viticulture: Implications on yield and quality of Pinot Gris grapevines. *Agronomy* 2021, 11, 1026.

[183] Tauseef, A., Hisamuddin,, Khalilullah, A., Uddin, I. Role of MgO nanoparticles in the suppression of *Meloidogyne incognita*, infecting cowpea and improvement in plant growth and physiology. *Experimental Parasitology* 2021, 220, 108045.

[184] Delfani, M., Firouzabadi, M. B., Farrokhi, N., Makarian, H. Some physiological responses of black-eyed pea to iron and magnesium nanofertilizers. *Communications in Soil Science and Plant Analysis* 2014, 45, 530–540.

[185] Anand, K. V., Anugraga, A. R., Kannan, M., Singaravelu, G., Govindaraju, K. Bio-engineered magnesium oxide nanoparticles as nano-priming agent for enhancing seed germination and seedling vigour of green gram (*Vigna radiata* L.). *Material Letters* 2020, 271, 127792.

[186] Salcido-Martinez, A., Sanchez, E., Licon-Trillo, L. P., Perez-Alvarez, S., Palacio-Marquez, A., Amaya-Olivas, N. I., Preciado-Rangel, P. Impact of the foliar application of magnesium nano fertilizer on physiological and biochemical parameters and yield in green beans. *Notulae Botanicae Horti Agrobotanici Cluj-Napoca* 2020, 48, 2167–2181.

[187] Kanjana, D. Foliar application of magnesium oxide nanoparticles on nutrient element concentrations, growth, physiological, and yield parameters of cotton. *Journal of Plant Nutrition* 2020, 43, 3035–3049.

[188] Jayarambabu, N., Kumari, B. S., Rao, K. V., Prabhu, Y. T. Enhancement of growth in maize by biogenic synthesized MgO nanoparticles. *International Journal of Pure and Applied Zoology* 2016, 4, 262–270.

[189] Luo, X. Q., Zhang, C. L., Xu, W. H., Peng, Q., Jiao, L. C., Deng, J. B. Effects of nanometer magnesium hydroxide on growth, cadmium (Cd) uptake of chinese cabbage (*Brassica campestris*) and soil Cd form. *International Journal of Agriculture and Biology* 2020, 24, 1006–1016.

[190] Shinde, S., Paralikar, P., Ingle, A. P., Rai, M. Promotion of seed germination and seedling growth of *Zea mays* by magnesium hydroxide nanoparticles synthesized by the filtrate from *Aspergillus niger*. *Arabian Journal of Chemistry* 2020, 13, 3172–3182.

[191] Tombuloglu, H., Tombuloglu, G., Slimani, Y., Ercan, I., Sozeri, H., Baykal, A. Impact of manganese ferrite ($MnFe_2O_4$) nanoparticles on growth and magnetic character of barley (*Hordeum vulgare* L.). *Environmental Pollution* 2018, 243, 872–881.

[192] Jin, H., Sun, Y. J., Sun, Z. J., Yang, M., Gui, R. Zero-dimensional sulfur nanomaterials: Synthesis, modifications and applications. *Coordination Chemistry Reviews* 2021, 438, 213913.

[193] Kurmanbayeva, M., Sekerova, T., Tileubayeva, Z., Kaiyrbekov, T., Kusmangazinov, A., Shapalov, S., Madenova, A., Burkitbayev, M., Bachilova, N. Influence of new sulfur-containing fertilizers on performance of wheat yield. *Saudi Journal of Biological Sciences* 2021, 28, 4644–4655.

[194] Burkitbayev, M., Bachilova, N., Kurmanbayeva, M., Tolenova, K., Yerezhepova, N., Zhumagul, M., Mamurova, A., Turysbek, B., Demeu, G. Effect of sulfur-containing agrochemicals on growth, yield, and protein content of soybeans (*Glycine max* (L.) Merr. *Saudi Journal of Biological Sciences* 2021, 28, 891–900.

[195] Esmaeil, M. A., Abd Elghany, S. H., Abdel Fattah, A. K., Arafat, A. K. Assessment of the effect of nano sulfur on some soil properties and maize productivity in saline soil. *Current Science International* 2020, 09, 656–665.

[196] Saad-Allah, K. M., Ragab, G. A. Sulfur nanoparticles mediated improvement of salt tolerance in wheat relates to decreasing oxidative stress and regulating metabolic activity. *Physiology and Molecular Biology* 2020, 26, 2209–2223.

[197] Ragab, G., Saad-Allah, K. Seed priming with greenly synthesized sulfur nanoparticles enhances antioxidative defense machinery and restricts oxidative injury under manganese stress in *Helianthus annuus* (L.) seedlings. *Journal of Plant Growth Regulation* 2021, 40, 1894–1902.

[198] Kaya, M., Karaman, R., Sener, A. Effects of nano sulfur (S) applications on yield and some yield properties of bread wheat. *Science Paper Series A. Agronomy* 2018, 61, 274–279.

[199] Kiferle, C., Martinelli, M., Salzano, A. M., Gonzali, S., Beltrami, S., Salvadori, P. A., Hora, K., Holwerda, H. T., Scaloni, A., Perata, P. Evidences for a nutritional role of iodine in plants. *Frontiers in Plant Science* 2021, 12, 616868.

[200] Maathuis, F. J. M. Sodium in plants: Perception, signalling, and regulation of sodium fluxes. *Journal of Experimental Botany* 2014, 65, 849–858.

[201] Hasanuzzaman, M., Hossain, M. A., Fujita, M. Selenium in higher plants: Physiological role, antioxidant metabolism and abiotic stress tolerance. *Journal of Plant Science* 2010, 5, 354–375.

[202] El-Ramady, H., Abdalla, N., Taha, H. S., Alshaal, T., El-Henawy, A., Faizy, S. E. D. A., et al. Selenium and nano-selenium in plant nutrition. *Environmental Chemistry Letters* 2016, 14, 123–147.

[203] Akeel, A., Jahan, A. Role of cobalt in plants: Its stress and alleviation. In Contaminants in Agriculture. Cham, Switzerland, Springer International Publishing, 2020, 339–357.

[204] Rout, G. R., Sahoo, S. Role of iron in plant growth and metabolism. *Reviews in Agricultural Science* 2015, 3, 1–24.

[205] Soliemanzadeh, A., Fekri, M. Effects of green iron nanoparticles on iron changes and phytoavailability in a calcareous soil. *Pedosphere* 2021, 31, 761–770.

[206] Haydar, M. S., Ghosh, S., Mandal, P. Application of iron oxide nanoparticles as micronutrient fertilizer in mulberry propagation. *Journal of Plant Growth Regulation* 2022, 41, 1726–1746.

[207] Farshchi, H. K., Azizi, M., Teymouri, M., Nikpoor, A. R., Jaafari, M. R. Synthesis and characterization of nanoliposome containing Fe^{2+} element: A superior nano-fertilizer for ferrous iron delivery to sweet basil. *Scientia Horticulturae* 2021, 283, 110110.

[208] Guha, T., Gopal, G., Das, H., Mukherjee, A., Kundu, R. Nanopriming with zero-valent iron synthesized using pomegranate peel waste: A "green" approach for yield enhancement in *Oryza sativa* L. cv. Gonindobhog. *Plant Physiology and Biochemistry* 2021, 163, 261–275.

[209] Guha, T., Mukherjee, A., Kundu, R. Nano-scale zero valent iron (nZVI) priming enhances yield, alters mineral distribution and grain nutrient content of *Oryza sativa* L. cv. Gobindobhog: A field study. *Journal of Plant Growth Regulation* 2022, 41, 710–733.

[210] Afzal, S., Sharma, D., Singh, N. K. Eco-friendly synthesis of phytochemical-capped iron oxide nanoparticles as nano-priming agent for boosting seed germination in rice (*Oryza sativa* L.). *Environmental Science and Pollution Research* 28, 40275–40287.

[211] Neto, M. E., Britt, D. W., Jackson, K. A., Coneglian, C. F., Inoue, T. T., Batista, M. A. Early growth of corn seedlings after seed priming with magnetite nanoparticles synthetised in easy way. *Acta Agriculturae Scandinavica, Section B* 2021, 71, 91–97.

[212] Li, M. S., Zhang, P., Adeel, M., Guo, Z. L., Chetwynd, A. J., Ma, C. X., Bai, T. H., Hao, Y., Rui, Y. K. Physiological impacts of zero valent iron, Fe_3O_4 and Fe_2O_3 nanoparticles in rice plants and their potential as Fe fertilizers. *Environmental Pollution* 2021, 269, 116134.

[213] Sega, D., Baldan, B., Zamboni, A., Varanini, Z. $FePO_4$ NPs are an efficient nutritional source for plants: Combination of nano-material properties and metabolic responses to nutritional deficiencies. *Frontiers in Plant Science* 2020, 11, 586470.

[214] Hasan, M., Rafique, S., Zafar, A., Loomba, S., Khan, R., Hassan, S. G., Khan, M. W., Zahra, S., Zia, M., Mustafa, G. Physiological and anti-oxidative response of biologically and chemically synthesized iron oxide: *Zea mays* a case study. *Heliyon* 2020, 6, e04595.

[215] Al-Amri, N., Tombuloglu, H., Slimani, Y., Akhtar, S., Barghouthi, M., Almessiere, M., Alshammari, T., Baykal, A., Sabit, H., Ercan, I. Size effect of iron (III) oxide nanomaterials on the growth, and their uptake and translocation in common wheat (*Triticum aestivum* L.). *Ecotoxicology & Environmental Safety* 2020, 194, 110377.

[216] Kasivelu, G., Selvaraj, T., Malaichamy, K., Kathickeyan, D., Shkolnik, D., Chaturvedi, S. Nano-micronutrients [γ-Fe_2O_3 (iron) and ZnO (zinc)]: Green preparation, characterization, agro-morphological characteristics and crop productivity studies in two crops (rice and maize). *New Journal of Chemistry* 2020, 44, 11373–11383.

[217] Yang, X. L., Alidoust, D., Wang, C. Y. Effects of iron oxide nanoparticles on the mineral composition and growth of soybean (*Glycine max* L.) plants. *Acta Physiologiae Plantarum* 2020, 42, 128.

[218] Shahrekizad, M., Ahangar, A. G., Mir, N. EDTA-coated Fe_3O_4 nanoparticles: A novel biocompatible fertilizer for improving agronomic traits of sunflower (*Helianthus annuus*). *Journal of Nanostructures* 2015, 5, 117–127.

[219] Iannone, M. F., Groppa, M. D., Zawoznik, M. S., Coral, D. F., Van Raap, M. B. F., Benavides, M. P. Magnetite nanoparticles coated with citric acid are not phytotoxic and stimulate soybean and alfalfa growth. *Ecotoxicology & Environmental Safety* 2021, 211, 111942.

[220] Palchoudhury, S., Jungjohann, K. L., Weerasena, L., Arabshahi, A., Gharge, U., Albattah, A., Miller, J., Patel, K., Holler, R. A. Enhanced legume root growth with pre-soaking in α-Fe_2O_3 nanoparticle fertilizer. *RSC Advances* 2018, 8, 24075–24083.

[221] Hu, J., Guo, H. Y., Li, J. L., Wang, Y. Q., Xiao, L., Xing, B. S. Interaction of γ-Fe_2O_3 nanoparticles with *Citrus maxima* leaves and the corresponding physiological effects via foliar application. *Journal of Nanobiotechnology* 2017, 15, 51.

[222] Wang, Y. Q., Hu, J., Dai, Z. Y., Li, J. L., Huang, J. In vitro assessment of physiological changes of watermelon (*Citrullus lanatus*) upon iron oxide nanoparticles exposure. *Plant Physiology and Biochemistry* 2016, 108, 353–360.

[223] Wang, Y. Q., Wang, S. X., Xu, M. X., Xiao, L., Dai, Z. Y., Li, J. L. The impacts of γ-Fe_2O_3 and Fe_3O_4 nanoparticles on the physiology and fruit quality of muskmelon (*Cucumis melo*) plants. *Environmental Pollution* 2019, 249, 1011–1018.

[224] Rui, M. M., Ma, C. X., Hao, Y., Guo, J., Rui, Y. K., Tang, X. L., Zhao, Q., Fan, X., Zhang, Z. T., Hou, T. Q. Iron oxide nanoparticles as a potential iron fertilizer for peanut (*Arachis hypogaea*). *Frontiers in Plant Science* 2016, 7, 815.

[225] Hassan, M., El-Feky, S. A. Green synthesis of magnetic nano particles using novel plant extracts and its impact as fertilizers on *Helianthus annuus* and *Vicia faba*. *Journal of Biosciences* 2019, 16, 2165–2178.

[226] Bahrami, M. K., Movafeghi, A., Mahdavinia, G. R., Hassanpouraghdam, M. B., Gohari, G. Effects of bare and chitosan-coated Fe_3O_4 magnetic nanoparticles on seed germination and seedling growth of *Capsicum annuum* L. *Biointerface Research in Applied Chemistry* 2018, 8, 3552–3559.

[227] Duran, N. M., Medina-Llamas, M., Cassanji, J. G. B., de Lima, R. G., De Almeida, E., Macedo, W. R., Mattia, D., De Carvalho, H. W. P. Bean seedling growth enhancement using magnetite nanoparticles. *Journal of Agricultural and Food Chemistry* 2018, 66, 5746–5755.

[228] Maksymiec, W. Effect of copper on cellular processes in higher plants. *Photosynthetica* 1997, 33, 321–342.

[229] Yurela, I. Copper in Plants. *Brazilian Journal of Plant Physiology* 2005, 17, 145–156.

[230] Mottaleb, S. A., Hassan, A. Z. A., El-Bahbohy, R., Mahmoud, A. W. M. Are copper nanoparticles toxic to all plants? A case study on onion (*Allium cepa* L.). *Agronomy-Basel* 2021, 11, 1006.

[231] Kohatsu, M. Y., Pelegrino, M. T., Monteiro, L. R., Freire, B. M., Pereira, R. M., Fincheira, P., Rubilar, O., Tortella, G., Batista, B. L., De Jesus, T. A. Comparison of foliar spray and soil irrigation of biogenic CuO nanoparticles (NPs) on elemental uptake and accumulation in lettuce. *Environmental Science and Pollution Research* 2021, 28, 16350–16367.

[232] Leonardi, M., Caruso, G. M., Carroccio, S. C., Boninelli, S., Curcuruto, G., Zimbone, M., Allegra, M., Torrisi, B., Ferlito, F., Miritello, M. Smart nanocomposites of chitosan/alginate nanoparticles loaded with copper oxide as alternative nanofertilizers. *Environmental Science Nano* 2021, 8, 174–187.

[233] Wang, Y., Deng, C. Y., Cota-Ruiz, K., Peralta-Videa, J. R., Sun, Y. P., Rawat, S., Tan, W. J., Reyes, A., Hernandez-Viezcas, J. A., Niu, G. H. Improvement of nutrient elements and allicin content in green onion (*Allium fistulosum*) plants exposed to CuO nanoparticles. *Science of the Total Environment* 2020, 725, 138387.

[234] Jaskulska, I., Jaskulski, D. Effects of using nanoparticles of silver (AgNPs) and copper (CuNPs) in foliar fertilizers. *Przemysl Chemiczny* 2020, 99, 250–253.

[235] Jahagirdar, A. S., Shende, S., Gade, A., Rai, M. Bioinspired synthesis of copper nanoparticles and its efficacy on seed viability and seedling growth in mungbean (*Vigna radiata* L.). *Current Nanoscience* 2020, 16, 246–252.

[236] Rawat, S., Adisa, I. O., Wang, Y., Sun, Y. P., Fadil, A. S., Niu, G. H., Sharma, N., Hernandez-Viezcas, J. A., Peralta-Videa, J. R., Gardea-Torresdey, J. L. Differential physiological and biochemical impacts of nano vs micron Cu at two phenological growth stages in bell pepper (*Capsicum annuum*) plant. *Nanoimpact* 2019, 14. 100161.

[237] Wang, Y. J., Lin, Y. J., Xu, Y. W., Yin, Y., Guo, H. Y., Du, W. C. Divergence in response of lettuce (*var. ramosa* Hort.) to copper oxide nanoparticles/microparticles as potential agricultural fertilizer. *Environmental Pollution Bioavailability* 2019, 31, 80–84.

[238] Essa, H. L., Abdelfattah, M. S., Marzouk, A. S., Shedeed, Z., Guirguis, H. A., El-Sayed, M. M. H. Biogenic copper nanoparticles from *Avicennia marina* leaves: Impact on seed germination, detoxification enzymes, chlorophyll content and uptake by wheat seedlings. *PloS ONE* 2021, 16, e0249764.

[239] Yassir, S. H., Yassen, A. A. Effect of different concentrations of nano boron, copper, inoculation of *Anabena* sp. and magnetizing water on yield of garlic (*Allium sativum* L.). *Plant Archives* 2021, 21, 391–396.

[240] Broadley, M. R., White, P. J., Hammond, J. P., Zelko, I., Lux, A. Zinc in plants. *New Phytologist* 2007, 173, 677–702.

[241] Sheoran, P., Grewal, S., Kumari, S., Goel, S. Enhancement of growth and yield, leaching reduction in *Triticum aestivum* using biogenic synthesized zinc oxide nanofertilizer. *Biocatalysis & Agricultural Biotechnology* 2021, 32, 101938.

[242] Yusefi-Tanha, E., Fallah, S., Rostamnejadi, A., Pokhrel, L. R. Zinc oxide nanoparticles (ZnONPs) as a novel nanofertilizer: Influence on seed yield and antioxidant defense system in soil grown soybean (*Glycine max* cv. Kowsar). *Science of the Total Environment* 2020, 738, 140240.

[243] Sabir, S., Zahoor, M. A., Waseem, M., Siddique, M. H., Shafique, M., Imran, M., Hayat, S., Malik, I. R., Muzammil, S. Biosynthesis of ZnO nanoparticles using *Bacillus subtilis*: Characterization and nutritive significance for promoting plant growth in *Zea mays* L. *Dose-Response* 2020, 18, 1559325820958911.

[244] Neto, M. E., Britt, D. W., Lara, L. M., Cartwright, A., Dos Santos, R. F., Inoue, T. T., Batista, M. A. Initial development of corn seedlings after seed priming with nanoscale synthetic zinc oxide. *Agronomy-Basel* 2020, 10, 307.

[245] Garcia-Lopez, J. I., Nino-Medina, G., Olivares-Saenz, E., Lira-Saldivar, R. H., Barriga-Castro, E. D., Vazquez-Alvarado, R., Rodriguez-Salinas, P. A., Zavala-Garcia, F. Foliar application of zinc oxide nanoparticles and zinc sulfate boosts the content of bioactive compounds in habanero peppers. *Plants-Basel* 2019, 8, 254.

[246] Bala, R., Kalia, A., Dhaliwal, S. S. Evaluation of efficacy of ZnO nanoparticles as remedial zinc nanofertilizer for rice. *Journal of Soil Science and Plant Nutrition* 2019, 19, 379–389.

[247] Cordova, M. A., Perez-Leal, R., Munoz, R. M. Y. Efficiency of nanoparticle, sulfate, and zinc-chelate use on biomass, yield, and nitrogen assimilation in green beans. *Agronomy-Basel* 2019, 9, 128.

[248] Rossi, L., Fedenia, L. N., Sharifan, H., Ma, X. M., Lombardini, L. Effects of foliar application of zinc sulfate and zinc nanoparticles in coffee (*Coffea arabica* L.) plants. *Plant Physiology and Biochemistry* 2019, 135, 160–166.

[249] Sharma, D., Afzal, S., Singh, N. K. Nanopriming with phytosynthesized zinc oxide nanoparticles for promoting germination and starch metabolism in rice seeds. *Journal of Biotechnology* 2021, 336, 64–75.

[250] Lafmejani, Z. N., Jafari, A. A., Moradi, P., Moghadam, A. L. Application of chelate and nano-chelate zinc micronutrient on morpho-physiological traits and essential oil compounds of peppermint (*Mentha piperita* L.). *Journal of Medicinal Plants and By-products* 2021, 10, 21–28.

[251] Pullagurala, V. L. R., Adisa, I. O., Rawat, S., Kim, B., Barrios, A. C., Medina-Velo, I. A., Hernandez-Viezcas, J. A., Peralta-Videa, J. R., Gardea-Torresdey, J. L. Finding the conditions for the beneficial use of ZnO nanoparticles towards plants-A review. *Environmental Pollution* 2018, 241, 1175–1181.

[252] Shireen, F., Nawaz, M. A., Chen, C., Zhang, Q. K., Zheng, Z. H., Sohail, H., Sun, J. Y., Cao, H. S., Huang, Y., Bie, Z. L. Boron: Functions and approaches to enhance its availability in plants for sustainable agriculture. *International Journal of Molecular Sciences* 2018, 19, 1856.

[253] Meier, S., Moore, F., Morales, A., Gonzalez, M. E., Seguel, A., Merino-Gergichevich, C., Rubilar, O., Cumming, J., Aponte, H., Alarcon, D., Mejias, J. Synthesis of calcium borate nanoparticles and its use as a potential foliar fertilizer in lettuce (*Lactuca sativa*) and zucchini (*Cucurbita pepo*). *Plant Physiology and Biochemistry* 2020, 151, 673–680.

[254] Rios, J. J., Yepes-Molina, L., Martinez-Alonso, A., Carvajal, M. Nanobiofertilization as a novel technology for highly efficient foliar application of Fe and B in almond trees. *Royal Society Open Science* 2020, 7, 200905.

[255] Kandil, E. E., Abdelsalam, N. R., Abd El Aziz, A. A., Ali, H. M., Siddiqui, M. H. Efficacy of nanofertilizer, fulvic acid and boron fertilizer on sugar beet (*Beta vulgaris* L.) yield and quality. *Sugar Technology* 2020, 22, 782–791.

[256] Vishekaii, Z. R., Soleimani, A., Fallahi, E., Ghasemnezhad, M., Hasani, A. The impact of foliar application of boron nano-chelated fertilizer and boric acid on fruit yield, oil content, and quality attributes in olive (*Olea europaea* L.). *Scientia Horticulturae* 2019, 257, 108689.

[257] Wang, S. L., Nguyen, A. D. Effects of Zn/B nanofertilizer on biophysical characteristics and growth of coffee seedlings in a greenhouse. *Research on Chemical Intermediates* 2018, 44, 4889–4901.

[258] Songkhum, P., Wuttikhun, T., Chanlek, N., Khemthong, P., Laohhasurayotin, K. Controlled release studies of boron and zinc from layered double hydroxides as the micronutrient hosts for agricultural application. *Applied Clay Science* 2018, 152, 311–322.

[259] Davarpanah, S., Tehranifar, A., Davarynejad, G., Abadia, J., Khorasani, R. Effects of foliar applications of zinc and boron nano-fertilizers on pomegranate (*Punica granatum* cv. Ardestani) fruit yield and quality. *Scientia Horticulturae* 2016, 210, 57–64.

[260] Tartoura, E. A., Seif El-Deen, U. M., El-Adawy, A. Y. Effect of irrigation intervals and foliar applications with some nano-fertilizers on growth and productivity of globe artichoke plant: A–Vegetative growth and chemical content in leaves. *Journal of Plant Production* 2021, 12, 209–216.

[261] Tartoura, E. A., El-Deen UM, S., El-Adawy, A. Y. Effect of irrigation intervals and foliar applications with some nano-fertilizers on growth and productivity of globe artichoke plant: B–Yield and its quality. *Journal of Plant Production* 2021, 12, 217–224.

[262] Wassel, A. E. M. M., Saied, H. H. M., Salama, M. I. A. E. Response of Flame seedless grapevines to spray boron prepared by nanotechnology. *Minia Journal of Agricultural Research and Development* 2020, 40, 197–205.

[263] Arnon, D. I., Stout, P. R. Molybdenum as an essential element for higher plants. *Plant Physiology*. 1939, 14, 599–602.

[264] Kaiser, B. N., Gridley, K. L., Brady, J. N., Phillips, T., Tyerman, S. D. The role of molybdenum in agricultural plant production. *Annals of Botany* 2005, 96, 745–754.

[265] Rana, M. S., Bhantana, P., Imran, M., Saleem, M. H., Moussa, M. G., Khan, Z., Khan, I., Akam, M., Abbas, M., Binyamin, R., Afzak, J., Syaifudin, M., Din, I. U., Younas, M., Ahmad, I., Shah, M. A., Hu, C. X. Molybdenum potential vital role in plants metabolism for optimizing the growth and development. *Annals of Environmental Science and Technology* 2020, 4, 032–044.

[266] Osman, S. A., Salama, D. M., Abd El-Aziz, M. E., Shaaban, E. A., Abd Elwahed, M. S. The influence of MoO$_3$-NPs on agro-morphological criteria, genomic stability of DNA, biochemical assay, and production of common dry bean (*Phaseolus vulgaris* L.). *Plant Physiology and Biochemistry*. 2020, 151, 77–87.

[267] Sharma, P. K., Raghubanshi, A. S., Shah, K. V. Examining the uptake and bioaccumulation of molybdenum nanoparticles and their effect on antioxidant activities in growing rice seedlings. *Environmental Science and Pollution Research* 2020, 28, 13439–13453.

[268] Li, Y. D., Liu, Y. L., Yang, D. S., Jin, Q., Wu, C. L., Cui, J. H. Multifunctional molybdenum disulfide-copper nanocomposite that enhances the antibacterial activity, promotes rice growth and induces rice resistance. *Journal of Hazardous Materials* 2020, 394, 122551.

[269] Abbasifar, A., ValizadehKaji, B., Iravani, M. A. Effect of green synthesized molybdenum nanoparticles on nitrate accumulation and nitrate reductase activity in spinach. *Journal of Plant Nutrition* 2020, 43, 13–27.

[270] Aubert, T., Burel, A., Esnault, M. A., Cordier, S., Grasset, F., Cabello-Hurtado, F. Root uptake and phytotoxicity of nanosized molybdenum octahedral clusters. *Journal of Hazardous Materials* 2012, 219, 111–118.

[271] Yang, J. H., Song, Z. Y., Ma, J., Han, H. Y. Toxicity of molybdenum-based nanomaterials on the soybean-rhizobia symbiotic system: Implications for nutrition. *ACS Applied Nano Materials* 2020, 3, 5773–5782.

[272] Millaleo, R., Reyes-Diaz, M., Ivanov, A. G., Mora, M. L., Alberdi, M. Manganese as essential and toxic element for plants: Transport, accumulation and resistance mechanisms. *Journal of Soil Science and Plant Nutrition* 2010, 10, 476–494.

[273] Ye, Y. Q., Medina-Velo, I. A., Cota-Ruiz, K., Moreno-Olivas, F., Gardea-Torresdey, J. L. Can abiotic stresses in plants be alleviated by manganese nanoparticles or compounds? *Ecotoxicology & Environmental Safety* 2019, 184, 10967.

[274] Shebl, A., Hassan, A. A., Salama, D. M., Abd El-Aziz, M. E., Abd Elwahed, M. S. A. Green synthesis of nanofertilizers and their application as a foliar for *Cucurbita pepo* L. *Journal of Nanomaterials* 2019, 2019, 3476347.

[275] Shebl, A., Hassan, A. A., Salama, D. M., Abd El-Aziz, M. E., Abd Elwahed, M. S. A. Template-free microwave-assisted hydrothermal synthesis of manganese zinc ferrite as a nanofertilizer for squash plant (*Cucurbita pepo* L). *Heliyon* 2020, 6, e03596.

[276] Pradhan, S., Patra, P., Mitra, S., Dey, K. K., Jain, S., Sarkar, S., Roy, S., Palit, P., Goswami, A. Manganese nanoparticles: Impact on non-nodulated plant as a potent enhancer in nitrogen metabolism and toxicity study both in vivo and in vitro. *Journal of Agricultural and Food Chemistry* 2014, 62, 8777–8785.

[277] Dimkpa, C. O., Singh, U., Adisa, I. O., Bindraban, P. S., Elmer, W. H., Gardea-Torresdey, J. L., White, J. C. Effects of manganese nanoparticle exposure on nutrient acquisition in wheat (*Triticum aestivum* L.). *Agronomy-Basel* 2018, 8, 158.

[278] Liu, R. Q., Zhang, H. T., Lal, R. Effects of stabilized nanoparticles of copper, zinc, manganese, and iron oxides in low concentrations on lettuce (*Lactuca sativa*) seed germination: Nanotoxicants or nanonutrients? *Water, Air, & Soil Pollution* 2016, 227, 42.

[279] Kasote, D. M., Lee, J. H. J., Jayaprakasha, G. K., Patil, B. S. Manganese oxide nanoparticles as safer seed priming agent to improve chlorophyll and antioxidant profiles in watermelon seedlings. *Nanomaterials* 2021, 11, 1016.

[280] Ye, Y. Q., Cota-Ruiz, K., Hernandez-Viezcas, J. A., Valdes, C., Medina-Velo, I. A., Turley, R. S., Peralta-Videa, J. R., Gardea-Torresdey, J. L. Manganese nanoparticles control salinity-modulated molecular responses in *Capsicum annuum* L. through priming: A sustainable approach for agriculture. *ACS Sustainable Chemistry & Engineering* 2020, 3, 1427–1436.

[281] Tian, H., Ghorbanpour, M., Kariman, K. Manganese oxide nanoparticle-induced changes in growth, redox reactions and elicitation of antioxidant metabolites in deadly nightshade (*Atropa belladonna* L.). *Industrial Crops and Products* 2018, 126, 403–414.

[282] Brown, P. H., Welch, R. M., Cary, E. E. Nickel: A micronutrient essential for higher plants. *Plant Physiology* 1987, 85, 801–803.

[283] Liu, G. D. A new essential mineral element – Nickel. *Plant Nutrition and Fertilizer Science* 2001, 7, 101–103.

[284] Ragsdale, S. Nickel-based enzyme systems. *Journal of Biological Chemistry* 2001, 284, 18571–18575.

[285] Uddin, S., Bin Safdar, L., Iqbal, J., Yaseen, T., Laila, S., Anwar, S., Abbasi, B. A., Saif, M. S., Quraishi, U. M. Green synthesis of nickel oxide nanoparticles using leaf extract of *Berberis balochistanica*: Characterization, and diverse biological applications. *Microscopy Research and Technique* 2021, 84, 2004–2016.

[286] Tombuloglu, H., Ercan, I., Alshammari, T., Tombuloglu, G., Slimani, Y., Almessiere, M., Baykal, A. Incorporation of micro-nutrients (nickel, copper, zinc, and iron) into plant body through nanoparticles. *Journal of Soil Science and Plant Nutrition* 2020, 20, 1872–1881.

[287] Zotikova, A. P., Astafurova, T. P., Burenina, A. A., Suchkova, S. A., Morgalev Yu, N. Morphophysiological features of wheat (*Triticum aestivum* L.) seedlings upon exposure to nickel nanoparticles. *Sel'skokhozyaistvennaya Biologiya* 2018, 53, 578–586.

[288] Krishnamoorthy, V., Rajiv, S. Potential seed coatings fabricated from electrospinning hexaaminocyclotriphosphazene and cobalt nanoparticles incorporated polyvinylpyrrolidone for sustainable agriculture. *ACS Sustainable Chemistry & Engineering* 2017, 5, 146–152.

[289] Motevalli, S., Hassani, S. B., Ghalamboran, M. R., Chahardeh, H. R. Increase of diosgenin in fenugreek seedlings by cobalt nanoparticles. *Rhizosphere* 2021, 8, 100335.

[290] Jahani, M., Khavari-Nejad, R. A., Mahmoodzadeh, H., Saadatmand, S. Effect of foliar application of cobalt dioxide nanoparticles on growth, photosynthetic pigments, oxidative indicators, non-enzymatic antioxidants and compatible osmolytes in canola (*Brassica napus* L.). *Acta Biologica Cracoviensia: Series Botanica* 2019, 61, 29–42.

[291] Jahani, M., Khavari-Nejad, R. A., Mahmoodzadeh, H., Saadatmand, S. Effects of cobalt oxide nanoparticles (Co_3O_4 NPs) on ion leakage, total phenol, antioxidant enzymes activities and cobalt accumulation in *Brassica napus* L. *Notulae Botanicae Horti Agrobotanici Cluj-Napoca* 2020, 48, 1260–1275.

[292] Polischuk, S. D., Churilov, G. I., Borychev, S. N., Byshov, N. V., Nazarova, A. A. Nanopowders of cuprum, cobalt and their oxides used in the intensive technology for growing cucumbers. *International Journal of Nanotechnology* 2018, 15, 352–369.

[293] Tombuloglu, H., Slimani, Y., AlShammari, T. M., Tombuloglu, G., Almessiere, M. A., Sozeri, H., Baykal, A., Ercan, I. Delivery, fate and physiological effect of engineered cobalt ferrite nanoparticles in barley (*Hordeum vulgare* L.). *Chemosphere* 2021, 265, 129138.

[294] Ngo, Q. B., Dao, Y. H., Nguyen, H. C., Tran, X. T., Van Nguyen, T., Khuu, T. D., Huynh, T. H. Effects of nanocrystalline powders (Fe, Co and Cu) on the germination, growth, crop yield and product quality of soybean (Vietnamese species DT-51). *Advances in Natural Sciences: Nanoscience and Nanotechnology* 2014, 5, 015016.

[295] Luyckx, M., Hausman, J. F., Lutts, S., Guerriero, G. Silicon and plants: Current knowledge and technological perspectives. *Frontiers in Plant Science* 2016, 8, 411.

[296] Souri, Z., Khanna, K., Karimi, N., Ahmad, P. Silicon and plants: Current knowledge and future prospects. *Journal of Plant Growth Regulation* 2021, 40, 906–925.

[297] Rastogi, A., Tripathi, D. K., Yadav, S., Chauhan, D. K., Zivcak, M., Ghorbanpour, M., El-Sheery, N. I., Brestic, M. Application of silicon nanoparticles in agriculture.*3 Biotech* 2019, 9, 90.

[298] Neu, S., Schaller, J., Dudel, E. G. Silicon availability modifies nutrient use efficiency and content, C:N: P stoichiometry, and productivity of winter wheat (*Triticum aestivum* L.). *Scientific Reports* 2017, 7, 40829.

[299] Mukarram, M., Khan, M. M., Corpas, F. J. Silicon nanoparticles elicit an increase in lemongrass (*Cymbopogon flexuosus* (Steud.) Wats) agronomic parameters with a higher essential oil yield. *Journal of Hazardous Materials* 2021, 412, 125254.

[300] Bhat, J. A., Rajora, N., Raturi, G., Sharma, S., Dhiman, P., Sanand, S., Shivaraj, S. M., Sonah, H., Deshmukh, R. Silicon nanoparticles (SiNPs) in sustainable agriculture: Major emphasis on the practicality, efficacy and concerns. *Nanoscale Advances* 2021, 3, 4019–4028.

[301] Avestan, S., Ghasemnezhad, M., Esfahani, M., Barker, A. V. Effects of nanosilicon dioxide on leaf anatomy, chlorophyll fluorescence, and mineral element composition of strawberry under salinity stress. *Journal of Plant Nutrition* 2021, 44, 3005–3019.

[302] Afshari, M., Pazoki, A., Sadeghipour, O. Foliar-applied silicon and its nanoparticles stimulate physio-chemical changes to improve growth, yield and active constituents of coriander (*Coriandrum sativum* L.) essential oil under different irrigation regimes. *Silicon* 2021, 13, 4177–4188.

[303] Gomez-Vera, P., Blanco-Flores, H., Francisco, A. M., Castillo, J., Tezara, W. Silicon dioxide nanofertilizers improve photosynthetic capacity of two Criollo cocoa clones (*Theobroma cacao* L.). *Experience Agriculture* 2021, 57, 85–102.

[304] Mathur, P., Roy, S. Nanosilica facilitates silica uptake, growth and stress tolerance in plants. *Plant Physiology and Biochemistry* 2020, 157, 114–127.

[305] Asadpour, S., Madani, H., Mohammadi, G. N., Heravan, I. M., Abad, H. H. S. Improving maize yield with advancing planting time and nano-silicon foliar spray alone or combined with zinc. *Silicon* 2022, 14, 201–209.

[306] Lu, X. H., Sun, D. Q., Zhang, X. M., Hu, H. G., Kong, L. X., Rookes, J. E., Xie, J. H., Cahill, D. M. Stimulation of photosynthesis and enhancement of growth and yield in *Arabidopsis thaliana* treated with amine-functionalized mesoporous silica nanoparticles. *Plant Physiology and Biochemistry* 2020, 156, 566–577.

[307] Felisberto, G., Prado, R. D., De Oliveira, R. L. L., Felisberto, P. A. D. Are nanosilica, potassium silicate and new soluble sources of silicon effective for silicon foliar application to soybean and rice plants? *Silicon* 2021, 13, 3217–322.

[308] Ahmad, B., Khan, M. M. A., Jaleel, H., Shabbir, A., Sadiq, Y., Uddin, M. Silicon nanoparticles mediated increase in glandular trichomes and regulation of photosynthetic and quality attributes in *Mentha piperita* L. *Journal of Plant Growth Regulation* 2020, 39, 346–357.

[309] Kheyri, N., Norouzi, H. A., Mobasser, H. R., Torabi, B. Effects of silicon and zinc nanoparticles on growth, yield, and biochemical characteristics of rice. *Journal of Agronomy* 2019, 111, 3084–3090.

[310] Li, Y. J., Li, W., Zhang, H. R., Liu, Y. L., Ma, L., Lei, B. F. Amplified light harvesting for enhancing Italian lettuce photosynthesis using water soluble silicon quantum dots as artificial antennas. *Nanoscale* 2020, 12, 155–166.

[311] El-Ramady, H. R., Domokos-Szabolcsy, E., Abdalla, N. A., Alshaal, T. A., Shalaby, T. A., Sztrik, A., Prokisch, J., Fari, M. Selenium and nano-selenium in agroecosystems. *Environmental Chemistry Letters* 2014, 12, 495–510.

[312] Li, D., Zhou, C. R., Zhang, J. B., An, Q. S., Wu, Y. L., Li, J. Q., Pan, C. P. Nanoselenium foliar applications enhance the nutrient quality of pepper by activating the capsaicinoid synthetic pathway. *Journal of Agricultural and Food Chemistry* 2020, 68, 9888–9895.

[313] Babajani, A., Iranbakhsh, A., Ardebili, Z. O., Eslami, B. Differential growth, nutrition, physiology, and gene expression in *Melissa officinalis* mediated by zinc oxide and elemental selenium nanoparticles. *Environmental Science and Pollution Research* 2019, 26, 24430–24444.

[314] Zahedi, S. M., Hosseini, M. S., Meybodi, N. D. H., da Silva, J. A. T. Foliar application of selenium and nano-selenium affects pomegranate (*Punica granatum* cv. Malase Saveh) fruit yield and quality. *South African Journal of Botany* 2019, 124, 350–358.

[315] Hussein, H. A. A., Darwesh, O. M., Mekki, B. B. Environmentally friendly nano-selenium to improve antioxidant system and growth of groundnut cultivars under sandy soil conditions. *Biocatalysis & Agricultural Biotechnology* 2019, 18, 101080.

[316] Nurminsky, V. N., Perfileva, A. I., Kapustina, I. S., Graskova, I. A., Sukhov, B. G., Trofimov, B. A. Growth-stimulating activity of natural polymer-based nanocomposites of selenium during the germination of cultivated plant seeds. *Doklady Biochemistry and Biophysics* 2020, 495, 296–299.

[317] Shedeed, S. I., Fawzy, Z. F., El-Bassiony, A. M. Nano and mineral selenium foliar application effect on pea plants (*Pisum sativum* L.). *Journal of Biosciences* 2018, 15, 645–654.

[318] Khan, M. N., Mobin, M., Abbas, Z. K., AlMutairi, K. A., Siddiqui, Z. H. Role of nanomaterials in plants under challenging environments. *Plant Physiology and Biochemistry* 2017, 110, 194–209.

[319] Rajput, V., Minkina, T., Kumari, A., Harish, Singh, V. K., Verma, K. K., Mandzhieva, S., Sushkova, S., Srivastava, S., Keswani, C. Coping with the challenges of abiotic stress in plants: New dimensions in the field application of nanoparticles. *Plants-Basel* 2021, 10, 1221.

[320] Singh, A., Tiwari, S., Pandey, J., Lata, C., Singh, I. K. Role of nanoparticles in crop improvement and abiotic stress management. *Journal of Biotechnology* 2021, 337, 5770.

[321] Zhao, L. J., Lu, L., Wang, A., Zhang, H., Huang, M., Wu, H. H., Xing, B. S., Wang, Z. U., Ji, R. Nano-biotechnology in agriculture: Use of nanomaterials to promote plant growth and stress tolerance. *Journal of Agricultural and Food Chemistry* 2020, 68, 1935–1947.

[322] Zulfiqar, F., Ashraf, M. Nanoparticles potentially mediate salt stress tolerance in plants. *Plant Physiology and Biochemistry* 2021, 160, 257–268.

[323] Ali, M. A., Ahmed, T., Wu, W. G., Hossain, A., Hafeez, R., Masum, M. I., Wang, Y. L., An, Q. L., Sun, G. C., Li, B. Advancements in plant and microbe-based synthesis of metallic nanoparticles and their antimicrobial activity against plant pathogens. *Nanomaterials* 2020, 10, 1146.

[324] Kalia, A., Abd-Elsalam, K. A., Kuca, K. Zinc-based nanomaterials for diagnosis and management of plant diseases: Ecological safety and future prospects. *Journal of Fungi* 2020, 6, 222.

[325] Hernandez-Diaz, J. A., Garza-Garcia, J. J. O., Zamudio-Ojeda, A., Leon-Morales, J. M., Lopez-Velazquez, J. C., Garcia-Morales, S. Plant-mediated synthesis of nanoparticles and their antimicrobial activity against phytopathogens. *Journal of Sol-Gel Science and Technology* 2021, 101, 1270–1287.

[326] Ashkavand, P., Zarafshar, M., Tabari, M., Mirzaie, J., Nikpour, A., Bordbar, S. K., Struve, D., Striker, G. G. Application of SiO$_2$ nanoparticles as pretreatment alleviates the impact of drought on the physiological performance of *Prunus mahaleb* (*Rosaceae*). *Boletín de la Sociedad Argentina de Botánica* 2018, 53, 207–219.

[327] Behboudi, F., Sarvestani, Z. T., Kassaee, M. Z., Sanavi, S. A. M. M., Sorooshzadeh, A. Improving growth and yield of wheat under drought stress via application of SiO$_2$ nanoparticles. *Journal of Agricultural Science & Technology* 2018, 20, 1479–1492.

[328] Ghassemi, A., Farahvash, F. Effect of nano-zinc foliar application on wheat under drought stress. *Fresenius Environmental Bulletin* 2018, 27, 5022–5026.

[329] Semida, W. M., Abdelkhalik, A., Mohamed, G. F., Abd El-Mageed, T. A., Abd El-Mageed, S., Rady, M. M., Ali, E. F. Foliar application of zinc oxide nanoparticles promotes drought stress tolerance in eggplant (*Solanum melongena* L.). *Plants-Basel* 2021, 10, 421.

[330] Ahmadian, K., Jalilian, J., Pirzad, A. Nano-fertilizers improved drought tolerance in wheat under deficit irrigation. *Agricultural Water Management* 2021, 244, 106544.

[331] Astaneh, N., Bazrafshan, F., Zare, M., Amiri, B., Bahrani, A. Nano-fertilizer prevents environmental pollution and improves physiological traits of wheat grown under drought stress conditions. *Scientia Agropecuaria* 2021, 12, 41–47.

[332] Gonzalez-Garcia, Y., Cardenas-Alvarez, C., Cadenas-Pliego, G., Benavides-Mendoza, A., Cabrera-de-la-fuente, M., Sandoval-Rangel, A., Valdes-Reyna, J., Juarez-Maldonado, A. Effect of three nanoparticles (Se, Si and Cu) on the bioactive compounds of bell pepper fruits under saline stress. *Plants-Basel* 2021, 10, 217.

[333] Zahedi, S. M., Abdelrahman, M., Hosseini, M. S., Hoveizeh, N. F., Tran, L. S. P. Alleviation of the effect of salinity on growth and yield of strawberry by foliar spray of selenium-nanoparticles. *Environmental Pollution* 2019, 253, 246–258.

[334] Maswada, H. F., Djanaguiraman, M., Prasad, P. V. V. Seed treatment with nano-iron (III) oxide enhances germination, seeding growth and salinity tolerance of sorghum. *Journal of Agronomy and Crop Science* 2018, 204, 577–587.

[335] Wang, Z. S., Li, H., Li, X. N., Xin, C. Y., Si, J. S., Li, S. D., Li, Y. J., Zheng, X. X., Li, H. W., Wei, X. H., Zhang, Z. W., Kong, L. G., Wang, F. H. Nano-ZnO priming induces salt tolerance by promoting photosynthetic carbon assimilation in wheat. *Archives of Agronomy and Soil Science* 2020, 6, 1259–1273.

[336] Farouk, S., Al-Amri, S. M. Exogenous zinc forms counteract NaCl-induced damage by regulating the antioxidant system, osmotic adjustment substances, and ions in canola (*Brassica napus* L. cv. Pactol) plants. *Journal of Soil Science and Plant Nutrition* 2019, 19, 887–899.

[337] Djanaguiraman, M., Belliraj, N., Bossmann, S. H., Prasad, P. V. V. High-temperature stress alleviation by selenium nanoparticle treatment in grain sorghum. *ACS Omega* 2018, 3, 2479–2491.

[338] Tripathi, D. K., Singh, S., Singh, V. P., Prasad, S. M., Dubey, N. K., Chauhan, D. K. Silicon nanoparticles more effectively alleviated UV-B stress than silicon in wheat (*Triticum aestivum*) seedlings. *Plant Physiology and Biochemistry* 2017, 110, 70–81.

[339] Al-Harbi, H. F. A., Abdelhaliem, E., Araf, N. M. Modulatory effect of zincoxide nanoparticles on gamma radiation-induced genotoxicity in *Vicia faba* (*Fabaceae*). *Genetics & Molecular Research* 2019, 18, GMR18232.

[340] Zhou, P. F., Adeel, M., Shakoor, N., Guo, M., Hao, Y., Azeem, I., Li, M. S., Liu, M. Y., Rui YK Application of nanoparticles alleviates heavy metals stress and promotes plant growth: An overview. *Nanomaterials* 2021, 11, 26.

[341] Rizwan, M., Ali, S., Ali, B., Adrees, M., Arshad, M., Hussain, A., Rehman, M. Z. U., Warris, A. A. Zinc and iron oxide nanoparticles improved the plant growth and reduced the oxidative stress and cadmium concentration in wheat. *Chemosphere* 2019, 214, 269–277.

[342] Mohammadi, H., Hatami, M., Feghezadeh, K., Ghorbanpour, M. Mitigating effect of nano-zerovalent iron, iron sulfate and EDTA against oxidative stress induced by chromium in *Helianthus annuus* L. *Acta Physiologiae Plantarum* 2018, 40, 69.

[343] Liu, J., Dhungana, B., Cobb, G. P. Copper oxide nanoparticles and arsenic interact to alter seedling growth of rice (*Oryza sativa japonica*). *Chemosphere* 2018, 206, 330–337.

[344] Ali, S., Rizwan, M., Hussain, A., Rehman, M. Z. U., Ali, B., Yousaf, B., Wijaya, L., Alyemeni, M. N., Ahmad, P. 2019. Silicon nanoparticles enhanced the growth and reduced the cadmium accumulation in grains of wheat (*Triticum aestivum* L.). *Plant Physiology and Biochemistry*. 140, 1–8.

[345] Hussain, A., Rizwan, M., Ali, Q., Ali, S. Seed priming with silicon nanoparticles improved the biomass and yield while reduced the oxidative stress and cadmium concentration in wheat grains. *Environmental Science and Pollution Research* 2019, 26, 7579–7588.

[346] Cui, J. H., Liu, T. X., Li, F. B., Yi, J. C., Liu, C. P., Yu, H. Y. Silica nanoparticles alleviate cadmium toxicity in rice cells: Mechanisms and size effects. *Environmental Pollution* 2017, 228, 363–369.

Noman Shakoor[+], Muhammad Arslan Ahmad[+], Muhammad Adeel[+*],
Muzammil Hussain, Imran Azeem, Muhammad Zain, Yuanbo Li,
Ming Xu, Yukui Rui[*]

Chapter 4
Iron-based nanomaterials are emerging nanofertilizers to fulfil iron deficiency

Abstract: Iron (Fe) is an essential micronutrient for all organisms because it plays a significant role in metabolic processes (e.g., photosynthesis, respiration, and DNA synthesis) and many metabolic pathways are activated by Fe. Globally, Fe deficiency is a serious health concern due to the consumption of low-iron diets (or plant-based diets). Fe deficiency in humans induces severe diseases such as anemia. Considering the importance of Fe for the development of both humans and plants, this chapter reports some of the recent advancements in nanofertilizer for improving crop quality and subsequent human intake to alleviate the Fe deficiency. Furthermore, synthesis methods of Fe-based NPs, e.g., chemically and bio-synthesis, are discussed in details alongside advanced characterization techniques. Exposure of Fe-based nanomaterials via controlled-release systems can enhance crops' growth and nutritional content. Our findings indicated that nano-enabled technologies could be beneficial for reducing the dependence on chemical fertilizers and solving the Fe deficiency problem. The application of Fe-based nanofertilizers can prove to be a sound and sustainable approach to achieve the goal of increasing micronutrient content and crop yield. Fe-based nanofertilizers can bring nutrient-rich crops with economic advantages if the products are environmentally and economically sustainable.

Note: [+]These authors contributed equally to this work.

[*]**Corresponding author: Muhammad Adeel**, BNU-HKUST Laboratory of Green Innovation, Advanced Institute of Natural Sciences, Beijing Normal University at Zhuhai, 18 Jinfeng Road, Tangjiawan, Zhuhai 519087, Guangdong, China, e-mail: Chadeel969@gmail.com
[*]**Corresponding author: Yukui Rui**, Beijing Key Laboratory of Farmland Soil Pollution Prevention and Remediation and College of Resources and Environmental Sciences, China Agricultural University, Beijing 100193, China, e-mail: ruiyukui@163.com
Noman Shakoor, Imran Azeem, Yuanbo Li, Beijing Key Laboratory of Farmland Soil Pollution Prevention and Remediation and College of Resources and Environmental Sciences, China Agricultural University, Beijing 100193, China
Muhammad Arslan Ahmad, Muzammil Hussain, College of Life Science and Oceanography, Shenzhen University, Shenzhen 518071, China
Ming Xu, BNU-HKUST Laboratory of Green Innovation, Advanced Institute of Natural Sciences, Beijing Normal University at Zhuhai, 18 Jinfeng Road, Tangjiawan, Zhuhai 519087, Guangdong, China
Muhammad Zain, Department of Botany, University of Lakki Marwat, Lakki Marwat 28420, Pakistan

https://doi.org/10.1515/9781501523229-004

Keywords: iron-based nanoparticles, synthesis methods, plant health, nutritional quality

4.1 Introduction

Iron (Fe) is one of the most essential trace elements for all living organisms based on its involvement in many metabolic processes [1]. The oxidation state is key mechanism for functional conversion, and Fe involves multiple functions in the animals (oxygen storage and transport, cell cycle control, and neuron signaling) [2], and in the plants (photosynthesis, DNA synthesis, hormone synthesis, and elimination of reactive oxygen species) [3, 4]. Deficiency of Fe is one of the prime factors causing the disability and death, affecting more than two billion people around the world [5]. The World Health Organization (WHO) has reported that 52% pregnant women and 48% children are anemic. Additionally, by 2050, 9.8 billion people will need 70% more food products, posing a serious threat to food security [6].

Fe is micronutrient that is absorbed by plants roots in the form of Fe^{3+} and Fe^{2+}, and its sufficiency ranges from 50 to 250 ppm for optimum plant growth. Fe plays a vital role in photosynthesis, respiration, chlorophyll biosynthesis, catalyst agent, and enzymatic activity, e.g., nitrogenase. When Fe concentration is less than 50 ppm deficiency symptoms appear firstly on leaves (chlorosis and necrotic) [7]. There is a greater requirement for effective and integrated management of food production to overcome the adverse effects of Fe deficiency and improve agricultural yields.

Nanotechnology is an emerging tool to address the challenges that will restrict efforts to maintain and achieve global food security [8]. In recent years, nanotechnology has gained attention in agricultural field due to its efficiency to delivering the macro- and micro-nutrients nutrients, and ability to enhance soil and plant health [9]. Fe-based NPs could potentially be a promising fertilizer by promoting the Fe contents in the plants and improving its absorption efficiency, with lowest collateral damage in the soil environment [10–12]. Recently Li et al. documented that the exposure of Fe-based NPs (Fe_3O_4 and nanoscale zero-valent iron (nZVI)) in hydroponic media significantly improved the rice growth (26–30%) under Fe-deficient condition by enhancing the phytohormones levels and alleviating the oxidative stress [13]. Furthermore, Rui et al. reported that Fe-based NPs (Fe_2O_3 NPs) as an iron fertilizer increased the physiological growth of peanut (*Arachis hypogaea*) by regulating the phytohormones levels and antioxidant enzymatic activities [10]. Another study reported that Fe-based NPs enhanced the Fe contents in leaves, numbers of chloroplast and grana, and doubled the pepper (*Capsicum annuum*) growth [14]. Nevertheless, a number of studies documented the adverse effects of Fe-based NPs at elevated concentrations. For example, Fe-based NPs (Fe_3O_4, γ-Fe_2O_3, and α-Fe_2O_3) significantly reduced (14–31%) the chlorophyll content in *Citrus maxima*. As well, *Citrus maxima* showed various physiological and molecular responses to different source of Fe-based NPs in hydroponic media [15].

Fe-based NP toxicity is related to the oxidation state [16] and Fe_3O_4-based NPs at 200 mg L^{-1} significantly inhibited the physiological growth (fresh root and leaf biomass decreased by 52% and 48%) of *Eichhornia crassipes* and significantly increased (18-fold) malondialdehyde (MDA) level, although Fe^{3+} ions and Fe_3O_4 bulk particles showed non-significant adverse impacts [17]. Aforementioned studies indicated that negative effects may be short duration of growth and that did not highlight the real case scenario. Bio-availability and fate of Fe-based NPs varied greatly depending on exposure medium and duration. A few full life-cycle studies reported to observe the long-term effects of Fe-based NPs on the quality and yield of edible portion of crops or vegetables that are needed to understand the safety of Fe-based NPs' incorporation in agricultural application. Recently, Fakharzadeh et al. conducted a field-based study and reported that chelated Fe-based NP fertilizer 27% enhanced the biological yield with improving the quality of rice by increasing protein contents (13%) [18]. Shakoor et al. documented that Fe-based NPs' enhanced biomass of cherry radish with enhancing the Fe concentration and amino acids content in fruiting body [19].

Nevertheless, there are critical gaps regarding Fe-based NPs' impacts on food quality and nutritional levels (proteins content, amino acids and fatty acids) with agriculture production. In this chapter, we investigate whether these Fe-NPs-based fertilizer could alter the nutritional level of crop grains and vegetables, as well as discuss about the effect of Fe-based NPs on the crop protein and vitamins contents. Our findings could significantly evaluate the NP impacts on crop grain nutritional quality and shed light on the use of NPs as a novel fertilizer.

4.2 Synthesis and characterization of Fe-based nanomaterials

Fe-based NPs are formed in a variety of polymorphs with various crystalline and stoichiometric structures [20]. Different applications of Fe oxide NPs require specific nanoparticle properties [21], and the most ubiquitous synthesis method for magnetite (Fe_3O_4) NPs is the coprecipitation of ferrous (Fe^{2+}) and ferric (Fe^{3+}) ions through the addition of an alkaline solution [22]. Additionally, natural/green Fe-based NPs are continuously gaining interest among scientists as a potential alternative to conventional NPs [23]. According to the reports the physicochemical properties of Fe-based NPs, namely, shape, size, crystal structure, stability, chemical composition, surface area, energy, and roughness, significantly regulate their magnetic properties and, consequently, their uptake, fate, toxicity, and efficiency in the development of crops [24–26]. As well as improving characterization techniques, it is also necessary to increase the accuracy of nanoparticle properties evaluation.

4.2.1 Synthesis methods for iron oxide nanoparticles

A top-down approach and a bottom-up approach are well-established methods for synthesizing nanoparticles. Mechanical action is generally used to break, crush, or fractionate bulk materials into smaller pieces to produce nanoparticles [27, 28]. Such synthesis methods contain mechanical crushing, grinding, or milling, sputtering, laser ablation, electron beam deposition, offering an alternative eco-friendly way despite the required high time and power consumption [29–31]. As opposed to top-bottom approaches, bottom-up approaches rely on chemical reactions between specific atoms, ions, or molecules. Based on these principles, synthesis routes can be divided further based on the type of process involved, such as physical, which corresponds to top-down processes, chemical (e.g., coprecipitation, sol-gel, thermal decomposition, emulsions and micro emulsions, hydrothermal, and microwave-assisted methods), and biological (which utilizes plants or microbes for the generation of nanoparticles) as shown in Figure 4.1 [27, 28, 32, 33].

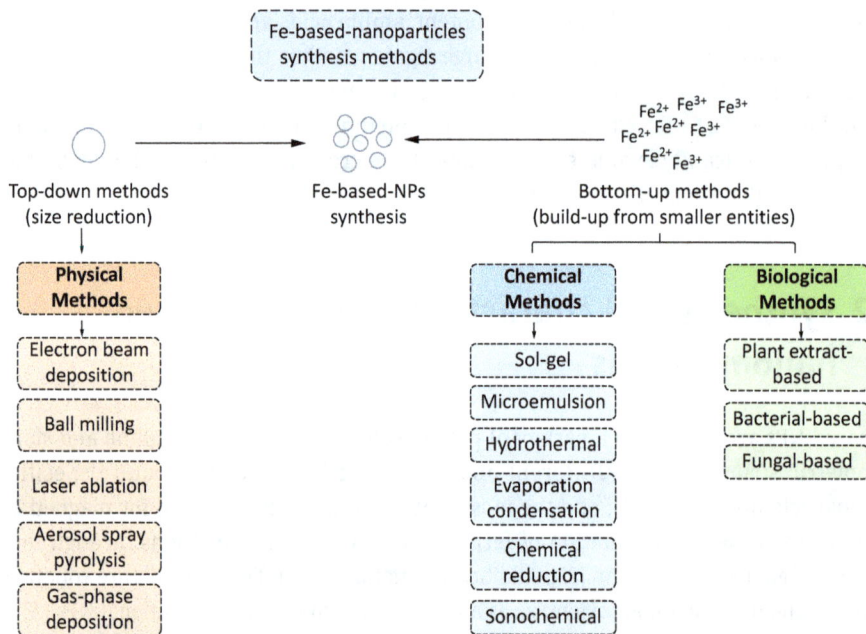

Figure 4.1: The main methods for the synthesis of Fe-based NPs [32].

In addition, Figure 4.2 compares the most widely used methods for synthesizing Fe nanoparticles. Coprecipitation accounts for the largest percentage of investigated routes, followed by chemical methods. Definitely, coprecipitation involves two possible pathways, partial oxidation of ferrous salts or the aging of a stoichiometric mixture of ferric and ferrous salts by adding an alkaline solution, which leads to the nucleation and growth of Fe_3O_4 nanoparticles and ultimately results in Fe_3O_4-based NPs.

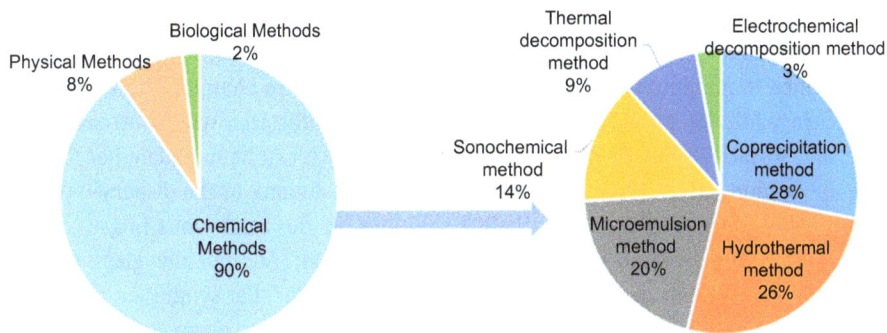

Figure 4.2: The most commonly used methods for synthesizing Fe nanoparticles [34].

While coprecipitation is the simplest, most time-efficient, and most safe method, as it uses only a small number of toxic solvents, it has substantial disadvantages in terms of reproducibility and control over nanoparticle properties [35, 36]. In order to enhance the characteristics of Fe nanoparticles, novel synthesis routes are required. A number of previously nonconventional methods have been implemented in recent years as a means of modulating chemical properties and potentially providing a plethora of alternatives. As a result, the following sections will describe the most recent progress in producing Fe nanoparticles using microwaves, microfluidics, and green synthesis processes.

4.2.1.1 Microwave-assisted NP synthesis method

It is based on the efficient heating of matter through microwave dielectric heating, which is the application of microwave irradiation to chemical reactions. It is based on the ability of a particular material (such as reagents or solvents) to absorb microwave energy and turn it into heat. Ionic conduction and dipolar polarization are the two main mechanisms that trigger heating in liquid phase synthesis of nanomaterials [37]. Microwave-assisted synthesis is becoming increasingly popular due to its numerous advantages. A particular advantage of this technique is the ease with which nanoparticles can be made at substantially reduced cost and energy consumption [38–40]. Aside from the economic benefits, microwave-assisted methods have gained scientific interest due to their ability to be tuned to produce magnetite nanoparticles of the desired size and shape with narrow distributions and high reproducibility, phase purity, and yield [35, 39, 40]. Due to the uniform heating and nucleation, rapid kinetics and crystallization, and phase selectivity, this is possible [39, 40].

4.2.1.2 Microfluidic-based NP synthesis method

In microfluidic systems, nanoparticles are synthesized in micro reactors with dimensions less than 1 mm. Lab-on-a-chip or tubular micro reactors are two types of micro reactors [41]. The field of microfluidics combines principles of fluid dynamics, chemistry, and material science to manipulate small fluid volumes within micro channels with precision and accuracy [42, 43]. Magnetite nanoparticles can be synthesized with controlled size, shape, and surface chemistry by using microfluidics, which can be modulated according to the application [42, 43]. Using this method, precursor solutions, or the dispersed phase, will intersect channels containing alkaline solutions, or the continuous phase, where nanoparticles will form and be further transported [27, 44]. Additionally, glass, metals, silicon, or polymers can be used in the microfluidic platforms that synthesize magnetite nanoparticles, but they must resist fluids introduced into the micro channels [27].

4.2.1.3 Green synthesis methods

The green synthesis method refers to the synthesis of NPs by using environment-friendly biological stabilizing and reducing agents present in microorganisms, plants, or biosurfactants. Metal ions present in the reaction mixture are reduced to metal atoms by reducing agents. As metal atoms cluster, they become stabilized and form NPs. Biotechnology and nanotechnology have combined to create a highly advanced field of NP synthesis using microorganisms such as bacteria, algae, fungi, viruses, yeasts, and plants [45] as shown in Figure 4.3. In this context, the synthesis of Fe-based nanoparticles, which are based on green technologies that use plant extracts and microorganisms as well, is constantly emerging as an effective, safe, renewable, and environmentally friendly alternative to other methods that do not involve complex protocols or complex procedures [45–48]. As a result of using green synthesis methods, nanoparticles formed are much more stable, since no chemicals are used to enhance the biocompatibility, antimicrobial activity, reactivity, and nontoxicity of the NPs [46–48]. A method such as this is possible due to the resistance mechanisms developed by plants and microorganisms and in order to survive the highly toxic environments generated by high levels of metal concentrations in the environment.

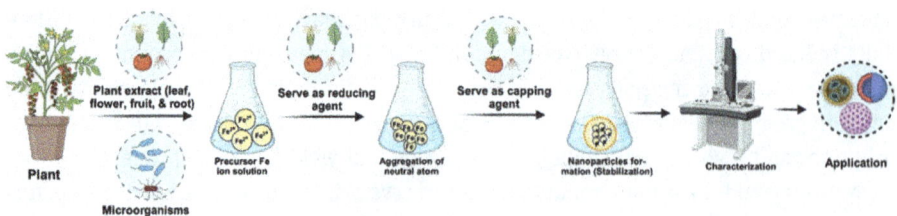

Figure 4.3: Illustration of green synthesis method for production of iron nanomaterials.

There is a general consensus that plant-based synthesis of NPs is more advantageous since it demonstrates higher kinetics, increased reduction and stabilization yields, and is easier to scale up at large scales [46, 47, 49]. Nanoparticles can be formed either intracellularly or inside the plant by specific biomolecules (e.g., ketones, aldehydes, phenols, flavones, amino acids, proteins, polysaccharides, terpenoids, tannins, vitamins, saponins), or extracellularly by plant extracts/phytochemicals. The nature and properties of phytochemicals are responsible for their shape, size, stability, and reactivity variations [47, 50]. Plant parts have been investigated based on their phytochemical production, including leaves, stems, petals, roots, flowers, fruits, peels, and seeds [49]. Starting materials are collected, washed, dried, weighed, ground into a fine powder, boiled in water or methanol/ethanol, centrifuged, and filtered. Afterward, the extract is mixed with the precursor salt solutions, e.g., $FeCl_3 \cdot 6H_2O$, $FeSO_4 \cdot 7H_2O$, $(FeNO_3)_3 \cdot 9H_2O$, $FeCl_3$, $FeSO_4$, $FeCl_2 \cdot 4H_2O$, $FeSO_4 \cdot 5H_2O$, or $FeCl_2$, of variable molarities. Finally, the mixture is heated and vigorously stirred in order to achieve a color change and intensification of the solution depending on the type of Fe salts used. The obtained Fe oxide NP pellets are further washed and dried [51].

4.3 Advanced iron oxide nanoparticles characterization techniques

A variety of different techniques are used to evaluate the physicochemical properties of Fe oxide nanoparticles, depending on the parameter that needs to be determined at the time [52–54]. An overview of iron oxide nanoparticles and their characterization is presented in Table 4.1, while actual iron nanomaterial has been shown in Figure 4.4.

It is very significant to evaluate the size and distribution of the NPs, since they can influence other properties and determine how the final product will perform within the application envisioned [52, 55]. A macroscale size measurement might seem trivial, but a nanoscale size determination might be interpreted differently depending on the characterization technique. The effective size of a nanoparticle in a matrix or solvent can be correlated with its atomic structure or with its diffusion/sedimentation-dependent effective size, as well as its mass/electron distribution weighted effective size [56, 57]. A nanoparticle's shape also plays an important role in its behavior, as it can further cause toxic effects due to cell damage. It is commonly used to evaluate the morphology and shape of nanoparticles using electron microscopy techniques [55, 58]. The crystal structure and chemical composition of Fe-based NPs are also important characterization steps [83]. Crystallinity, crystal structure, and phases can be evaluated by X-ray diffraction [59]. By using electron energy loss spectroscopy, to distinguish the iron oxide phases present within the NPs, it is also possible to distinguish iron(II) from iron(III) [60]. BET analysis is a straightforward technique for determining surface area [56].

Table 4.1: An overview of Fe-based NPs' characteristics and associated characterization methods. Adapted from previous published studies [52, 53, 55].

NPs' properties	Characterization techniques
Size	XRD, TEM, NTA, DLS, SAXS, HRTEM, AFM, SEM
Size distribution	NTA, DLS, SAXS
Shape	TEM, SEM, HRTEM, STEM, AFM
Crystal structure	XRD, SAED, HRTEM, STEM
Elemental/chemical composition	XRD, ICP-MS, SEM-EDX, EELS, XPS
Surface area/specific surface area	NMR, BET analysis
Surface charge	Zeta potential
Magnetic properties	SQUID, VSM, MFM

XRD: X-ray diffraction; TEM: transmission electron microscopy; DLS: dynamic light scattering; NTA: nanoparticle tracking analysis; HRTEM: high-resolution TEM; SAXS: small-angle X-ray scattering; SEM: scanning electron microscopy; AFM: atomic force microscopy; STEM: scanning transmission electron microscope; SAED: selected area electron diffraction; EDX: energy-dispersive X-ray spectroscopy; ICP-MS: inductively coupled plasma mass spectrometry; XPS: X-ray photoelectron spectroscopy; EELS: electron energy loss spectroscopy; BET: Brunauer–Emmett–Teller; NMR: nuclear magnetic resonance; VSM: vibrating sample magnetometry; SQUID: superconducting quantum interference device; MFM: magnetic force microscopy.

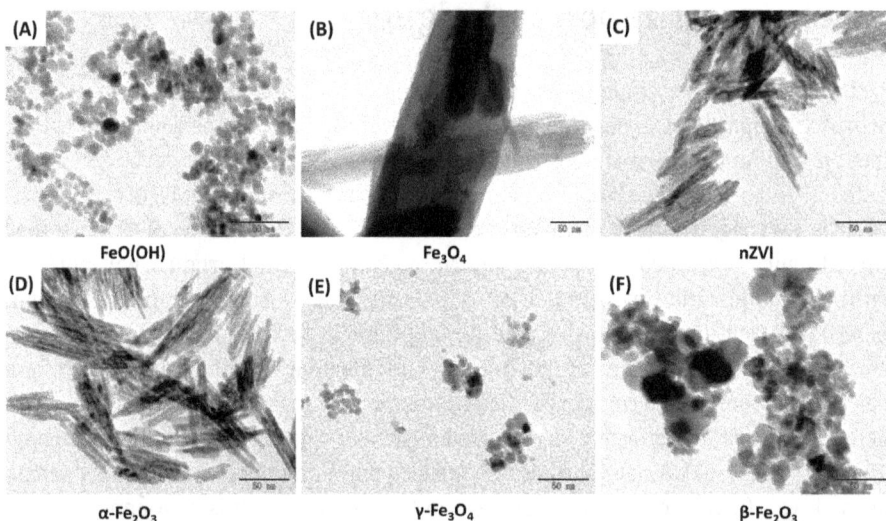

Figure 4.4: Characterization of chemically synthesized Fe-based nanoparticles (Fe-based-NPs) by Transmission electron microscope (TEM). (A) FeO(OH), (B) Fe_3O_4, (C) nZVI, (D) α-Fe_2O_3, (E) γ-Fe_3O_4, and (F) β-Fe_2O_3.

By applying a voltage to the samples, a zeta potentiometer is generally used to measure the surface charge of nanoparticles. Results are given as zeta potential, which is the difference between the potential of the particle surrounding stationary layers and the potential of the solution [61, 62]. The zeta potentials of colloidally stable suspensions usually fall below −15 mV and above +15 mV, as these conditions promote electrostatic repulsion, which keeps the nanoparticles from aggregating [56, 57, 63].

4.4 Nano iron–based fertilization

Recently Shakoor et al. conducted a full life cycle study of cherry radish and documented that Fe-based NPs (FeO(OH), Fe_3O_4, nZVI, α-Fe_2O_3, β-Fe_2O_3, and γ-Fe_3O_4) at 50 mg kg^{-1} showed nonsignificant impacts on the physiological parameters of charry radish. However, with 100 mg kg^{-1} of Fe_3O_4, FeO(OH), and nZVI-based NPs, the fresh and dry biomass of cherry radish increased by 115–244%, 243–314%, and 110–232%, respectively, compared to the control. Fresh and dry biomass of cherry radish leaves under different Fe-based NPs at 50 mg kg^{-1} showed nonsignificant impacts except nZVI (increased 45%). In contrast, Fe_3O_4 NPs, FeO(OH), and nZVI at 100 mg kg^{-1} enhanced the cherry radish leaves fresh and dry biomass by 185–232%, 135–115% and 97–112%, respectively. Furthermore, 200 mg kg^{-1} of Fe-based NPs did not influence the fresh and dry biomass of cherry radish leaves compared to the control [19]. Another study reported that foliar application of both FeO-based NPs (50 and 60 mg kg^{-1}) with poultry manure significantly enhanced the fresh root and dry biomass of red radish (*Raphanus sativus*) as compared to nontreated plants [64]. Furthermore, plant biomass was increased by 38% when *Arabidopsis thaliana* was exposed to 500 mg kg^{-1} of nZVI [65]. Whereas a concentration of nZVI higher than 200 mg L^{-1} significantly reduced the plant growth and biomass of *Populous nigra* [66, 67]. The effects of nZVI vary by plant species and exposure medium (soil or water). For example, nZVI was more immobilized in water than in soil [68, 69].

It has been reported that peanut biomass (shoots and roots) was less toxic to Fe_2O_3-based NPs at 50 and 500 mg kg^{-1} [70, 71]. Fe-NPs may have been mixed with poultry manure, leading to a difference in findings. This may be because the exposure of α- and γ-Fe_2O_3 NPs caused root cell wall wrinkling and deformation, whereas the exposure of Fe_3O_4-based NPs revealed that root and mitochondrial wall structures remained unchanged. Therefore, Fe-based NPs at 100 mg kg^{-1} are the most effective for increasing plant biomass. Rui et al. documented that Fe_2O_3-based NPs increased the physiological parameters of *Arachis hypogaea* (plant height, root length, biomass, and SPAD value) compared to conventional fertilizer (ethylenediaminetetraacetic acid-Fe) as well as Fe-based NPs upregulating the phytohormones that promoted the growth [72]. More studies are required for a better understanding of the impact of Fe-based NPs as fertilizer in a different medium and concentration.

4.5 Bioaccumulation and dietary intake of iron from Fe-based NPs

In vegetables, the edible portion contains a significant amount of Fe, which is essential to the human diet. Increasing Fe-based nanoparticle concentration in radish leads to increased Fe absorption. For example, soils amended with four different concentrations of Fe-based NPs showed higher accumulation of iron in cherry radish as the total estimated daily intake (EDI) values were boosted from 4.58 mg kg^{-1} day^{-1} (noninoculated soil) to 8.74–9.34 mg kg^{-1} day^{-1} in the inoculated soils with FeO(OH) and Fe$_3$O$_4$ at 100 mg kg^{-1} [19]. Similarly, various concentration levels of Ferric EDTA (C$_{10}$H$_{13}$FeN$_2$O$_8$) increased Fe content in radish sprouts, and Fe accumulation was 6.50–5.4% higher than control in low to high concentration levels [73]. The Fe content of treated radish sprouts was 76.4–156.7% higher than what is permissible for daily consumption. Various factors can contribute to this deviation, including the design of the experiment, the growth medium, the concentration level, the applied treatments, and the variety of radish used. In light of these results, Fe-based NPs are unarguably more suitable to enrich fruits and vegetable with Fe to meet the recommendation of EDI in the vegetarian diet.

4.6 Impact of Fe-based NPs on the nutritional content of plants

4.6.1 Impact of Fe nanoparticles on the protein content of plant/crop

A variety of Fe-based NPs differentially alter the content of proteins and N in fruits and vegetables. For example, previous studies documented that the exposure of Fe$_3$O$_4$ had the highest N and crude protein contents in cherry radish, while the lowest results were found for α-Fe$_2$O$_3$. Further, the concentration of Fe-based NPs also affects N content and crude protein content. Cherry radish N and protein contents were improved at 100 mg kg^{-1} as compared to other applied concentrations (0, 50, 200, and 500 mg kg^{-1}). Another previous study reported that Fe$_3$O$_4$ and γ-Fe$_2$O$_3$ application to muskmelon increased the soluble protein content at concentrations ranging from 100–400 mg L^{-1} compared to the control [74, 75]. A comparison of moringa leaves treated with different concentrations of Fe$_3$O$_4$ NPs (20–60 ppm) showed a significant increase in crude protein and some nutrients (N, P, K and K/Na) [76]. A study by Rui et al. found that an increase in soluble proteins could account for decreased amino acids [70]. Chicken manure was also shown to increase the crude protein content of red radish storage roots (5.88%) when ZnO- and FeO-NPs were applied (50 and 60 mg kg^{-1}). An unbalanced redox system is caused by NP-induced oxidative stress, which damages proteins and nitrogen macro-molecules in cells [77, 78]. A more in-depth study of the mechanisms underpinning the regulation of plant nutrients by Fe-based nanoparticles and concentration levels is needed.

4.6.2 Fe-NPs' impact effect on the contents of the vitamin (C)

Free radical scavenger vitamin C (VC) facilitates plant metabolism by donating reducing equivalents and enhancing certain enzymes' activity [79]. A previous study observed that in cherry radish, at 100 mg kg^{-1} exposure level, Fe_3O_4 (48.26%) and FeO (OH) (45.26%) showed the highest increase in VC contents than control. Furthermore, following control, the lowest VC content was observed under γ-Fe_3O_4 than other applied Fe-based NPs. Another study showed that melatonin (17.4 mg L^{-1}) application on cherry radish increased VC contents relative to control [80]. Wang et al. also reported that Fe_3O_4 NPs increase the VC content at 100 mg L^{-1} in muskmelon, while γ-Fe_2O_3 NPs increase VC content at 200 mg L^{-1} compared to the control [81].

4.6.3 Effects of Fe-based NPs on amino acids

Various Fe-based NPs significantly altered the amino acid composition of fruits and vegetables as shown in Figure 4.5. When Fe_3O_4 nanoparticles were sprayed on *Zea mays*, the glutamate and cystine contents were the highest. Cysteine content in wheat crops was also found to increase by approximately 20% when exposed to Fe_2O_3 NPs compared with controls but did not differ in significant manner between the exposure doses [82, 83]. Wang et al. documented that muskmelon exposed to γ-Fe_2O_3 increased (13.4%) the total amino acids content, however, Fe_3O_4 significantly decreased (15.5%) the content of the amino acid at 200 mg L^{-1} [81]. Another study reported that Fe_2O_3-based NPs reduce the leucine, arginine, glutamate, glycine, and aspartate concentrations in peanut grains [70]. A similar study reported that TiO_2-NPs with a dose of 250 mg kg^{-1} did not significantly affect peanut amino acid content whereas at 500 mg kg^{-1}, arginine content reduced by 25% [70]. Moreover, the concentration of nonessential amino acids in cherry radish increased when exposed to Fe_3O_4 at 100 mg kg^{-1} but decreased when exposed to Fe-based nanoparticles at higher concentrations [19]. Aspartate and leucine content that is involved in the citric acid cycle and glycolytic pathway were altered by the presence of engineered nanoparticles. As a component of many metabolic processes in plants, amino acids play an important role in protecting them from abiotic stresses [84]. Proline plays a crucial role in plant cell walls as a scavenger of reactive oxygen species [85]. In plants, histidine production correlates with metal accumulation (Cu, Zn, Ni). When cucumbers were exposed to CuO_2-NPs (400 and 800 mg kg^{-1}) it significantly reduced the lysine (55–61%) and methionine contents (13–25%) as compared to controls

[86, 87]. Fe-based NPs alter essential amino acid contents in fruits and vegetables. In cherry radishes, most Fe-based NPs' treatments at different concentrations significantly decreased leucine and isoleucine content except for FeO(OH) and Fe_3O_4 NPs' treatments. However, FeO(OH) and Fe_3O_4 at lower exposure concentrations (50 and 100 mg kg^{-1}) significantly affected valine and phenylalanine contents [19]. In response to various Fe-based NPs at different exposure concentrations, further studies are needed to understand better the mechanisms that control amino acid contents.

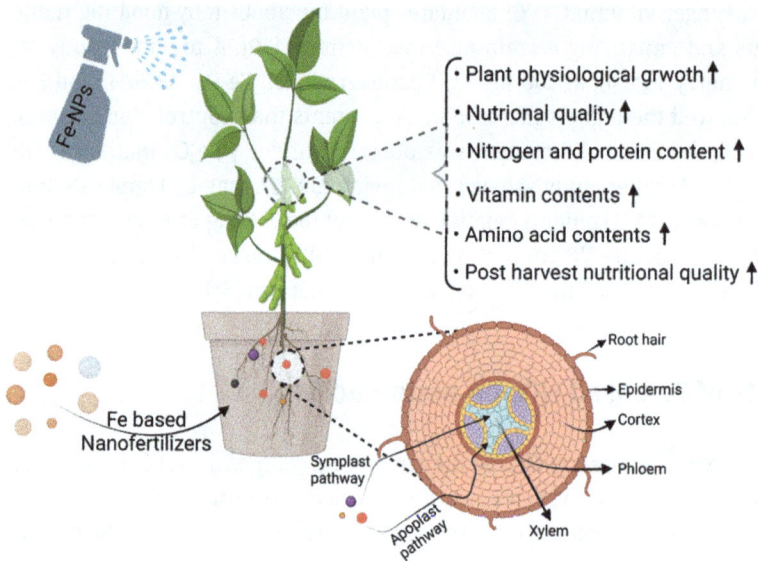

Figure 4.5: Impact of Fe-based NPs on the crops exposed through soil medium or foliar application.

4.7 Conclusion and future outlooks

Humans suffer from iron deficiency because they do not consume enough nutrient-diverse diets. In this chapter, we report on recent advancements in nanofertilizers for improving crop quality. A controlled-release system that exposes Fe-based nanomaterials to crops can enhance their nutritional content and growth. A critical literature review indicates that most Fe-based nanofertilizers are designed to deliver nutrients. A nano-enabled technology could aid in reducing the use of chemical fertilizers and solving the Fe deficiency problem. Nanofertilizers based on Fe are an effective and sustainable means of increasing crop yields and micronutrient content. A sustainable environmental and economical approach can make Fe-based nanofertilizers useful for producing nutrient-rich crops.

References

[1] Saini, R. K., Nile, S. H., Keum, Y.-S. Food science and technology for management of iron deficiency in humans: A review. *Trends in Food Science & Technology* 2016, 53, 13–22.

[2] Zhou, Z. D., Tan, E.-K. Iron regulatory protein (IRP)-iron responsive element (IRE) signaling pathway in human neurodegenerative diseases. *Molecular Neurodegeneration* 2017, 12, 1–12.

[3] Kobayashi, T., Nishizawa, N. K. Annual Review of Plant Biology, Vol. 63. In Merchant, S. S., Ed., Annual Reviews, 2012, 131–152.

[4] Tiwari, A., et al. Photodamage of iron-sulphur clusters in photosystem I induces non-photochemical energy dissipation. *Nature Plants* 2016, 2, doi: 10.1038/nplants.2016.35.

[5] Lopez, A., Cacoub, P., Macdougall, I. C., Peyrin-Biroulet, L. Iron deficiency anaemia. *Lancet [Internet]* 2016, 387, 907–916.

[6] Lowry, G. V., Avellan, A., Gilbertson, L. M. J. N. Opportunities and challenges for nanotechnology in the agri-tech revolution 2019, 14, 517.

[7] Jeyasubramanian, K., et al. Enhancement in growth rate and productivity of spinach grown in hydroponics with iron oxide nanoparticles. *RSC Advances* 2016, 6, 15451–15459, doi: 10.1039/C5RA23425E.

[8] Elmer, W., White, J. C. The future of nanotechnology in plant pathology 2018, 56, 111–133, doi: 10.1146/annurev-phyto-080417-050108.

[9] Lowry, G. V., Avellan, A., Gilbertson, L. M. Opportunities and challenges for nanotechnology in the agri-tech revolution. *Nature Nanotechnology* 2019, 14, 517.

[10] Rui, M., et al. Iron oxide nanoparticles as a potential iron fertilizer for peanut (Arachis hypogaea). *Frontiers Plant Science* 2016, 7, doi: 10.3389/fpls.2016.00815.

[11] Li, J., Hu, J., Xiao, L., Wang, Y., Wang, X. Interaction mechanisms between alpha-Fe2O3, gamma-Fe2O3 and Fe3O4 nanoparticles and Citrus maxima seedlings. *Science of the Total Environment* 2018, 625, 677–685, doi: 10.1016/j.scitotenv.2017.12.276.

[12] Liu, R., Lal, R. Potentials of engineered nanoparticles as fertilizers for increasing agronomic productions. *Science of the Total Environment* 2015, 514, 131–139, doi: 10.1016/j.scitotenv.2015.01.104.

[13] Li, M., et al. Physiological impacts of zero valent iron, Fe3O4 and Fe2O3 nanoparticles in rice plants and their potential as Fe fertilizers. *Environmental Pollution* 2020, 116134.

[14] Yuan, J., et al. New insights into the cellular responses to iron nanoparticles in Capsicum annuum. *Scientific Reports* 2018, 8, doi: 10.1038/s41598-017-18055-w.

[15] Li, J. L., Hu, J., Xiao, L., Wang, Y. Q., Wang, X. L. Interaction mechanisms between alpha-Fe2O3, gamma-Fe2O3 and Fe3O4 nanoparticles and Citrus maxima seedlings. *Science of the Total Environment* 2018, 625, 677–685, doi: 10.1016/j.scitotenv.2017.12.276.

[16] Lei, C., Zhang, L. Q., Yang, K., Zhu, L. Z., Lin, D. H. Toxicity of iron-based nanoparticles to green algae: Effects of particle size, crystal phase, oxidation state and environmental aging. *Environmental Pollution* 2016, 218, 505–512, doi: 10.1016/j.envpol.2016.07.030.

[17] Ding, Y., Bai, X., Ye, Z., Ma, L., Liang, L. Toxicological responses of Fe3O4 nanoparticles on Eichhornia crassipes and associated plant transportation. *Science of the Total Environment* 2019, 671, 558–567, doi: 10.1016/j.scitotenv.2019.03.344.

[18] Fakharzadeh, S., et al. Using nanochelating technology for Biofortification and Yield increase in Rice. *Scientific Reports* 2020, 10, 1–9.

[19] Shakoor, N., et al. Exposure of cherry radish (*Raphanus sativus* L. var. *Radculus Pers*) to iron-based nanoparticles enhances its nutritional quality by trigging the essential elements. *NanoImpact* 2022, 25, 100388, doi: 10.1016/j.impact.2022.100388.

[20] Chircov, C., Vasile, B. S. Iron oxide nanoparticles 2022.

[21] Akbarzadeh, A., Samiei, M., Davaran, S. Magnetic nanoparticles: Preparation, physical properties, and applications in biomedicine. *Nanoscale Research Letters* 2012, 7, 144, doi: 10.1186/1556-276X-7-144.

[22] LaGrow, A. P., et al. Unravelling the growth mechanism of the co-precipitation of iron oxide nanoparticles with the aid of synchrotron X-Ray diffraction in solution. *Nanoscale* 2019, 11, 6620–6628, doi: 10.1039/c9nr00531e.

[23] Guo, H., Barnard, A. S. Naturally occurring iron oxide nanoparticles: Morphology, surface chemistry and environmental stability. *Journal of Materials Chemistry A* 2013, 1, 27–42, doi: 10.1039/C2TA00523A.

[24] Ortega, G., Reguera, E. Materials for Biomedical Engineering. Elsevier, 2019, 397–434.

[25] Gatoo, M. A., et al. Physicochemical properties of nanomaterials: Implication in associated toxic manifestations. *BioMed Research International* 2014, 2014, 498420, doi: 10.1155/2014/498420.

[26] Hui, B. H., Salimi, M. N. Production of iron oxide nanoparticles by co-precipitation method with optimization studies of processing temperature, pH and stirring rate. *IOP Conference Series: Materials Science and Engineering* 2020, 743, 012036, doi:10.1088/1757-899x/743/1/012036.

[27] Niculescu, A. G., Chircov, C., Grumezescu, A. M. Magnetite nanoparticles: Synthesis methods – A comparative review. *Methods* 2022, 199, 16–27, doi: 10.1016/j.ymeth.2021.04.018.

[28] Singh, J. P., et al. Sonochemical Reactions. London, UK, IntechOpen, 2020.

[29] Priyadarshana, G., Senaratne, A., De Alwis, A., Karunaratne, V., Kottegoda, N. Synthesis of magnetite nanoparticles by top-down approach from a high purity ore. *Journal of Nanomaterials* 2015, 2015, 1–8, doi: 10.1155/2015/317312.

[30] Aryal, S., Park, H., Leary, J. F., Key, J. Top-down fabrication-based nano/microparticles for molecular imaging and drug delivery. *International Journal of Nanomedicine* 2019, 14, 6631–6644, doi: 10.2147/ijn.S212037.

[31] Baig, N., Kammakakam, I., Falath, W. Nanomaterials: A review of synthesis methods, properties, recent progress, and challenges. *Materials Advances* 2021, 2, 1821–1871, doi: 10.1039/D0MA00807A.

[32] Mihai, A. D., Chircov, C., Grumezescu, A. M., Holban, A. M. Magnetite nanoparticles and essential oils systems for advanced antibacterial therapies. *International Journal of Molecular Sciences* 2020, 21, doi: 10.3390/ijms21197355.

[33] Rashid, H., et al. Synthesis and characterization of magnetite nano particles with high selectivity using in-situ precipitation method. *Separation Science and Technology* 2020, 55, 1207–1215, doi: 10.1080/01496395.2019.1585876.

[34] Ansari, S., et al. Magnetic iron oxide nanoparticles: Synthesis, characterization and functionalization for biomedical applications in the central nervous system. *Materials* 2019, 12, doi: 10.3390/ma12030465.

[35] Wallyn, J., Anton, N., Vandamme, T. F. Synthesis, principles, and properties of magnetite nanoparticles for in vivo imaging applications-a review. *Pharmaceutics* 2019, 11, doi: 10.3390/pharmaceutics11110601.

[36] Liu, S., Yu, B., Wang, S., Shen, Y., Cong, H. Preparation, surface functionalization and application of Fe_3O_4 magnetic nanoparticles. *Advances in Colloid Interface Science* 2020, 281, 102165, doi: 10.1016/j.cis.2020.102165.

[37] Saleh, T. A., Majeed, S., Nayak, A., Bhushan, B. Advanced Nanomaterials for Water Engineering, Treatment, and Hydraulics. IGI Global, 2017, 40–57.

[38] Wu, W., Jiang, C. Z., Roy, V. A. Designed synthesis and surface engineering strategies of magnetic iron oxide nanoparticles for biomedical applications. *Nanoscale* 2016, 8, 19421–19474, doi: 10.1039/c6nr07542h.

[39] Morsali, A., Hashemi, L. Advances in Inorganic Chemistry, Vol. 76, Elsevier, 2020, 33–72.

[40] Gupta, D., Jamwal, D., Rana, D., Katoch, A. Applications of Nanocomposite Materials in Drug Delivery, Vol. 26, Elsevier, 2018, 619–632.

[41] Thomée, E. Microfluidic nanoparticle synthesis: A short review. *Elveflow* 2021.

[42] Chircov, C., et al. Synthesis of magnetite nanoparticles through a lab-on-chip device. *Materials* 2021, 14, doi: 10.3390/ma14195906.

[43] Chircov, C., Bîrcă, A. C., Grumezescu, A. M., Andronescu, E. Biosensors-on-Chip: An up-to-date review. *Molecules* 2020, 25, doi: 10.3390/molecules25246013.

[44] Abedini-Nassab, R., Pouryosef Miandoab, M., Şaşmaz, M. Microfluidic synthesis, control, and sensing of magnetic nanoparticles: A review. *Micromachines (Basel)* 2021, 12, doi: 10.3390/mi12070768.

[45] Kaur, M., Gautam, A., Guleria, P., Singh, K., Kumar, V. Green synthesis of metal nanoparticles and their environmental applications. *Current Opinion in Environmental Science & Health* 2022, 29, 100390, doi: 10.1016/j.coesh.2022.100390.

[46] Ahmad, S., et al. Green nanotechnology: A review on green synthesis of silver nanoparticles – An ecofriendly approach. *International Journal of Nanomedicine* 2019, 14, 5087–5107, doi: 10.2147/ijn. S200254.

[47] Dikshit, P. K., et al. Green synthesis of metallic nanoparticles: Applications and limitations. *Catalysts* 2021, 11, 902, doi: 10.3390/catal11080902.

[48] Naikoo, G. A., et al. Bioinspired and green synthesis of nanoparticles from plant extracts with antiviral and antimicrobial properties: A critical review. *Journal of Saudi Chemical Society* 2021, 25, 101304, doi: https://doi.org/10.1016/j.jscs.2021.101304.

[49] Jadoun, S., Arif, R., Jangid, N. K., Meena, R. K. Green synthesis of nanoparticles using plant extracts: A review. *Environmental Chemistry Letters* 2021, 19, 355–374, doi: 10.1007/s10311-020-01074-x.

[50] El Shafey, A. M. Green synthesis of metal and metal oxide nanoparticles from plant leaf extracts and their applications: A review. *Green Processing Synthesis* 2020, 9, 304–339, doi: 10.1515/gps-2020-0031.

[51] Yadwade, R., Kirtiwar, S., Ankamwar, B. A review on green synthesis and applications of iron oxide nanoparticles. *Journal of Nanoscience & Nanotechnology* 2021, 21, 5812–5834, doi: 10.1166/ jnn.2021.19285.

[52] Ali, A., et al. Review on recent progress in magnetic nanoparticles: Synthesis, characterization, and diverse applications. *Frontiers in Chemistry* 2021, 9, 629054, doi: 10.3389/fchem.2021.629054.

[53] Kaushal, P., Verma, N., Kaur, K., Sidhu, A. Green synthesis: An eco-friendly route for the synthesis of iron oxide nanoparticles. *Frontiers in Nanotechnology* 2021, 3, 655062, doi: 10.3389/ fnano.2021.655062.

[54] Ealia, S. A. M., Saravanakumar, M. P. A review on the classification, characterisation, synthesis of nanoparticles and their application. *IOP Conference Series Materials Science and Engineering* 2017, 263, 032019, doi:10.1088/1757-899X/263/3/032019.

[55] Mourdikoudis, S., Pallares, R. M., Thanh, N. T. K. Characterization techniques for nanoparticles: Comparison and complementarity upon studying nanoparticle properties. *Nanoscale* 2018, 10, 12871–12934, doi: 10.1039/c8nr02278j.

[56] Modena, M. M., Rühle, B., Burg, T. P., Wuttke, S. Nanoparticle characterization: What to measure?. *Advances in Materials* 2019, 31, 1901556, doi: 10.1002/adma.201901556.

[57] Bélteky, P., et al. Are smaller nanoparticles always better? Understanding the biological effect of size-dependent silver nanoparticle aggregation under biorelevant conditions. *International Journal of Nanomedicine* 2021, 16, 3021–3040, doi: 10.2147/ijn.S304138.

[58] Boselli, L., et al. Classification and biological identity of complex nano shapes. *Communications Materials* 2020, 1, 35, doi: 10.1038/s43246-020-0033-2.

[59] Khan, I., Saeed, K., Khan, I. Nanoparticles: Properties, applications and toxicities. *Arabian Journal of Chemistry* 2019, 12, 908–931, doi: 10.1016/j.arabjc.2017.05.011.

[60] Hilal, N., Ismail, A. F., Matsuura, T., Oatley-Radcliffe, D. Membrane Characterization. Elsevier, 2017.

[61] Kumar, C. V., Pattammattel, A. Introduction to Graphene, 2017, 155–186.

[62] Nasrollahzadeh, M., Shafiei, N., Soleimani, F., Nezafat, Z., Bidgoli, N. S. S. Biopolymer-Based Metal Nanoparticle Chemistry for Sustainable Applications, Vol. 1, 2021, 317.

[63] Anu Mary Ealia, S., Saravanakumar, M. P. A review on the classification, characterisation, synthesis of nanoparticles and their application. *IOP Conference Series: Materials Science and Engineering* 2017, 263, 032019, doi:10.1088/1757-899x/263/3/032019.

[64] Mahmoud, A. W. M., Abdelaziz, S. M., El-Mogy, M. M., Abdeldaym, E. A. J. A. Effect of foliar ZnO and FeO nanoparticles application on growth and nutritional quality of red radish and assessment of their accumulation on human health. *Agriculture* 2019, 65, 16–29.

[65] Yoon, H., Kang, Y. G., Chang, Y. S., Kim, J. H. Effects of zerovalent iron nanoparticles on photosynthesis and biochemical adaptation of soil-grown Arabidopsis Thaliana. Nanomaterials 2019, 9, doi: 10.3390/nano9111543.

[66] Ma, X., Gurung, A., Deng, Y. Phytotoxicity and uptake of nanoscale zero-valent iron (nZVI) by two plant species. *Science of the Total Environment* 2013, 443, 844–849, doi: 10.1016/j.scitotenv.2012.11.073.

[67] Javed, R., Ahmad, M. A., Gul, A., Ahsan, T., Cheema, M. Comprehensive Analytical Chemistry, Vol. 94. In Verma, S. K., Das, A. K., Eds. Elsevier, 2021, 303–329.

[68] Gil-Diaz, M., Alvarez, M. A., Alonso, J., Lobo, M. C. Effectiveness of nanoscale zero-valent iron for the immobilization of Cu and/or Ni in water and soil samples. *Scientific Reports* 2020, 10, 15927, doi: 10.1038/s41598-020-73144-7.

[69] Ahmad, M. A., et al. Influence of calcium and magnesium elimination on plant biomass and secondary metabolites of Stevia rebaudiana Bertoni. *Biotechnology & Applied Biochemistry* 2021, 1, 1–9, doi: 10.1002/bab.2263.

[70] Rui, M., et al. Metal oxide nanoparticles alter peanut (Arachis hypogaea L.) physiological response and reduce nutritional quality: A life cycle study. *Environmental Science-Nano* 2018, 5, 2088–2102, doi: 10.1039/c8en00436f.

[71] Pang, L.-J., et al. Engineered nanomaterials suppress the soft rot disease (Rhizopus Stolonifer) and slow down the loss of nutrient in sweet potato. *Nanomaterials* 2021, 11, 2572, doi: 10.3390/nano11102572.

[72] Rui, M., et al. Iron oxide nanoparticles as a potential iron fertilizer for peanut (*Arachis hypogaea*). *Frontiers in Plant Science* 2016, 7, 815, doi: 10.3389/fpls.2016.00815.

[73] Przybysz, A., Wrochna, M., Malecka-Przybysz, M., Gawronska, H., Gawronski, S. W. Vegetable sprouts enriched with iron: Effects on yield, ROS generation and antioxidative system. *Scientia Horticulturae* 2016, 203, 110–117, doi: 10.1016/j.scienta.2016.03.017.

[74] Tanveer, M., Hasanuzzaman, M., Wang, L. Lithium in environment and potential targets to reduce lithium toxicity in plants. *Journal of Plant Growth Regulation* 2019, 38, 1574–1586, doi: 10.1007/s00344-019-09957-2.

[75] Ahmad, M. A., et al. Appraisal of comparative therapeutic potential of undoped and nitrogen-doped titanium dioxide nanoparticles. *Molecules* 2019, 24, 3916, doi: 10.3390/molecules24213916.

[76] Tawfik, M. M., Mohamed, M. H., Sadak, M. S., Thalooth, A. T. Iron oxide nanoparticles effect on growth, physiological traits and nutritional contents of Moringa oleifera grown in saline environment. *Bulletin of the National Research Centre* 2021, 45, 177, doi: 10.1186/s42269-021-00624-9.

[77] Ma, C., White, J. C., Zhao, J., Zhao, Q., Xing, B. Uptake of engineered nanoparticles by food crops: Characterization, mechanisms, and implications. *Annual Reviews of Food Science and Technology* 2018, 9, 129–153, doi: 10.1146/annurev-food-030117-012657.

[78] Farooq, T., et al. Nanotechnology and plant viruses: An emerging disease management approach for resistant pathogens. *ACS Nano* 2021, 15, 6030–6037, doi: 10.1021/acsnano.0c10910.

[79] Paciolla, C., et al. Vitamin C in plants: From functions to biofortification. *Antioxidants (Basel)* 2019, 8, 519, doi: 10.3390/antiox8110519.

[80] Jia, C. H., et al. Application of melatonin-enhanced tolerance to high-temperature stress in cherry radish (Raphanus sativus L. var. radculus pers). *Journal of Plant Growth Regulation* 2020, 39, 631–640, doi: 10.1007/s00344-019-10006-1.

[81] Wang, Y., et al. The impacts of gamma-Fe_2O_3 and Fe_3O_4 nanoparticles on the physiology and fruit quality of muskmelon (*Cucumis melo*) plants. *Environmental Pollution* 2019, 249, 1011–1018, doi: 10.1016/j.envpol.2019.03.119.

[82]   Wang, Y., et al. Effect of metal oxide nanoparticles on amino acids in wheat grains (*Triticum aestivum*) in a life cycle study. *Journal of Environmental Management* 2019, 241, 319–327, doi: 10.1016/j. jenvman.2019.04.041.

[83]   Ahmad, M. A., Javed, R., Adeel, M., Rizwan, M., Yang, Y. PEG 6000-stimulated drought stress improves the attributes of in vitro growth, steviol glycosides production, and antioxidant activities in Stevia rebaudiana Bertoni 2020, 9, 1552, doi: 10.3390/plants9111552.

[84]   Thakur, S., et al. Plant-driven removal of heavy metals from soil: Uptake, translocation, tolerance mechanism, challenges, and future perspectives. *Environmental Monitoring and Assessment* 2016, 188, 206, doi: 10.1007/s10661-016-5211-9.

[85]   Sharma, S. S., Dietz, K. J. The relationship between metal toxicity and cellular redox imbalance. *Trends in Plant Science* 2009, 14, 43–50, doi: 10.1016/j.tplants.2008.10.007.

[86]   Zhao, L., et al. (1)H NMR and GC-MS based metabolomics reveal nano-Cu altered cucumber (Cucumis sativus) fruit nutritional supply. *Plant Physiology and Biochemistry* 2017, 110, 138–146, doi: 10.1016/j.plaphy.2016.02.010

[87]   Hussain, M. et al. Nano-enabled plant microbiome engineering for disease resistance. *Nano Today* 2023, 48, 101752, doi: https://doi.org/10.1016/j.nantod.2023.101752.

Mehdi Rahmati*, Mehdi Kousehlou

Chapter 5
Nanoparticles as soil amendments

Abstract: Soils not only form the thin skin of the land surface, but also play an important role in controlling water, matter, and energy cycles. They store the largest amount of organic carbon on the land surface and, through respiration, generate a carbon flux similar in magnitude to the carbon absorbed through photosynthesis. They are also the primary source of water transpired by plants and provide important ecosystem services to society, including supporting and regulating services. Therefore, soil conservation has always been of great interest to researchers from various disciplines, especially soil scientists. Nanoparticles have emerged as a promising soil amendment technology for combating soil degradation and improving its quality. Nanoparticles such as nanoclays, carbon nanotubes, and metal oxide nanoparticles have been shown to improve soil properties such as water retention, nutrient availability, and soil stability. This chapter summarizes recent advances in the use of nanoparticles as soil amendments to mitigate soil degradation and enhance its quality and health. It highlights the potential of nanoparticles as a sustainable and eco-friendly approach to prevent soil degradation and improve soil quality, and their potential to enhance crop productivity and promote sustainable agriculture. It also discusses the challenges and potential risks associated with the use of nanoparticles as soil amendments and the need for further research to understand their long-term effects on the environment and human health. Overall, the use of nanoparticles as soil amendments represents a promising way to reduce soil degradation and improve soil quality in agriculture.

Keywords: biodiversity, nanotechnology, soil degradation, soil quality, water management

5.1 Introduction

As the world population grows, the demand for food and agricultural products is increasing rapidly, so much so that the Food and Agriculture Organization (FAO) predicts that 200 million tons of food and crops will be needed annually in 2050 [1]. However, due to factors such as climate change, limited resources of water and land, and increasing soil, water, and air pollution, it will be very difficult to meet the huge demand for food during this

---

*Corresponding author: Mehdi Rahmati,** Department of Soil Science and Engineering, University of Maragheh, P.O. Box 55136-553, Maragheh, Iran; Institute of Bio- and Geosciences: Agrosphere (IBG-3), Forschungszentrum Jülich GmbH, Jülich, Germany, e-mail: mehdirmti@gmail.com
Mehdi Kousehlou, Department of Soil Science, College of Water and Soil Sciences, Gorgan University of Agricultural Sciences and Natural Resources, Gorgan, Iran

https://doi.org/10.1515/9781501523229-005

period. As a result, scientists are using various tools, including nanotechnology, to improve soil quality and create healthy soils that can, in part, ensure food security. Nanotechnology has the potential to revolutionize the agricultural sector. Nanoscience is concerned with the study of materials in the order of about 1 to 100 nm and the production and application of nanoscale control instruments and systems [2]. Man-made nanoparticles are released into the environment primarily in the form of metal oxides (e.g., zinc oxide (ZnO) and titanium dioxide (TiO_2)), semiconductor nanocrystals (e.g., quantum dots), carbon nanomaterials (e.g., carbon nanotubes), zero-valent metals (e.g., zero-valent iron), and nanopolymers. Nanotechnology has numerous applications in soil [3], as one of the most important resources for agricultural production. In fact, for sustainable agricultural production, it is very important to maintain the health and fertility of the soil. The nutrient and moisture content of the soil should always be desirable, and the concentration of pollutants (heavy metals, toxins, etc.) should be reduced to a minimum. Nanotechnology can be useful for all the above purposes, which are discussed below.

5.2 Nanotechnology in plant nutrition and soil fertility

The use of nanotechnology in development and production of nano-fertilizers opens new opportunities for the agriculture sector to increase the efficiency of nutrient use and minimize the cost of environmental protection. Nano-fertilizers are suitable substitutes for conventional soluble fertilizers due to their gradual and slow release of nutrients. Nano-fertilizers are nutrient carriers with a size of 30 to 40 nm and can transport nutrient ions well due to their high specific values [4]. In general, the advantages of using these nano-fertilizers can be summarized as follows:
1. making smart fertilizers so that the nutrients release rate of fertilizers matches the absorption pattern of plants;
2. making fertilizers with low-consumption nano-sized nutrients that increase the solubility and dispersion of these nutrients in the soil and improve the plant's ability to absorb these elements [5];
3. reducing nutrient loss through leaching [6, 7];
4. increasing plant resistance to environmental stresses due to proper nutrition and availability of nutrients [6, 7]; and
5. being economically viable due to reduced fertilizer use [5].

5.3 Nanohydrogels as soil modifiers

Hydrogels have an empty lattice interstitial space and can therefore serve as nanoreactors to produce nanoparticles. These materials respond to environmental stimuli such as pH, light, electric field, etc. [8]. One of the roles of hydrogels, especially nanohydrogels, is to provide an environment for plant growth where nutrients and water can be stored for long periods of time and gradually released to the plant. In general, there are the following possibilities for the use of nanohydrogels:

1. to increase water absorption through nanoscale pores;
2. to increase the water holding capacity; and
3. to store nutrients and gradually release them to the plants [9].

The use of nanohydrogels or superabsorbents results in rapid water absorption and conservation which in consequence increases the efficiency of water uptake during extensive rainfall and, in the case of irrigation practices, extends irrigation intervals. Superabsorbents also improve soil aeration through a continuous change of its volume (swelling during water gain and shrinkage during water loss). The anionic types of superabsorbents are of particular importance in agriculture because they not only absorb cations useful for plant growth but also prevent their stabilization and leaching out of the soil by improving cation exchange capacity (CEC) of the soil. In some cases, if these hydrogels are properly applied, majority of the water stored in the hydrogel can be available for plant uptake [10]. When these adsorbents are mixed with the soil, water absorption turns them into amorphous gelatinous masses that can absorb and release water over a long period of time. Therefore, these superadsorbents act as a source of water and soluble nutrients. More specifically, hydrogel superabsorbents coated with silver nanocomposites were prepared for agricultural purposes.

5.4 Nanoparticles as soil amendment

Due to improper soil management, the soil gradually loses its quality and fertility and becomes inferior and less healthy. Therefore, the use of special additives such as nano amendments, which can improve soil quality and prevent its degradation, seems unavoidable. However, there is little information on the potential effects of nanoparticles on the physical, chemical, and biological indicators of soil quality and soil health. In the following subsections, we will briefly report on a few of the current uses of nanoparticles affecting soil physical, chemical, and biological properties.

5.4.1 Impacts on soil physical properties

In most areas of the world with sandy soils, water scarcity is a major problem, leading to a decline in the production of agricultural products, which in turn drives up world prices for these products. Therefore, several companies have developed new nanomaterials to alleviate problems such as water scarcity in irrigation [11]. Nanoclays are among those nanomaterials suitable for such an aim. In an example case, the use of nanoclay in hot and dry sandy soils in Egypt saved two-thirds of normal irrigation water use, i.e., a reduction in irrigation use to one-third resulting in a 416% increase in crop yields [12]. Nanoclays are better suited for sandy soils, where they act as binders, stabilize the sand, and can retain up to 25% more water in those coarse-grained soils. Nanoclays are also useful for increasing water availability in desert soils and supporting revegetation in these areas. This is because they form cohesive masses in the soil, storing water and nutrients and making them available to plants. However, the treatment of soils with platinum nanoparticles revealed that these nanoparticles accelerated soil drying, which contributed to increased evapotranspiration and potentially affected other soil processes [13], which in turn may affect soil water availability for plants. Reclamation of deserts with nanoclay can also lead to a reduction in wind erosion and formation of soil grains in those soils. The high specific surface area of nanoparticles causes a very strong surface energy between soil particles which can increase the flocculation in soil [14]. This property of nanoparticles can be used as a binding agent between soil particles and the formation of soil aggregates.

Various macroscopic and microscopic properties of soils can also be affected by nanomaterials such as metal oxide nanoparticles [15]. The accumulation of nanoparticles in soil may lead to the blockage of some flow paths and the reduction of hydraulic conductivity. In fact, soil has pores in the range of micrometers to nanometers. When these pores are filled with nanoparticles, the properties of the soil are also likely to change. Thus, some properties such as shear strength, bulk density, fluid resistance, etc. may improve, while compressibility, permeability, plastic index, etc. may decrease. An increase in soil shear strength and bulk density, a decrease in permeability of the soil, and an improvement in soil granularity due to the application of calcium chloride ($CaCl_2$), calcium oxide (CaO), and potassium nitrate (KNO_3) nanoparticles [16] as well as an increase in the average size and number of grains and a reduction in hydraulic conductivity due to application of iron oxide (Fe_3O_4) and copper oxide (CuO) nanoparticles [15] as well as a reduction in hydraulic conductivity due to the application of nano-alumina (aluminum oxide, Al_2O_3, nanoparticles), nano-copper [16], and carbon nanotubes and nanofibers [17] is already reported. However, contradictory to the increase in soil bulk density due to $CaCl_2$, CaO, and KNO_3 nanoparticles [16], application of nano-alumina [18], carbon nanotubes, and carbon nanofibers [17], as well as zinc and silica nanoparticles [19] resulted in a decrease in soil bulk density. It is assumed that the reduction of soil permeability after the addition of nanoparticles is due to the filling of soil pores with these nanoparticles. The use of nanoparticles can also reduce the attenuation ratio and increase the shear modulus of the soil, which consequently reduces the liquefaction sensitivity of the non-adherent soil [16]. According to Saad Kheir et al. [19], the use of zinc and silica nanoparticles increased soil hydraulic conductivity and decreased soil compaction.

Soil plasticity and liquid limits are also among the soil properties that can be affected by the application of nanoparticles. Conflicting results exist for the effects of nanoparticles application on soil plasticity index (subtraction of liquid and plasticity limits), where an increase in plasticity index due to the application of multi-carbon nanotubes and carbon nanofibers [20] and a slight decrease due to the addition of nano-alumina and a sharp decrease due to the addition of nano-copper [18] are re-ported. Taha and Alsharef [17] showed that the liquid limit increased slightly with the increase of carbon nanotubes doses, while it did not change for carbon nanofibers. In contrast, the changes in plastic indices were clear with the addition of carbon nano-tubes, but there was no specific relationship with applied doses (there was no increas-ing or decreasing trend). However, in soils treated with carbon nanofibers, the changes in plastic indices insignificantly increased.

The use of nanomaterials can also affect the swelling and shrinking behavior of soils. Thus, the use of nano-alumina and nano-copper reduced both shrinkage and swelling of the soil [18]. As reported by Taha and Taha [18], the application of nano-alumina and nano-copper also modified and reduced soil cracking without signifi-cantly affecting the hydraulic conductivity of the soil.

It seems that the outcome of soil treatment by nanoparticles is highly dependent on soil type. For example, clayey soil most probably can contain more nanoparticles after being treated (adding and washing) with nanoparticles compared to sandy soil being treated in a similar way so probably resulting in different effects. The reason seems to be the absorption of nanoparticles by the ions in the clays [15]. Taha and Taha [21] also showed that nanomaterials are more suitable for soils with high clay content (bentonite).

The effects of nanoparticles on soil porosity probably are also dependent on the type of applied nanoparticles as some research reports no change in soil porosity by the addition of Fe_3O_4 and CuO nanoparticles [15] but some others report a reduction in soil porosity due to the application of multi-carbon nanotubes and carbon nanofib-ers [20]. In general, the accumulation of nanoparticles in the soil is likely to clog pores and reduce porosity [20] if the concentration of applied nanoparticles is high enough. The accumulation of nanoparticles in the soil can also lead to clogging of some flow paths and a reduction in hydraulic conductivity [21], as previously reported. There-fore, the use of nanomaterials is a possible method to decrease the hydraulic conduc-tivity of the soil without increasing the soil suction [21]. The reduction of soil bulk density and hydraulic conductivity is also reported by the application of multi-carbon nanotubes and carbon nanofibers with a stronger effect for carbon nanofibers than for multi-walled carbon nanotubes [20]. Taha and Taha [21] reported that mixing the soil with the optimal concentration of nanomaterials decreased the water content of the studied soils. However, at a given suction in the characteristic curve, the soil water content was lower for samples containing nano-alumina than for nano-copper. Bayat et al. [22] also showed that magnesium oxide (MgO) nanoparticles resulted in lower soil bulk density compared to nano-iron. Their measurements showed that nano-magnesium increased the void ratio and improved the soil structure and in-creased the porosity, while nano-iron accumulated in the soil pores due to its high flocculation ability and decreased the porosity by clogging the pores. They showed that none of these nano-oxides had any effect on the compaction curve under compression. The comparison of the average tensile stability of soil grains showed that nano-iron acts as a bridge between soil particles by strengthening the bonds between iron and soil particles and increasing the strength and tensile stability of soil grains against soil suction.

Few studies also show that the clogging effect of nanoparticles and thus their effects on soil hydraulic conductivity and soil porosities may vary depending on the size of nanoparticles. In this regard, according to Bahmani et al. [23], as the size of nanoparticles increases, the clogging effect of nanoparticles seems to decrease and consequently the hydraulic conductivity of soil increases.

The effects of nanoparticles on soil properties are probably also dependent on the concentration of nanoparticles used. For example, Alsharef et al. [20] showed that any increase in the concentration of nanocarbons up to a certain amount increases the maximum dry density of the soil, but thereafter any increase in the concentration of nanocarbons decreases the maximum dry density. Taha and Taha [18] also showed that increasing nano-alumina above a certain level can lead to the accumulation and linkage of nano-alumina, which in turn increases voids and water content of the soil and decreases soil bulk density. Taha and Taha [21] also showed that the hydraulic conductivity of the dense samples decreased with increasing content of nanomaterials, which was due to the clogging of pores and interaction between particles by the nanomaterials.

5.4.2 Impacts on soil chemical properties

It is easy to expect that the application of nanoparticles can also change the chemical properties of soil. Therefore, the effects of nanoparticle application on soil chemical properties were also of interest to researchers. For example, pH is one of the soil chemical properties that can be affected by the accumulation of different types of nanoparticles such as Zn, Ag, Au, Cu, etc. It seems that depending on nanoparticles type, soil pH can increase [20] or decrease [24, 25] due to the use of nanoparticles.

Another important chemical property of soil whose fate after nanoparticle application is of interest to soil scientists is CEC. The CEC value indicates the number of negatively charged sites in the soil, which indicates the ability of the soil to exchangeable the cations with soil solution. This parameter is important because it is a good indicator of the soil's ability to absorb nutrients, thus indicating soil quality and productivity. Ben-Moshe et al. [15] examined the changes in the adsorption capacity of nickel adsorption isotherms for soil samples with and without CuO and Fe_3O_4 nanoparticles showing that no change in isotherms was observed when nanoparticles were added, indicating that the adsorption capacity of the soil was not affected by nanoparticles. The SEM images show that the nanoparticles accumulated on the surface of the soil grains. It is likely that the change in surface area is relatively small and therefore has little effect on the absorption process. Saad Kheir et al. [19] showed that the use of zinc and silica nanoparticles resulted in a slight decrease in soil salinity at the mean values of ECe and an increase in organic matter compared to the control.

5.4.3 Impacts on soil biological properties

The bacterial community of soil and its functions are among the soil properties most sensitive to nanoparticle application. Therefore, some research in this area are aimed at investigating the effects of nanoparticle application on the biological properties of soil. For example, it is already reported that CuO and Fe_3O_4 nanoparticles affected the composition of the bacterial community in soil [15]. In contrast, it seems fullerene nanoparticles had little effect on the microbial population and microbial activities in the soil [26]. Multi-walled carbon nanotubes decreased enzyme activity and microbial C and N biomass in soil [27]. Xu et al. [28] reported that TiO_2 and CuO nanomaterials reduced soil microbial biomass by studying the soil of rice fields and caused a change in the chemical behavior of microbial biomass in rice soil by affecting its structure. You et al. [29] investigated the effect of ZnO, CeO_2, TiO_2, and Fe_3O_4 nanoparticles on enzyme activities (invertase, urease, catalase, and phosphatase) and bacterial communities of saline and alkaline soils, revealing the effects on soil enzyme activity and changes in the soil bacterial community and threats to biological nitrogen fixation. High concentrations of Fe_3O_4 nanoparticles generally reduce the matter of microscopic organisms in the soil. The TiO_2 nanoparticles reduce the amount of beneficial microscopic soil organisms and enzyme activities and prevent the effects on microbial performance and diversity. Zinc oxide nanoparticles (ZnO) and CeO_2 affect *Azotobacter* and phosphorus- and potassium-solubilizing bacteria by inhibiting enzymatic activity [30].

Luo et al. [31] showed that the addition of $nCeO_2$ and nCr_2O_3 nanoparticles increased the nitrate, organic matter, and total nitrogen content of the soil, and the nitrate content was higher in the soil with nanoparticles than in the soil without nanoparticles. By increasing the concentration of $nCeO_2$ nanoparticles, soil nitrate was also increased, but in soil with nCr_2O_3 with increasing nanoparticles, a decreasing trend was observed in nitrate content of total ergogenic, dissolved organic nitrogen, and dissolved organic carbon.

Copper ions released from copper nanoparticles are lethal to both pathogenic and beneficial bacteria [32, 33]. Concha-Guerrero et al. [34] showed that copper oxide nanoparticles are highly toxic to indigenous soil bacteria, as the authors observed and analyzed the formation of cavities, holes, cell loss, and collapse in isolated cells of soil bacteria. In general, inorganic nanoparticles appear to exhibit higher toxicity to soil microorganisms than organic nanoparticles [35].

Ge et al. [36] examined the effects of TiO_2 and ZnO nanoparticles on bacterial communities and found that although both nanoparticles decreased microbial biomass and extractable DNA, their response curve was different, so that it was linear for TiO_2 and exponential for ZnO, indicating the difference in bioavailability and environmental behavior of these nanomaterials. The effect of ZnO on soil DNA was more and stronger than TiO_2, so at a dose of 0.5 mg/g of both nanoparticles, the soil DNA extraction was much lower for ZnO than TiO_2. Their results showed that the bacterial toxicity of ZnO was higher than that of TiO_2, which was due to the greater solubility of ZnO. Other effects of these nanoparticles were that they decreased substrate-induced respiration. To investigate the effects of different doses of ZnO and TiO_2 nanoparticles on microbial activity and organic matter mineralization, Ge et al. [36] measured baseline respiration and found no significant effect on baseline respiration at any concentration of nanomaterials.

Kumpiene et al. [37] showed that the application of zero-valent iron to stabilize chromium and copper and arsenic in soils contaminated with chromate–copper–arsenate increased the activity of many soil enzymes and reduced microbial toxicity. The study conducted by Samarajeewa et al. [38] showed that soil exposure to silver nanoparticles negatively affects microbial growth, so silver can cause toxicity to microorganisms through several functions. In this experiment, basal respiration and respiration caused by the addition of substrate were also measured, and the changes in basal respiration were different, which showed an increasing trend in the first days compared with the control but decreased with time. In contrast, the respiration caused by the addition of substrate was lower. The activity of the enzyme dehydrogenase decreased completely, while the activity of the enzyme β-glucosidase decreased but the rate of decrease was lower than that of dehydrogenase. Their results also showed that microbial diversity decreased significantly after the addition of silver nanoparticles.

The TiO_2 nanoparticles decrease the number of functional soil bacteria and enzymatic activity and have adverse effects on microbial activity, number, and diversity [39, 40]. Data by Maliszewska [41] showed that Au bio-nanoparticles up to 33 mg/kg have no effect on the soil process and cannot be classified as harmful nanoparticles. However, antibacterial activities were observed in clinical isolates (*Bacillus subtilis*, *Escherichia coli*, *Klebsiella pneumoniae*, *Pseudomonas aeruginosa*, *Salmonella typhi*, and *Staphylococcus aureus*) when soils were treated with ZnO nanoparticles synthesized by biological and chemical methods. Huang et al. [24] investigated the effect of zinc nano-oxide on lead bioavailability and soil microbial properties and reported that at a concentration of Zn nanoparticles greater than 10 mg/kg, diethylenetriaminepentaacetic acid (DTPA)-extractable lead decreased considerably between 10.6 and 21.3% over 60 experimental days. Their results also showed that zinc nano-oxide improved the microbial diversity and abundance of some metal-resistant bacteria after 60 days. The results of Xiaohong et al. [42] showed that silver nanoparticles inhibited the enzyme activity of the nitrogen cycle, the number of nitrifying bacteria, and the gross nitrification rate, and the extent of nitrification was negatively correlated with the surfaces of silver nanoparticles. Verma et al. [25] found that the application of Zn nanoparticles resulted in a decrease in microbial carbon biomass, but a significant increase in available phosphorus and zinc. Feng et al. [43] conducted an experiment to investigate the effect of zinc oxide nanoparticles on nitrogen oxide emissions in agricultural soils and reported that ZnO nanoparticles increased N_2O emissions through enhanced nitrification and nitrate removal and contributed to global warming.

Soil pH was found to affect the toxicity of nanoparticles to microorganisms and nematodes. The effect of organic matter and soil pH on the toxicity of ZnO nanoparticles to Folsomia candida was observed [44]. The more toxic effect of Zn nanoparticles compared to Zn dissolution on the bacterial community in soil was also observed [45].

5.5 Nanoparticles as soil remediators

Nanoscience and nanotechnology have the potential to have a profound impact on the environment as they can interact with polymers, minerals, pollutants, and nutrients. To increase the speed and efficiency of removing pollutants from the environment, researchers have produced modified iron nanoparticles, such as nanoparticles with catalysts. For example, they are used to absorb metals and anionic pollutants such as arsenic, chromium, lead, mercury, selenium, copper, uranium, and heavy metals [46]. An example of the use of this technique is the use of neutral iron as a reduction chemical. In general, the use of nanoparticles, specifically nanoscale zero-valent iron, for in situ modification of soil pollution caused by heavy metals and organic pollutants has raised concerns about the performance, migration, and transformation of them under the influence of soil physical and chemical conditions due to its excellent activity potential, low cost, and low toxicity. However, the use of nanomaterials has some limitations. For example, some materials can passivate rapidly, accumulate readily [47], be sensitive to geochemical conditions [48], and potentially endanger the environment and human health. In the next subsections, we will briefly discuss the use of nanoparticles for the removal of heavy metals, ionic pollutants, and organic pollutants.

5.5.1 Removal of heavy metals

Nano-oxides are used as important remediation agents for soil contamination mainly due to their high reactivity and large specific surface area [49]. As for the removal of chromium and related substances, studies have focused on the use of nanoparticles to remove chromium from groundwater and soil [50]. Various studies have been conducted for the removal of arsenic, one or both of its related species, e.g., As(III) and As(V) [51], some other studies focused on the removal of several polluting heavy metals such as lead [52], copper [53], various zinc species [54] from water and nickel ions from wastewater [55], and removal of cadmium ion dots. Carbon oxide and hydroxylated nanotubes are also used to absorb metals such as copper [56], silver [57, 58], cadmium [56], zinc [59], and americium [60].

Evidence shows the use of zero-valent iron nanoparticles for the remediation of soil contamination. The availability, easy accessibility, and affordable cost of those nanoparticles have led to its widespread use. Despite the high reactivity of zero-valent iron nanoparticles, the tendency of these materials to aggregate due to their magnetic properties leads to external surfaces, which increases their reactivity in solution and decreases their sedimentation rate [61]. To prevent the aggregation of zero-valent iron nanoparticles, several solutions have been proposed, including the use of nanoparticles composed of two metals instead of one, called bimetallic particles. In this context, metals that have a more positive redox potential are used, such as palladium, silver, and copper [62].

In particular, copper oxide (CuO) nanoparticles have been used to absorb (trivalent and tetravalent) arsenic [63], and zero-valent iron nanoparticles have been used to modify pollutants by nucleoid radiation [64]. The study by Naderi Peikam and Jalali [65] on the application of Al_2O_3, SiO_2, and TiO_2 nanoparticles for the remediation of metal-contaminated soils showed that although both calcareous and non-calcareous soils were affected by nanoparticles, nanoparticles decreased the exchangeable fraction of cadmium and zinc more effectively in non-calcareous soils and the change in nickel fraction was minimal, unlike cadmium and zinc in non-calcareous soils. In general, SiO_2 nanoparticles were very effective in immobilizing metals in calcareous soils. In non-calcareous soils, the maximum decrease in the mobile fraction of cadmium and zinc was in the presence of Al_2O_3 nanoparticles. Overall, the results showed that nanoparticles can reduce metals in exchangeable, oxide, carbonate, and organic fractions.

The application of zero-valent iron nanoparticles in the rhizosphere soils contaminated with heavy metals (Pb, Cd, As, and Zn) showed that the use of these nanoparticles increased the durability of As and Zn in the studied soils, thus increasing the biomass production for modified soils and significantly reducing the absorption of Pb, Cd, As, and Zn. Examination of the solid phase of soils showed that absorption of oxides, iron oxide and manganese, and the formation of secondary Fe–As phases are the main mechanism of immobilization of heavy metals after using zero-valent iron nanoparticles [66].

Baragaño et al. [67] showed that using graphene oxide and zero-valent iron nanoparticles in soils contaminated with arsenic-containing metals showed that the immobility of lead increased significantly with increasing doses of the nanoparticles. The availability of cadmium was reduced only at the highest dose of graphene oxide nanoparticles, while the availability of zinc was significantly reduced only at the highest doses of graphene oxide and zero-valent iron nanoparticles. In contrast, the availability of copper differed between the different doses of both nanoparticles. For copper, a significant decrease was observed with increasing doses of graphene oxide, while zero-valent iron treatment had no effect on copper availability. For arsenic, different effects were observed such that arsenic availability significantly increased by increasing the dose of graphene oxide nanoparticles, while arsenic availability significantly decreased by increasing the dose of zero-valent iron nanoparticles.

Kumpiene et al. [37] used zero-valent iron nanoparticles to stabilize chromium, copper, and arsenic in soils contaminated with chromium-containing copper arsenate (CCA). The results of this study showed that the use of 1% zero-valent iron nanoparticles reduced the dissolved forms of As and Cr in the soil and led to a decrease in the concentration of these elements in the leachate (98% and 45% for As and Cr, respectively), in the pore water of the soil (99% and 94%, respectively), and in the plant branches (84% and 95%, respectively). The bioavailability of arsenic also decreased dramatically. However, after stabilizing the CCA-contaminated soil with Fe^0, copper was available at high concentrations and may even cause some toxicity in the treated soil. In addition, Fe^0 reduced the leaching of As, Cr, and Cu by 99%, 93%, and 57%, respectively.

In studying the stabilization of lead and zinc by using zero-valent iron stabilized in both acidic and alkaline soils and its effect on some soil properties, it was found that the application of this nanoparticle significantly increased the availability of these two metals in exchangeable components and reduced carbonation and increased the residual content in the soil. The nanoparticles of zero-valent iron were more effective than zinc in binding lead [68].

5.5.2 Removal of anionic pollutants

Of several inorganic pollutants whose removal has been studied, nitrate has received the most attention. Zero-valent iron nanoparticles were used for nitrate removal. Another pollutant is phosphorus. Despite the widespread use of nanoparticles in various fields, there is little information on the immobilization of phosphorus by these nanoparticles, especially in soil [59]. Taghipour and Jalali [69] studied the effect of titanium dioxide and alumina nanoparticles on phosphorus release kinetics and fractionation. They showed that the addition of nanoparticles to the soil reduces phosphorus release over time. Nano-calcium sulfate is also used to reduce the loss of orthophosphates in soil [70]. There are several studies on the effect of nano-calcium sulfate to retain phosphorus in soil. Nano-calcium sulfate has advantages over conventional calcium sulfate including greater specific surface area, higher solubility, better contact with fertilizer and soil, better dispersibility, and lower consumption. Nano-calcium sulfate is expected to increase complexation (precipitation and adsorption) between orthophosphate and calcium and reduce phosphorus losses in fields [70]. By using palladium/iron nanoparticles at a concentration of 6.25 g/liter, the concentration of chlorinated compounds in the solution was reduced below the detection range of the instrument. The use of iron nanoparticles removed more than 99% of the chlorinated compounds in the solution within 24 h.

5.5.3 Removal of organic pollutants

Several studies have focused on the removal of trichloroethylene and polychlorinated biphenyls [71]. Carbon nanotubes have been used to absorb various organic compounds such as dioxin [72], polyaromatic hydrocarbons, diphenyl ethers [60], chlorobenzene and chlorophenol, and thiamethoxam pesticides [53]. Zero-valent iron oxide nanoparticles are used to modify organic pollutants as well as to remove chlorinated solvents [53], pesticides and herbicides, and dichlorodiphenyltrichloroethane [73]. Some pesticides that are stable in an aerobic environment are degraded much more readily and rapidly under anaerobic (regenerative) conditions.

5.6 Potential dangers of nanoparticles

As mentioned earlier, there is sufficient evidence that nanoparticles have toxic effects on soil microorganisms. Studies have shown that the toxicity of nanoparticles, especially metal nanoparticles, depends on the physical properties of the particles such as size, shape, surface coating, biocompatibility, and reactivity, as well as the method used to produce nanoparticles [74]. The possible mechanisms of nanoparticle toxicity include membrane disruption, oxidative damage to proteins, DNA damage, disruption of electron transport, formation of reactive oxygen species, and release of toxic compounds [49]. Researchers also found that the small size (high surface-to-volume ratio) and solubility of metal nanoparticles may increase their toxicity to bacteria. Studies on the toxicity of silver and zinc oxide nanoparticles to nitrifying bacteria and *Escherichia coli* showed that the antibacterial effect of these nanoparticles results from the release of silver and zinc ions [75]. Nanoparticles can bind to the membrane of bacteria through electrostatic interaction and disruption of membrane structure [76]. In addition, the close contact between nanoparticles and roots can lead to the absorption of nanoparticles by roots and cause the transfer of nanoparticles to plants [77]. Many studies have also shown the side effects and toxicity of nanoparticles on plants [78]. The processes by which metal oxide nanoparticles exert toxic effects on living organisms include direct adsorption of the particle, adsorption on biological surfaces, the release of metal ions, and transport and release of contaminants [79]. When metal oxide nanoparticles are directly ingested by organisms, they can cause damage to cell surfaces by disrupting biological processes [80]. Although studies have shown that plants can directly uptake and transport metal oxide nanoparticles in aquatic systems [15], there is little information on the uptake of nanoparticles by plants in the soil environment. Therefore, there is concern that nanoparticles may transfer contaminants to living organisms through surface complexation [81]. Although there is increasing concern about the unknown toxicity of nanoparticles, the lack of an appropriate experimental method to validate the toxicity is a major obstacle in cases of poisoning by nanocomposites. Therefore, the study of the effects of nanoparticle toxicity is still in the initial stage, and it is not possible to draw a definite conclusion based on the available studies.

5.7 Conclusion

Considering the advantages of using nanotechnology in the field of soil science, it seems that a significant market volume for nanotechnology-based products and services will soon be achieved. Undoubtedly, by using nanotechnology as an emerging advanced technology in the agricultural sector, some good results can be achieved, including ensuring food security and sustainable and environmentally friendly agricultural development in a worldwide perspective. Despite the diversity of products brought to the market, this technology, like any other new technology, needs to undergo a further scientific and industrial evaluation to disseminate it in soil science and in the development of agriculture as a whole, especially to make it popular among farmers, and to determine the side effects of its application on soil microorganisms. In general, it can be stated that:

1. The production of fertilizers containing nano-sized trace elements increases the solubility and dispersion of soil nutrients and improves the efficiency of adsorption of these nutrients by plants.
2. The use of nanomodifiers can help improve soil aeration and increase water retention in the soil.
3. Nanoparticles can be used to remove soil pollutants.
4. Nanoparticles can be used to control soil erosion and stabilize sand in arid and desert regions.
5. Studies should consider the toxicity of nanoparticles, the concentration and behavior of nanoparticles in the environment, and the toxicity to soil microorganisms.

References

[1]   Sekhon, B. S. Nanotechnology in agri-food production: An overview. *Nanotechnology Science and Applications* 2014, 7, 31–53.
[2]   Richards, L. A. (ed.). *Diagnosis and improvement of saline and alkali soils. No. 60.* US Government Printing Office, 1954.
[3]   Lal, R. Soil science and the carbon civilization. *Soil Science Society of America Journal* 2007, 71(5): 1425–1437.
[4]   Subramanian, K.S., Manikandan, A., Thirunavukkarasu, M., Rahale, C.S. (2015). Nano-fertilizers for Balanced Crop Nutrition. In Nanotechnologies in Food and Agriculture, Rai, M., Ribeiro, C., Mattoso, L., Duran, N. (eds). Springer, Cham.
[5]   Liu, R., Lal, R. Potentials of engineered nanoparticles as fertilizers for increasing agronomic productions. *Science of the Total Environment* 2015, 514, 131–139.
[6]   Johnston, C. T. Probing the nanoscale architecture of clay minerals. *Clay Minerals* 2010, 45(3): 245–279.

[7] Chen, H., Yada, R. Nanotechnologies in agriculture: New tools for sustainable development. *Trends in Food Science and Technology* 2011, 22(11): 585–594.

[8] Ganji, F., Vasheghani, F. E. Hydrogels in controlled drug delivery systems. *Iranian Polymer Journal* 2009, 18(1), 63–88.

[9] James, E. A., Richards, D. The influence of iron source on the water-holding properties of potting media amended with water-absorbing polymers. *Scientia Horticulturae* 1986, 28(3): 201–208.

[10] Johnson, M. S., Veltkamp, C. J. Structure and functioning of water-storing agricultural polyacrylamides. *Journal of the Science of Food and Agriculture* 1985, 36(9): 789–793.

[11] Wang, X., Lü, S., Gao, C., Xiubin, X., Wei, Y., Bai, X., Feng, C., Gao, N., Liu, M., Wu, L. Biomass-based multifunctional fertilizer system featuring controlled-release nutrient, water-retention and amelioration of soil. *RSC Advances* 2014, 4(35): 18382–18390.

[12] Olesen, K. P. Turning sandy soil to farmland: 66% water saved in sandy soil treated with NanoClay. *Desert Control Institute Inc* 2010, 1, 1–10.

[13] Komendova, R., Žídek, J., Berka, M., Jemelkova, M., Řezáčová, V., Conte, P., Kučerík, J. Small-sized platinum nanoparticles in soil organic matter: Influence on water holding capacity, evaporation and structural rigidity. *Science of the Total Environment* 2019, 694, 133822.

[14] Nanda, K. K., Maisels, A., Kruis, F. E., Fissan, H., Stappert, S. Higher surface energy of free nanoparticles. *Physical Review Letters* 2003, 91(10): 106102.

[15] Ben-Moshe, T., Frenk, S., Dror, I., Minz, D., Berkowitz, B. Effects of metal oxide nanoparticles on soil properties. *Chemosphere* 2013, 90(2): 640–646.

[16] Taipodia, J., Dutta, J., Dey, A. K. 2011. Effect of nano particles on properties of soil. *Proceedings of Indian Geotechnical Conference.*

[17] Taha, M. R., Alsharef, J. M. A. Performance of soil stabilized with carbon nanomaterials. *Chemical Engineering Transaction* 2018, 63, 757–762.

[18] Taha, M. R., Taha, O. M. E. Influence of nano-material on the expansive and shrinkage soil behavior. *Journal of Nanoparticle Research* 2012, 14, 1–13.

[19] Saad Kheir, A. M., Abouelsoud, H. M., Hafez, E. M., Ali, O. A. M. Integrated effect of nano-Zn, nano-Si, and drainage using crop straw–filled ditches on saline sodic soil properties and rice productivity. *Arabian Journal of Geosciences* 2019, 12, 1–8.

[20] Alsharef, J., Taha, M. R., Firoozi, A. A., Govindasamy, P. Potential of using nanocarbons to stabilize weak soils. *Applied and Environmental Soil Science* 2016, 2016, 5060531.

[21] Taha, O. M. E., Taha, M. R. Soil-water characteristic curves and hydraulic conductivity of nanomaterial-soil-bentonite mixtures. *Arabian Journal of Geosciences* 2016, 9(1): 12.

[22] Bayat, H., Kolahchi, Z., Valaey, S., Rastgou, M., Mahdavi, S. Novel impacts of nanoparticles on soil properties: Tensile strength of aggregates and compression characteristics of soil. *Archives of Agronomy and Soil Science* 2018, 64(6): 776–789.

[23] Bahmani, S. H., Huat, B. B. K., Asadi, A., Farzadnia, N. Stabilization of residual soil using SiO2 nanoparticles and cement. *Construction and Building Materials* 2014, 64, 350–359.

[24] Huang, H., Chen, J., Liu, S., Shengyan, P. Impact of ZnO nanoparticles on soil lead bioavailability and microbial properties. *Science of the Total Environment* 2022, 806, 150299.

[25] Verma, Y., Singh, S. K., Jatav, H. S., Rajput, V. D., Minkina, T. Interaction of zinc oxide nanoparticles with soil: Insights into the chemical and biological properties. *Environmental Geochemistry and Health* 2022, 44, 221–234

[26] Nyberg, L., Turco, R. F., Nies, L. Assessing the impact of nanomaterials on anaerobic microbial communities. *Environmental Science and Technology* 2008, 42(6): 1938–1943.

[27] Chung, H., Son, Y., Yoon, T. K., Kim, S., Kim, W. The effect of multi-walled carbon nanotubes on soil microbial activity. *Ecotoxicology and Environmental Safety* 2011, 74(4): 569–575.

[28] Chen, X., Peng, C., Sun, L., Zhang, S., Huang, H., Chen, Y., Shi, J. Distinctive effects of TiO2 and CuO nanoparticles on soil microbes and their community structures in flooded paddy soil. *Soil Biology and Biochemistry* 2015, 86, 24–33.

[29] You, T., Liu, D., Chen, J., Yang, Z., Dou, R., Gao, X., Wang, L. Effects of metal oxide nanoparticles on soil enzyme activities and bacterial communities in two different soil types. *Journal of Soils and Sediments* 2018, 18, 211–221.

[30] Chai, H., Yao, J., Sun, J., Zhang, C., Liu, W., Zhu, M., Ceccanti, B. The effect of metal oxide nanoparticles on functional bacteria and metabolic profiles in agricultural soil. *Bulletin of Environmental Contamination and Toxicology* 2015, 94, 490–495.

[31] Luo, J., Song, Y., Liang, J., Jinxing, L., Islam, E., Tingqiang, L. Elevated CO2 mitigates the negative effect of CeO2 and Cr2O3 nanoparticles on soil bacterial communities by alteration of microbial carbon use. *Environmental Pollution* 2020, 263, 114456.

[32] Lofts, S., Criel, P., Janssen, C. R., Lock, K., McGrath, S. P., Oorts, K., Rooney, C. P., Smolders, E., Spurgeon, D. J., Svendsen, C. Modelling the effects of copper on soil organisms and processes using the free ion approach: Towards a multi-species toxicity model. *Environmental Pollution* 2013, 178, 244–253.

[33] Venkataraju, J. L., Sharath, R., Chandraprabha, M. N., Neelufar, E., Hazra, A., Patra, M. Synthesis, characterization and evaluation of antimicrobial activity of zinc oxide nanoparticles. *Journal of Biochemical Technology* 2014, 3(5): 151–154.

[34] Concha-Guerrero, S. I., Brito, E. M. S., Piñón-Castillo, H. A., Tarango-Rivero, S. H., Caretta, C. A., Luna-Velasco, A., Duran, R., Orrantia-Borunda, E. Effect of CuO nanoparticles over isolated bacterial strains from agricultural soil. *Journal of Nanomaterials* 2015, 2014, 206–206.

[35] Frenk, S., Ben-Moshe, T., Dror, I., Berkowitz, B., Minz, D. Effect of metal oxide nanoparticles on microbial community structure and function in two different soil types. *PLoS One* 2013, 8(12): e84441.

[36] Ge, Y., Schimel, J. P., Holden, P. A. Evidence for negative effects of TiO2 and ZnO nanoparticles on soil bacterial communities. *Environmental Science and Technology* 2011, 45(4): 1659–1664.

[37] Kumpiene, J., Ore, S., Renella, G., Mench, M., Lagerkvist, A., Maurice, C. Assessment of zerovalent iron for stabilization of chromium, copper, and arsenic in soil. *Environmental Pollution* 2006, 144(1): 62–69.

[38] Samarajeewa, A. D., Velicogna, J. R., Princz, J. I., Subasinghe, R. M., Scroggins, R. P., Beaudette, L. A. Effect of silver nano-particles on soil microbial growth, activity and community diversity in a sandy loam soil. *Environmental Pollution* 2017, 220, 504–513.

[39] Buzea, C., Pacheco, I. I., Robbie, K. Nanomaterials and nanoparticles: Sources and toxicity. *Biointerphases* 2007, 2(4): MR17–MR71.

[40] Solanki, A., Kim, J. D., Lee, K.-B. Nanotechnology for regenerative medicine: Nanomaterials for stem cell imaging. 2008, 3(4), 1–10.

[41] Maliszewska, I. Effects of the biogenic gold nanoparticles on microbial community structure and activities. *Annals of Microbiology* 2016, 66, 785–794.

[42] Xiaohong, L. I. U., Juan, W., Lingli, W. U., Zhang, L., Youbin, S. I. Impacts of silver nanoparticles on enzymatic activities, nitrifying bacteria, and nitrogen transformation in soil amended with ammonium and nitrate. *Pedosphere* 2021, 31(6): 934–943.

[43] Feng, Z., Yu, Y., Yao, H., Ge, C. Effect of zinc oxide nanoparticles on nitrous oxide emissions in agricultural soil. *Agriculture* 2021, 11(8): 730.

[44] Waalewijn-Kool, P. L., Ortiz, M. D., Nico, M., Straalen, V., Van gestel, C. A. M. Sorption, dissolution and pH determine the long-term equilibration and toxicity of coated and uncoated ZnO nanoparticles in soil. *Environmental Pollution* 2013, 178, 59–64.

[45] Read, D. S., Matzke, M., Gweon, H. S., Newbold, L. K., Heggelund, L., Ortiz, M. D., Lahive, E., Spurgeon, D., Svendsen, C. Soil pH effects on the interactions between dissolved zinc, non-nano-and nano-ZnO with soil bacterial communities. *Environmental Science and Pollution Research* 2016, 23, 4120–4128.

[46] Yang, K., Zhu, L., Xing, B. Adsorption of polycyclic aromatic hydrocarbons by carbon nanomaterials. *Environmental Science and Technology* 2006, 40(6): 1855–1861.

[47] Fan, G., Cang, L., Qin, W., Zhou, C., Gomes, H. I., Zhou, D. Surfactants-enhanced electrokinetic transport of xanthan gum stabilized nanoPd/Fe for the remediation of PCBs contaminated soils. *Separation and Purification Technology* 2013, 114, 64–72.

[48] Shahwan, T., Üzüm, Ç., Eroğlu, A. E., Lieberwirth, I. Synthesis and characterization of bentonite/iron nanoparticles and their application as adsorbent of cobalt ions. *Applied Clay Science* 2010, 47(3–4): 257–262.

[49] Klaine, S. J., Alvarez, P. J. J., Batley, G. E., Fernandes, T. F., Handy, R. D., Lyon, D. Y., Mahendra, S., McLaughlin, M. J., Lead, J. R. Nanomaterials in the environment: Behavior, fate, bioavailability, and effects. *Environmental Toxicology and Chemistry: An International Journal* 2008, 27(9): 1825–1851.

[50] Albadarin, A. B., Yang, Z., Mangwandi, C., Glocheux, Y., Walker, G., Ahmad, M. N. M. Experimental design and batch experiments for optimization of Cr (VI) removal from aqueous solutions by hydrous cerium oxide nanoparticles. *Chemical Engineering Research and Design* 2014, 92(7): 1354–1362.

[51] Goswami, A., Raul, P. K., Purkait, M. K. Arsenic adsorption using copper (II) oxide nanoparticles. *Chemical Engineering Research and Design* 2012, 90(9): 1387–1396.

[52] Esfahani, A. R., Firouzi, A. F., Sayyad, G., Kiasat, A., Alidokht, L., Khataee, A. R. Pb (II) removal from aqueous solution by polyacrylic acid stabilized zero-valent iron nanoparticles: Process optimization using response surface methodology. *Research on Chemical Intermediates* 2014, 40, 431–445.

[53] Zhou, Y.-T., Nie, H.-L., Branford-White, C., Zhi-Yan, H., Zhu, L.-M. Removal of Cu2+ from aqueous solution by chitosan-coated magnetic nanoparticles modified with α-ketoglutaric acid. *Journal of Colloid and Interface Science* 2009, 330(1): 29–37.

[54] Kržišnik, N., Mladenovič, A., Škapin, A. S., Škrlep, L., Ščančar, J., Milačič, R. Nanoscale zero-valent iron for the removal of Zn2+, Zn (II)–EDTA and Zn (II)–citrate from aqueous solutions. *Science of the Total Environment* 2014, 476, 20–28.

[55] Panneerselvam, P., Morad, N., Lim, Y. L. Separation of Ni (II) ions from aqueous solution onto maghemite nanoparticle (γ-Fe3O4) enriched with clay. *Separation Science and Technology* 2013, 48 (17): 2670–2680.

[56] Liang, P., Liu, Y., Guo, L., Zeng, J., Hanbing, L. Multiwalled carbon nanotubes as solid-phase extraction adsorbent for the preconcentration of trace metal ions and their determination by inductively coupled plasma atomic emission spectrometry. *Journal of Analytical Atomic Spectrometry* 2004, 19(11): 1489–1492.

[57] Chen, C., Wang, X. Adsorption of Ni (II) from aqueous solution using oxidized multiwall carbon nanotubes. *Industrial and Engineering Chemistry Research* 2006, 45(26): 9144–9149.

[58] Ding, Q., Liang, P., Song, F., Xiang, A. Separation and preconcentration of silver ion using multiwalled carbon nanotubes as solid phase extraction sorbent. *Separation Science and Technology* 2006, 41(12): 2723–2732.

[59] Liu, X.-M., Feng, Z.-B., Zhang, F.-D., Zhang, S.-Q., He, X.-S. Preparation and testing of cementing and coating nano-subnanocomposites of slow/controlled-release fertilizer. *Agricultural Sciences in China* 2006, 5(9): 700–706.

[60] Wang, X., Chen, C., Wenping, H., Ding, A., Xu, D., Zhou, X. Sorption of 243Am (III) to multiwall carbon nanotubes. *Environmental Science and Technology* 2005, 39(8): 2856–2860.

[61] Phenrat, T., Saleh, N., Sirk, K., Tilton, R. D., Lowry, G. V. Aggregation and sedimentation of aqueous nanoscale zerovalent iron dispersions. *Environmental Science and Technology* 2007, 41(1): 284–290.

[62] Vance, D. 2005. Nanotechnology for Hazardous Waste Site Remediation. *Technical Workshop, Washington, DC, October.*

[63] Martinson, C. A., Reddy, K. J. Adsorption of arsenic (III) and arsenic (V) by cupric oxide nanoparticles. *Journal of Colloid and Interface Science* 2009, 336(2): 406–411.

[64] El Asri, S., Laghzizil, A., Saoiabi, A., Alaoui, A., El Abassi, K., M'hamdi, R., Coradin, T. A novel process for the fabrication of nanoporous apatites from Moroccan phosphate rock. *Colloids and Surfaces A: Physicochemical and Engineering Aspects* 2009, 350(1–3): 73–78.

[65] Naderi Peikam, E., Jalali, M. Application of three nanoparticles (Al 2 O 3, SiO 2 and TiO 2) for metal-contaminated soil remediation (measuring and modeling). *International Journal of Environmental Science and Technology* 2019, 16, 7207–7220.

[66] Vítková, M., Puschenreiter, M., Komárek, M. Effect of nano zero-valent iron application on As, Cd, Pb, and Zn availability in the rhizosphere of metal (loid) contaminated soils. *Chemosphere* 2018, 200, 217–226.

[67] Baragaño, D., Forján, R., Welte, L., Gallego, J. L. R. Nanoremediation of As and metals polluted soils by means of graphene oxide nanoparticles. *Scientific Reports* 2020, 10(1): 1–10.

[68] Mar Gil-Díaz, M., Pérez-Sanz, A., Vicente, M. A., Lobo, M. C. Immobilisation of Pb and Zn in soils using stabilised zero-valent iron nanoparticles: Effects on soil properties. *Clean–Soil, Air, Water* 2014, 42(12): 1776–1784.

[69] Taghipour, M., Jalali, M. Effect of nanoparticles on kinetics release and fractionation of phosphorus. *Journal of Hazardous Materials* 2015, 283, 359–370.

[70] Chen, D., Szostak, P., Wei, Z., Xiao, R. Reduction of orthophosphates loss in agricultural soil by nano calcium sulfate. *Science of the Total Environment* 2016, 539, 381–387.

[71] Wang, Q., Jeong, S.-W., Choi, H. Removal of trichloroethylene DNAPL trapped in porous media using nanoscale zerovalent iron and bimetallic nanoparticles: Direct observation and quantification. *Journal of Hazardous Materials* 2012, 213, 299–310.

[72] Long, R. Q., Yang, R. T. Carbon nanotubes as superior sorbent for dioxin removal. *Journal of the American Chemical Society* 2001, 123(9): 2058–2059.

[73] Joo, S. H., Zhao, D. Destruction of lindane and atrazine using stabilized iron nanoparticles under aerobic and anaerobic conditions: Effects of catalyst and stabilizer. *Chemosphere* 2008, 70(3): 418–425.

[74] Suresh, A. K., Pelletier, D. A., Doktycz, M. J. Relating nanomaterial properties and microbial toxicity. *Nanoscale* 2013, 5(2): 463–474.

[75] Li, M., Zhu, L., Lin, D. Toxicity of ZnO nanoparticles to Escherichia coli: Mechanism and the influence of medium components. *Environmental Science and Technology* 2011, 45(5): 1977–1983.

[76] Thill, A., Zeyons, O., Spalla, O., Chauvat, F., Rose, J., Auffan, M., Marie Flank, A. Cytotoxicity of CeO2 nanoparticles for Escherichia coli. Physico-chemical insight of the cytotoxicity mechanism. *Environmental Science and Technology* 2006, 40(19): 6151–6156.

[77] Gardea-Torresdey, J. L., Rico, C. M., White, J. C. Trophic transfer, transformation, and impact of engineered nanomaterials in terrestrial environments. *Environmental Science and Technology* 2014, 48 (5): 2526–2540.

[78] Midander, K., Cronholm, P., Karlsson, H. L., Elihn, K., Möller, L., Leygraf, C., Wallinder, I. O. Surface characteristics, copper release, and toxicity of nano-and micrometer-sized copper and copper (II) oxide particles: A cross-disciplinary study. *Small* 2009, 5(3): 389–399.

[79] Bottero, J.-Y., Auffan, M., Rose, J., Mouneyrac, C., Botta, C., Labille, J., Masion, A., Thill, A., Chaneac, C. Manufactured metal and metal-oxide nanoparticles: Properties and perturbing mechanisms of their biological activity in ecosystems. *Comptes Rendus Geoscience* 2011, 343(2–3): 168–176.

[80] Ken, D., Poland, C. A., Schins, R. P. F. Possible genotoxic mechanisms of nanoparticles: Criteria for improved test strategies. *Nanotoxicology* 2010, 4(4): 414–420.

[81] Peralta-Videa, J. R., Zhao, L., Lopez-Moreno, M. L., De la rosa, G., Hong, J., Gardea-Torresdey, J. L. Nanomaterials and the environment: A review for the biennium 2008–2010. *Journal of Hazardous Materials* 2011, 186(1): 1–15.

Parasto Pouraziz, Davoud Koolivand*

Chapter 6
Utilization of nanoparticles in plant protection against biotic stresses

Abstract: Nanotechnology has emerged as a promising field for developing novel strategies to protect plants against biotic stresses caused by pests and diseases. Biotic stress is a major threat to crop productivity worldwide, and conventional methods of plant protection such as pesticides and fungicides have significant limitations. Nanoparticles (NPs) have unique physical and chemical properties that make them effective tools for combating biotic stresses. This review summarizes the recent advances in the utilization of NPs for plant protection against biotic stresses. It covers a range of topics including the mechanisms of action of NPs, their effects on plant growth and physiology, and their potential applications in agriculture. The review highlights the potential of NPs as an eco-friendly and sustainable approach for plant protection against biotic stresses as well as their potential to enhance the efficacy of conventional plant protection methods. The review also discusses the challenges and potential risks associated with the use of NPs in agriculture and the need for further research to understand their long-term effects on the environment and human health.

Keywords: plant disease, pathogen, nanoparticles, plant protection, virus

6.1 Introduction

Nanotechnology has been considered a widespread science in the recent decade and can make the connection between other sciences such as biology, chemistry, physics, and other sciences [1]. It can be relied on to make less use of harmful materials and products, so this science is a perfect alternative to destructive technologies [2]. Nanotechnology is also considered as a science of manipulating matter to the nanometer size and can be used in a wide range of materials such as organic matter (chitosan, proteins, liposomes, and aptamers) and inorganic matter including silver (Ag), gold (Au), platinum (Pt), aluminum oxide (Al_2O_3), iron oxide (Fe_2O_3), D2 cerium (CeO_2), graphene, silica (SiO_2), and zinc oxide (ZnO). Due to the wide range of available materials for utilizing nanomaterials, different synthesis methods have been considered based on the goal [3].

*Corresponding author: Davoud Koolivand, Department of Plant Protection, Faculty of Agriculture, University of Zanjane, Zanjan, Iran, e-mail: koolivand@znu.ac.ir
Parasto Pouraziz, Department of Plant Protection, Faculty of Agriculture, University of Zanjan, Zanjan, Iran

https://doi.org/10.1515/9781501523229-006

Nanoparticles (NPs) are used in various fields (agriculture, materials, manufacturing, mechanical engineering, medicine, pharmaceuticals, and the environment). Nanotechnology has a high potential to deal with global problems in the vintage of agricultural products. It has prominent roles in agriculture such as nanosensors – nano barcoding – seed germination – hormone transfer in plants, target gene transfer, water management, and plant disease management (Figure 6.1). Nanotechnology, which employs nanomaterials (particles of sizes between 1 and 100 nm), is an emerging field of research with enormous space in agriculture; thus, agriculture is ranked second on the list of nanotechnology applications; this science can help increase fertility and increase crop yield [1, 4].

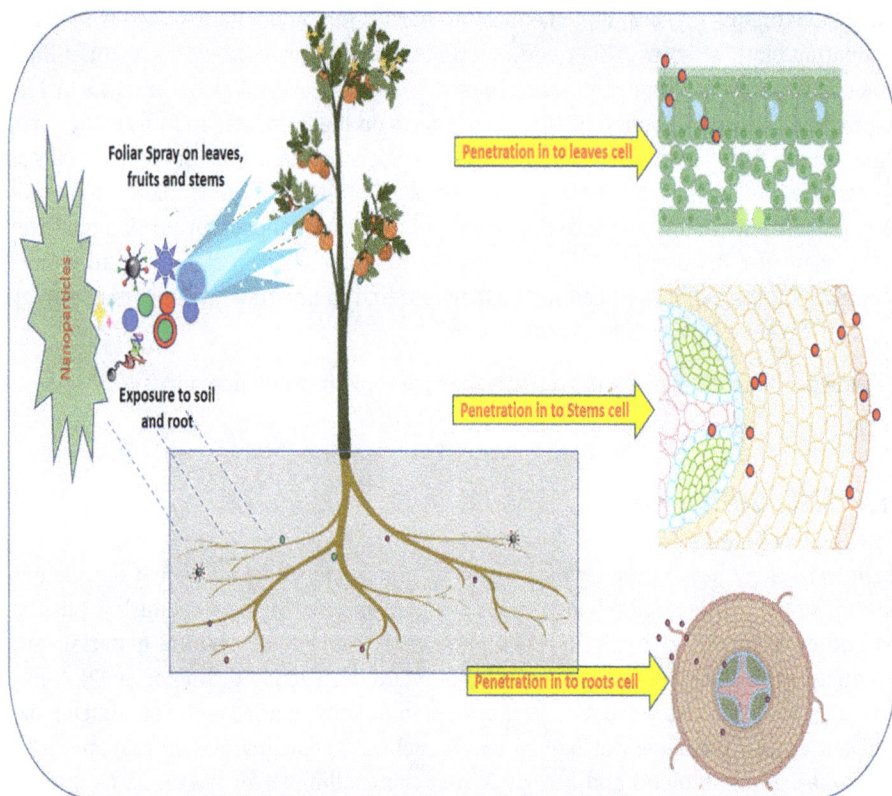

Figure 6.1: Internalization process of NPs in various parts of plants.

Plants are exposed to biological stresses caused by fungi, bacteria, viruses, phytoplasma, nematodes, rats, and weeds. These stresses and other factors can reduce crop production. Bio-stress can cause up to 40 or 50% of crop losses; immediate detection, like correct action against any of these stresses, can prevent the loss of agricultural products [3]. Existing strategies used against all these pathogens such as fungicides and insecticides or herbicides have positive and negative aspects. One of the disadvantages of these strategies is that they create resistance to the pathogen; on the contrary, the use of nanotechnology in agriculture can be a more appropriate and cleaner solution for the environment that the use of nanomaterials to protect the plant is only in the early stages of the disease [2]. At present, NPs can also be used as antifungal, antibacterial, and antiviral agents; since these agents cause damage to the plant, there are few protective and therapeutic strategies against them (especially plant viruses) [5]; therefore, using NPs can be a promising solution in the field of disease management in agriculture. In this way, some NPs act directly as antimicrobial agents, while others act more to alter the host's nutritional status and thus activate defense mechanisms. In addition, NPs can be used to diagnose plant diseases. This increase in diagnostic power will pave the way for the protection of agricultural products [3]. On the other hand, NPs can affect, directly and indirectly, the metabolic and physiological functions of plants in various ways [3].

6.2 The role of nanomaterials in plant protection

Several studies have shown that metal nanoparticles (MeNPs) are very effective against plant pathogens, insects, and pests [2]. These NPs' efficacy in plant pathogens is not apparent, but it can be said that these nanomaterials can disrupt protein production and membranes and cause impermeability; NPs can also create holes in pathogenic membranes as we can see under a TEM microscope [5]. It should be noted that NPs will be able to change some of the properties of the soil. The hypothesis of NPs acting against microbes can be explained by several obligations: (1) release of toxic ions such as Cd2C, Zn2C, and AgC that can bind to sulfur-containing proteins, and this accumulation of proteins in the membrane itself disrupts membrane and permeability; (2) they can act as genotoxic to the DNA genome; (3) can disrupt electron transfer and protein oxidation and cause membrane collapse [6].

6.3 Plant viruses

Nanomaterials have several advantages in agriculture such as reducing chemical material, water management, nutrient uptake, and increasing product through pest management (Figure 6.2). In the field of viral diseases for which there is no specific treatment, 900 species of plant viruses that infect more than 700 plant species have been reported so that NPs can prevent these diseases and the amount of damage caused by plant viruses [20]. The antiviral activity of MeNPs has been proven in vitro and in vivo experiments with numerous plants and is effective against positive- and negative-sense single-stranded RNA viruses [5]. The antiviral activity of ZnONPs and

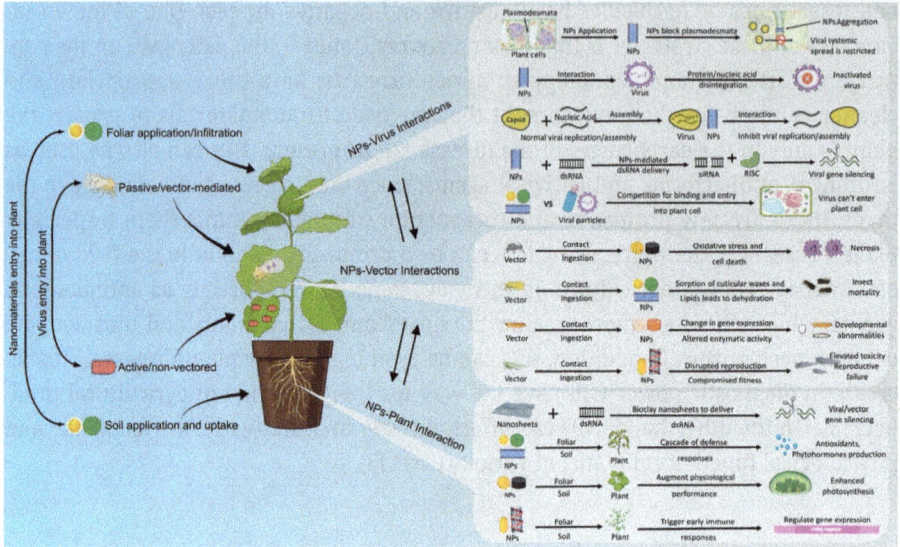

Figure 6.2: Mechanisms of tripartite interaction between NPs in plant-virus-vector systems.

SiO$_2$NPs against TMV tobacco mosaic virus has been tested in vitro due to direct inactivation of TMV by MeNPs due to interaction with coating glycoproteins, which directly damage TMV-coating proteins. These nanomaterials can be used against these viruses. MeNPs affect and interfere with the reproduction of viruses through various mechanisms [21]. NPs can significantly affect the control of the viral disease; For example, in tomatoes infected with ToMV and PVY virus, seven days after inoculation using silver nanoparticles (AgNPs), the severity of the disease is reduced, and this decrease is expected to be due to the attachment of NPs to viruses and their proliferative disorders. In other words, AgNP particles can bind to the ToMV and PVY coat protein. According to research, sometimes, NPs with other materials can have a synergistic effect. For example, on a plant infected with the TBSV virus, a mixture of AgNPs and salicylic acid has a synergistic effect. It also reduces the amount of infection when used alone [5, 22]. In greenhouse research, foliar application of AgNPs at concentrations of 50 mg/L on tomatoes reduced the severity of ToMV-induced disease and increased the levels of antioxidant enzymes PPO and POD and the total soluble protein content [23]. Graphene-based silver nanocomposites have strong antiviral power against the TBSV in lettuce, thus reducing virus concentration and disease severity [24]. According to research, foliar application of the plant and soil with nano SiO$_2$NPs against papaya ringspot virus (PRSV) in *Cucumis sativus* will significantly reduce the severity of the disease. In another example, they reduced the disease by using TiO$_2$NPs by spraying and washing the soil after 24 h against the bean strain virus (BBSV) in the bean plant [5]. In addition to preventing the virus from replicating, some NPs can also help the plant's defense system; for example, MeNPs, which prevent the virus from replicating, also activate defense mechanisms. MeNPs can impair cellular homeostasis by reducing or increasing oxidative stress depending on the type, concentration, and tissue type used [5]. In addition to possibly acting directly on plant viruses, NPs also indirectly affect plant disease treatment by stimulating overall plant growth such as increasing photosynthesis or inducing plant defense responses. Interestingly, NPs can be used as an RNAi delivery system for nucleic acids in viral disease management. It should be noted that RNAi is a biologically protected pathway that produces pathogen resistance [23] (Table 6.1).

Table 6.1: Antiviral activity of metal and metal and metal oxide nanoparticles.

Nanoparticle	Size	Plant species	Pathogen	Effect	Reference
Silver nanoparticles (AgNPs)	10–20 nm	*Cymopsis tetragonaloba*	Sunhemp rosette virus (SHRV)	Complete suppression of the disease	[7]
AgNPs	77 nm	*Vicia faba*	Bean yellow mosaic virus (BYMV)	Decrease in virus concentration, percentage of infection, and disease severity Reduction in lesions on infected leaves	[8]
Schiff-base nanosilver NPs		*Nicotiana tabacum*	Tobacco mosaic virus (TMV)	Reduction of the harm of TMV to tobacco	[9]
AgNPs	12 nm	*Solanum tuberosum*	Potato virus Y (PVY)	Resistance to virus infection	[10]
Graphene oxide-silver NPs (GO-AgNPs	3,050 nm	*Lactuca sativa*	Tomato bushy stunt virus (TBSV)	Decrease in virus concentration, infection percentage, and disease severity	[11]
AgNPs	12.6 ± 5 nm	*Solanum tuberosum* L. cv. Spunta	Tomato-spotted wilt virus (TSWV)	Decrease in TSWV infectivity and produces an inhibitory effect in local lesions	[12]
AgNPs		*Solanum lycopersicum*	Tomato mosaic virus (ToMV)	Reduction in disease severity and virus infection	[13]
AgNPs		*Solanum lycopersicum*	Potato virus Y (PVY)	Reduction in disease severity and virus infection	[13]
Gold NPs (AuNPs)		*Hordeum vulgare*	Barley yellow mosaic virus (BaYMV)	Dissociation of virus particle in vitro	[14]

[34] Beisl, S., Friedl, A., Miltner, A. Lignin from micro- to nanosize: Applications. *International Journal of Molecular Sciences* 2017, 18(11), Article 11, doi: https://doi.org/10.3390/ijms18112367.

[35] Matsakas, L., Karnaouri, A., Cwirzen, A., Rova, U., Christakopoulos, P. Formation of lignin nanoparticles by combining organosolv pretreatment of birch biomass and homogenization processes. *Molecules* 2018, 23(7), Article 7, doi: https://doi.org/10.3390/molecules23071822.

[36] Del Buono, D., Luzi, F., Puglia, D. Lignin nanoparticles: A promising tool to improve maize physiological, biochemical, and chemical traits. *Nanomaterials* 2021, 11(4), Article 4, doi: https://doi.org/10.3390/nano11040846.

[37] Del Buono, D., Luzi, F., Tolisano, C., Puglia, D., Di Michele, A. Synthesis of a lignin/zinc oxide hybrid nanoparticles system and its application by nano-priming in maize. *Nanomaterials* 2022, 12(3), Article 3, doi: https://doi.org/10.3390/nano12030568.

[38] Blackman, J. A., Binns, C. Chapter 1 Introduction. in Handbook of Metal Physics, Vol. 5. Elsevier, 2008, 1–16, doi: https://doi.org/10.1016/S1570-002X(08)00201-2.

[39] Mody, V., Siwale, R., Singh, A., Mody, H. Introduction to metallic nanoparticles. *Journal of Pharmacy and Bioallied Sciences* 2010, 2(4), Article 4, doi: https://doi.org/10.4103/0975-7406.72127.

[40] Das, C. K., Jangir, H., Kumar, J., Verma, S., Mahapatra, S. S., Philip, D., Srivastava, G., Das, M. Nano-pyrite seed dressing: A sustainable design for NPK equivalent rice production. *Nanotechnology for Environmental Engineering* 2018, 3(1), Article 1, doi: https://doi.org/10.1007/s41204-018-0043-1.

[41] Spanos, A., Athanasiou, K., Ioannou, A., Fotopoulos, V., Krasia-Christoforou, T. Functionalized magnetic nanomaterials in agricultural applications. *Nanomaterials* 2021, 11(11), Article 11, doi: https://doi.org/10.3390/nano11113106.

[42] Kasote, D. M., Lee, J. H. J., Jayaprakasha, G. K., Patil, B. S. Seed priming with iron oxide nanoparticles modulate antioxidant potential and defense-linked hormones in watermelon seedlings. *ACS Sustainable Chemistry & Engineering* 2019, 7(5), Article 5, doi: https://doi.org/10.1021/acssuschemeng. 8b06013.

[43] Afzal, S., Sharma, D., Singh, N. K. Eco-friendly synthesis of phytochemical-capped iron oxide nanoparticles as nano-priming agent for boosting seed germination in rice (Oryza sativa L.). *Environmental Science and Pollution Research* 2021, 28(30), doi: https://doi.org/10.1007/s11356-020-12056-5.

[44] Rizwan, M., Ali, S., Ali, B., Adrees, M., Arshad, M., Hussain, A., Zia Ur Rehman, M., Waris, A. A. Zinc and iron oxide nanoparticles improved the plant growth and reduced the oxidative stress and cadmium concentration in wheat. *Chemosphere* 2019, 214, 269–277, doi: https://doi.org/10.1016/j.che mosphere.2018.09.120.

[45] Sundaria, N., Singh, M., Upreti, P., Chauhan, R. P., Jaiswal, J. P., Kumar, A. Seed priming with iron oxide nanoparticles triggers iron acquisition and biofortification in wheat (Triticum aestivum L.) grains. *Journal of Plant Growth Regulation* 2019, 38(1), Article 1, doi: https://doi.org/10.1007/s00344-018-9818-7.

[46] Sturikova, H., Krystofova, O., Huska, D., Adam, V. Zinc, zinc nanoparticles and plants. *Journal of Hazardous Materials* 2018, 349, 101–110, doi: https://doi.org/10.1016/j.jhazmat.2018.01.040.

[47] Kim, I., Viswanathan, K., Kasi, G., Thanakkasaranee, S., Sadeghi, K., Seo, J. ZnO nanostructures in active antibacterial food packaging: Preparation methods, antimicrobial mechanisms, safety issues, future prospects, and challenges. *Food Reviews International* 2022, 38(4), Article 4, doi: https://doi.org/10.1080/87559129.2020.1737709.

[48] Singh, T. A., Das, J., Sil, P. C. Zinc oxide nanoparticles: A comprehensive review on its synthesis, anticancer and drug delivery applications as well as health risks. *Advances in Colloid and Interface Science* 2020, 286, 102317, doi: https://doi.org/10.1016/j.cis.2020.102317.

[49] Abdel Latef, A. A. H., Abu Alhmad, M. F., Abdelfattah, K. E. The possible roles of priming with ZnO nanoparticles in mitigation of salinity stress in lupine (Lupinus termis) plants. *Journal of Plant Growth Regulation* 2017, 36(1), doi: https://doi.org/10.1007/s00344-016-9618-x.

[50] Al-Salama, Y. Effect of seed priming with ZnO nanoparticles and saline irrigation water in yield and nutrients uptake by wheat plants. *LAFOBA2* 2022, 37, doi: https://doi.org/10.3390/ environsciproc2022016037.

[51] Li, Y., Liang, L., Li, W., Ashraf, U., Ma, L., Tang, X., Pan, S., Tian, H., Mo, Z. ZnO nanoparticle-based seed priming modulates early growth and enhances physio-biochemical and metabolic profiles of fragrant rice against cadmium toxicity. *Journal of Nanobiotechnology* 2021, 19(1), Article 1, doi: https://doi.org/10.1186/s12951-021-00820-9.

[52] Rai-Kalal, P., Jajoo, A. Priming with zinc oxide nanoparticles improve germination and photosynthetic performance in wheat. *Plant Physiology and Biochemistry* 2021, 160, 341–351, doi: https://doi.org/10.1016/j.plaphy.2021.01.032.

[53] Waqas Mazhar, M., Ishtiaq, M., Hussain, I., Parveen, A., Hayat Bhatti, K., Azeem, M., Thind, S., Ajaib, M., Maqbool, M., Sardar, T., Muzammil, K., Nasir, N. Seed nano-priming with zinc oxide nanoparticles in rice mitigates drought and enhances agronomic profile. *PLOS One* 2022, 17(3), Article 3, doi: https://doi.org/10.1371/journal.pone.0264967.

[54] Mahakham, W., Sarmah, A. K., Maensiri, S., Theerakulpisut, P. Nanopriming technology for enhancing germination and starch metabolism of aged rice seeds using phytosynthesized silver nanoparticles. *Scientific Reports* 2017, 7(1), Article 1, doi: https://doi.org/10.1038/s41598-017-08669-5.

[55] Khan, I., Raza, M. A., Awan, S. A., Shah, G. A., Rizwan, M., Ali, B., Tariq, R., Hassan, M. J., Alyemeni, M. N., Brestic, M., Zhang, X., Ali, S., Huang, L. Amelioration of salt induced toxicity in pearl millet by seed priming with silver nanoparticles (AgNPs): The oxidative damage, antioxidant enzymes and ions uptake are major determinants of salt tolerant capacity. *Plant Physiology and Biochemistry* 2020, 156, 221–232, doi: https://doi.org/10.1016/j.plaphy.2020.09.018.

[56] Acharya, P., Jayaprakasha, G. K., Crosby, K. M., Jifon, J. L., Patil, B. S. Nanoparticle-mediated seed priming improves germination, growth, yield, and quality of watermelons (citrullus lanatus) at multi-locations in texas. *Scientific Reports* 2020, 10(1), doi: https://doi.org/10.1038/s41598-020-61696-7.

[57] Wu, L., Huo, W., Yao, D., Li, M. Effects of solid matrix priming (SMP) and salt stress on broccoli and cauliflower seed germination and early seedling growth. *Scientia Horticulturae* 2019, 255, 161–168, doi: https://doi.org/10.1016/j.scienta.2019.05.007.

[58] Olsson, L. et al. Heat stress and heat waves. In Chadee, D. (Ed) Climate Change. Cambridge, United Kingdom: Cambridge University Press, 2014, 109–111.

[59] Hamada, A. M., Hamada, Y. M. Management of abiotic stress and sustainability. *Plant Life Under Changing Environment*, 2020, 883–916, doi: 10.1016/b978-0-12-818204-8.00041-2.

[60] M. S., A. et al. Priming with nanoscale materials for boosting abiotic stress tolerance in crop plants. *Journal of Agricultural and Food Chemistry* 2021, 69(35), 10017–10035, doi: 10.1021/acs.jafc.1c03673.

[61] Lowry, G. V., Avellan, A., Gilbertson, L. M. Opportunities and challenges for nanotechnology in the Agri-Tech Revolution. *Nature Nanotechnology* 2019, 14(6), 517–522, doi: 10.1038/s41565-019-0461-7.

[62] Zhao, L., et al. Nano-Biotechnology in agriculture: Use of nanomaterials to promote plant growth and stress tolerance. *Journal of Agricultural and Food Chemistry* 2020, 68(7), 1935–1947, doi: 10.1021/acs.jafc.9b06615.

[63] Harmeet Singh, J. S. K., Harrajdeep Kang, G. S., Jagroop Kaur, V. P. Abiotic stress and its amelioration in cereals and pulses: A review. *International Journal of Current Microbiology and Applied Sciences* 2017, 6(3), 10109–1045, doi: 10.20546/ijcmas.2017.603.120.

[64] Boonchai, C., et al. Rice overexpressing OSNUC1-s reveals differential gene expression leading to yield loss reduction after salt stress at the booting stage. *International Journal of Molecular Sciences* 2018, 19(12), 3936, doi: 10.3390/ijms19123936.

[65] Sahil, et al. Salicylic acid for vigorous plant growth and enhanced yield under harsh environment. *Plant Performance Under Environmental Stress* 2021, 99–127, doi: 10.1007/978-3-030-78521-5_5.

[66] Chandrasekaran, U., et al. Are there unidentified factors involved in the germination of Nanoprimed seeds? *Frontiers in Plant Science* 2020, 11, doi: 10.3389/fpls.2020.00832.

[67] Abbasi Khalaki, M., et al. Influence of nano-priming on seed germination and plant growth of forage and medicinal plants. *Plant Growth Regulation* 2020, 93(1), 13–28, doi: 10.1007/s10725-020-00670-9.

[68] Nile, S. H., et al. Nano-priming as emerging seed priming technology for sustainable agriculture— recent developments and future perspectives. *Journal of Nanobiotechnology* 2022, 20(1), doi: 10.1186/s12951-022-01423-8.

[69] Tripathi, D. K., et al. Silicon nanoparticles more effectively alleviated UV-B stress than silicon in wheat (triticum aestivum) seedlings. *Plant Physiology and Biochemistry* 2017, 110, 70–81, doi: 10.1016/j.plaphy.2016.06.026.

[70] Abdel Latef, A. A., et al. Titanium dioxide nanoparticles improve growth and enhance tolerance of broad bean plants under saline soil conditions. *Land Degradation & Development* 2017, 29(4), 1065–1073, doi: 10.1002/ldr.2780.

[71] Faizan, M., et al. Zinc oxide nanoparticle-mediated changes in photosynthetic efficiency and antioxidant system of tomato plants. *Photosynthetica* 2018, 56(2), 678–686, doi: 10.1007/s11099-017-0717-0.

[72] Shang, Y., et al. Applications of nanotechnology in plant growth and Crop Protection: A review. *Molecules* 2019, 24(14), 2558, doi: 10.3390/molecules24142558.

[73] Hussain, S., et al. Physiological and biochemical mechanisms of seed priming-induced chilling tolerance in rice cultivars. *Frontiers in Plant Science* 2016, 7, doi: 10.3389/fpls.2016.00116.

[74] Palocci, C., et al. Endocytic pathways involved in PLGA nanoparticle uptake by grapevine cells and role of cell wall and membrane in size selection. *Plant Cell Reports* 2017, 36(12), 1917–1928, doi: 10.1007/s00299-017-2206-0.

[75] Singh, A., et al. Role of nanoparticles in crop improvement and abiotic stress management. *Journal of Biotechnology* 2021, 337, 57–70, doi: 10.1016/j.jbiotec.2021.06.022.

[76] Sharma, D., Afzal, S., Singh, N. K. Nanopriming with phytosynthesized zinc oxide nanoparticles for promoting germination and starch metabolism in rice seeds. *Journal of Biotechnology* 2021, 336, 64–75, doi: 10.1016/j.jbiotec.2021.06.014.

[77] Savvides, A., et al. Chemical priming of plants against multiple abiotic stresses: Mission possible? *Trends in Plant Science* 2016, 21(4), 329–340, doi: 10.1016/j.tplants.2015.11.003.

[78] Shah, T., et al., Seed priming with titanium dioxide nanoparticles enhances seed vigor, leaf water status, and antioxidant enzyme activities in maize (Zea mays L.) under salinity stress. *Journal of King Saud University – Science* 2021, 33(1), 101207, doi: 10.1016/j.jksus.2020.10.004.

[79] Mishra, V., et al. Interactions of nanoparticles with plants. *Emerging Technologies and Management of Crop Stress Tolerance* 2014, 159–180, doi: 10.1016/b978-0-12-800876-8.00008-4.

[80] Gohari, G., et al. Titanium dioxide nanoparticles (TIO2 NPS) promote growth and ameliorate salinity stress effects on essential oil profile and biochemical attributes of Dracocephalum Moldavica. *Scientific Reports* 2020, 10(1), doi: 10.1038/s41598-020-57794-1.

[81] Jafari, L., et al. Improved marjoram (Origanum majorana L.) tolerance to salinity with seed priming using titanium dioxide (tio2). *Iranian Journal of Science and Technology, Transactions A: Science* 2021, 46(2), 361–371, doi: 10.1007/s40995-021-01249-3.

[82] Sardar, R., Ahmed, S., Yasin, N. A. Titanium dioxide nanoparticles mitigate cadmium toxicity in coriandrum sativum L. through modulating antioxidant system, stress markers and reducing cadmium uptake. *Environmental Pollution* 2022, 292, 118373, doi: 10.1016/j.envpol.2021.118373.

[83] Joshi, A., et al. Multi-walled carbon nanotubes applied through seed-priming influence early germination, root hair, growth and yield of bread wheat (*triticum aestivum L.*). *Journal of the Science of Food and Agriculture* 2018 [Preprint], doi: 10.1002/jsfa.8818.

[84] Wan, J., et al. Comparative physiological and metabolomics analysis reveals that single-walled carbon nanohorns and zno nanoparticles affect salt tolerance in *sophora alopecuroides*. *Environmental Science: Nano* 2020, 7(10), 2968–2981, doi: 10.1039/d0en00582g.

[85] Ali, Md. H., et al. Carbon nanoparticles functionalized with carboxylic acid improved the germination and seedling vigor in upland boreal forest species. *Nanomaterials* 2020, 10(1), 176, doi: 10.3390/nano10010176.

[86] Baz, H., et al. Water-soluble carbon nanoparticles improve seed germination and post-germination growth of lettuce under salinity stress. *Agronomy* 2020, 10(8), 1192, doi: 10.3390/agronomy10081192.

[87] Pourkhaloee, A., Haghighi, M., Saharkhiz, M., Jouzi, H., Mahdi, M. Carbon nanotubes can promote seed germination via seed coat penetration. *Seed Technology* 2011, 33, 155–169.

[88] Joshi, A., et al. Tracking multi-walled carbon nanotubes inside oat (Avena sativa L.) plants and assessing their effect on growth, yield, and mammalian (human) cell viability. *Applied Nanoscience* 2018, 8(6), 1399–1414, doi: 10.1007/s13204-018-0801-1.

[89] Cano, A. M., et al. Determination of uptake, accumulation, and stress effects in corn (Zea mays L.) grown in single-wall carbon nanotube contaminated soil. *Chemosphere* 2016, 152, 117–122, doi: 10.1016/j.chemosphere.2016.02.093.

[90] Husen, A., Siddiqi, K. S. Phytosynthesis of nanoparticles: Concept, controversy and application. *Nanoscale Research Letters* 2014, 9(1), doi: 10.1186/1556-276x-9-229.

[91] Xin, Q., et al. Antibacterial carbon-based nanomaterials. *Advanced Materials* 2018, 31(45), 1804838, doi: 10.1002/adma.201804838.

[92] Liu, Q., et al. Carbon nanotubes as molecular transporters for walled plant cells. *Nano Letters* 2009, 9(3), 1007–1010, doi: 10.1021/nl803083u.

[93] Jha, R., Singh, A., Sharma, P. K., Fuloria, N. K. Smart carbon nanotubes for drug delivery system: A comprehensive study. *Journal of Drug Delivery Science and Technology* 2020, 58, 101811.

Debasis Mitra*, Wiem Alloun, Shraddha Bhaskar Sawant,
Edappayil Janeeshma, Ankita Priyadarshini, Suchismita Behera,
Ansuman Senapati, Sucharita Satapathy, Pradeep K. Das Mohapatra,
Periyasamy Panneerselvam*

Chapter 9
Nanoparticles for the improved horticultural crop production

Abstract: Nanotechnology has a substantial potential for increasing agricultural output and hence enhancing future food safety through the utilization of green technology. In horticulture, it can be quite difficult to preserve the quality of harvested fruit and vegetables while minimizing deterioration. Nanofertilizers are slow-releasing, and that work very effective at the stage of enhancing flower fertility, pollination, and vegetative development of plant. Additionally, it has shown significant potential for enhancing production, extending shelf life, reducing postharvest damage, and enhancing crop quality for vegetables and fruit trees. It also serves as a special agrochemical carrier, enabling site-specific, controlled nutrient delivery while enhancing

Acknowledgments: Authors are thankful to ICAR – National Rice Research Institute, India; Raiganj University, India; University of Constantine 1, Algeria; Odisha University of agriculture and technology, India and MES KEVEEYAM College, India for support. D.M. and P.P. are grateful to the Department of Biotechnology, Government of India for giving grants (BT/PR36476/NNT/28/1723/2020) for this work.

*Corresponding author: Debasis Mitra, Department of Microbiology, Raiganj University, Raiganj, Uttar Dinajpur 733 134, West Bengal, India, e-mail: debasismitra3@raiganjuniversity.ac.in
*Corresponding author: Periyasamy Panneerselvam, ICAR – National Rice Research Institute, Cuttack 753006, Odisha, India, e-mail: panneerselvam.p@icar.gov.in
Wiem Alloun, Microbial Biotechnology and Bioprocesses, University of Constantine 1, RN79, Constantine, Algeria, e-mail: wiemalloun@gmail.com
Shraddha Bhaskar Sawant, Odisha University of agriculture and technology, Bhubaneswar 751003, Odisha, India, e-mail: sbsawant56@gmail.com
Edappayil Janeeshma, Department of Botany, MES KEVEEYAM College, Valanchery, Malappuram, 676552, Kerala, India, e-mail: edappayiljaneeshma@gmail.com
Ankita Priyadarshini, ICAR – National Rice Research Institute, Cuttack 753006, Odisha, India, e-mail: ankitapriyadarshini10697@gmail.com
Suchismita Behera, ICAR – National Rice Research Institute, Cuttack, Odisha 753006, India, e-mail: suchi.behera456@gmail.com
Ansuman Senapati, ICAR – National Rice Research Institute, Cuttack, Odisha 753006, India, e-mail: asenapati89@gmail.com
Sucharita Satapathy, ICAR – National Rice Research Institute, Cuttack 753006, Odisha, India, e-mail: sucharitasatapathy943@gmail.com
Pradeep K. Das Mohapatra, Department of Microbiology, Raiganj University, Raiganj, Uttar Dinajpur 733134, West Bengal, India, e-mail: pkdmvu@gmail.com

https://doi.org/10.1515/9781501523229-009

crop protection. Fruits and vegetables can also be transported and stored using anti-microbial nanomaterials, such as nanofilm on harvested products or packaging materials. Nanotools (like nanobiosensors) are useful in the advancement of high-tech agricultural practices by their direct and purposeful applications in the management and control of high inputs of fertilizers, herbicides, and pesticides. Additionally, they are frequently used to accurately determine the soil moisture, humidity, crop pest populations, pesticide residues, and nutrient. The availability of usable nanoparticles and field application safety assessments are required to ensure food and nutritional security for the world's ever-increasing population in a changing climate situation. Increased use of nanotechnology will result in climate-smart horticulture, reduced post-harvest losses, and improved overall crop quality.

Keywords: nanoparticles, formulation, horticultural crop production, nanofertilizers, plant growth

9.1 Nanomaterials as fertilizers

Nanoscale nutrients have singular physical and chemical properties compared to bulk materials, thus constituting a promising technology to enhance crop performance, improve soil fertility, decrease nutrient loss, keep the environment clean, and provide a suitable environment for microorganisms [1–3]. In agriculture, three types of nanotechnologies can be distinguished: nanofertilizers, nano-additives, and nano-coatings [4, 5]. Nanofertilizers (NFs) are developed using a variety of synthetic materials, including urea, ammonia, and organics [6]. Pretreatments, such as milling, crushing, or grinding, are applied to bulk ingredients to produce nanosized materials, which are then used to make effective NFs that release the required nutrients gradually over a long period of time. A few examples are urea combined with calcium cyanamide, urea combined with other biofertilizers, nano-emulsions created by combining colloids in nanoscale forms with emulsions, peat moss, ammonium humate, and other man-made substances [7]. The surface polymeric coating was chosen to protect the nutrients from homo- and hetero-aggregation, allowing the NPs to reach the rhizosphere and enhance crop productivity [8]. Top-down or bottom-up methods can be used to synthesize nanoscale particles [9]. The majority of nanofertilizers are produced using a difficult and expensive bottom-up method. The top-down strategy is an alternative method for high-volume, low-cost production. For instance, using planetary ball mills, high-energy ball milling is a top-down method [9, 10]. To get the best grinding parameters in a planetary ball mill, numerous tests are necessary [11]. Therefore, in the NF synthesis with different macro-/micronutrients based on nanotechnology, the prioritized delivery methods to the plants are the regulated release and progressive systems [12]. Eco-friendly cassava starch films as a controlled-release nutrient [13], chitosan nanoparticles as a sustainable delivery

system [14], and controlled-release phosphorus fertilizer based on biological macromolecules are a few examples of nanotechnology applications in sustainable agricultural
production [15]. Therefore, many researchers have developed slow-release fertilizers
that contain hydroxyapatite (HA) and urea to improve the delivery of nutrients to
plants [16–18]. Tarafder et al. [19] have effectively synthesized an HNF, which was
found to be in, and it is potential activity for the continuous, gradual release of urea and
nutrients, such as Ca^{2+}, $(PO_4)_3$, NO_2, NO_3, Cu^{2+}, Fe^{2+}, and Zn^{2+}, into the soil. The slow-
release investigation of the formulated HNF was studied over 14 days. Bioassays per-
formed on this nanofertilizer demonstrated maximum nutrient resource utilization *Abelmoschus esculentus* (L.) and increased yields. Within a short period, HNF was found to
boost the efficiency of nutrient uptake of Cu^{2+}, Fe^{2+}, and Zn^{2+} compared to commercial
fertilizer. According to the results of this study, HNF offers many benefits as a fertilizer,
including a delayed and sustained nanoscale nutrient release, low dosage (50 mg/week),
cheap cost, and insignificant detectable soil contamination [19]. In another study performed by Murgueitio-Herrera et al. [20], the foliar application of the formulated nanofertilizer at a concentration of 270 ppm resulted in a significant enhancement of the test
plants' growth. The formula contained nanoparticles with a diameter of 9.5 nm, 7.8 nm,
10.5 nm for ZnO, FeO, and MnO, respectively, aiming the facilitation the adsorption of
NPs by the plants' stomata. Indeed, treatment of Andean lupin with 270 ppm iron and
zinc led to increases in the height of 6%, root size of 19%, chlorophyll content index of
3.5%, and leaf area of 300% compared to the control, which was sprayed with distilled
water. The ZnO MnO-NPs show increases in root size of 10.3%, dry biomass of 55.1%,
chlorophyll content of 7.1%, and leaf area of 25.6% in cabbage at a concentration of 270
ppm. Similarly, Dimkpa et al. [21] reported that the coating efficiency of Zn nanoparticles
reached 80–100% using the coating agents N-acetyl cysteine (NAC) and sodium salicylate
(SAL), NAC and urea, or SAL and urea. Sorghum performance and soil retention of Zn,
N, and P were affected by soil treatment with nanofertilizers at 6.4 (rate-1) and 2.1 (rate-
2) mg Zn/ kg soil compared with the control (Zn-sulfate). In contrast to Zn-sulfate, SAL-
urea-Zn, NAC-SAL-Zn, and NAC-urea-Zn nanofertilizers elicited rate-dependent significant ($P < 0.05$) effects. Moreover, chlorophyll contents were significantly higher with
SAL-urea-Zn rate-1. Compared to Zn-sulfate, NAC-SAL-Zn, and SAL-urea-Zn rate-1 considerably increased shoot biomass. These results suggest that urea can be coated with Zn
nanoparticles to make applying nanoscale fertilizers in agriculture easier without sacrific-
ing plant performance or nutrient delivery.

Applying conventional fertilizers over an extended period harms soil and plant ecosystems and microbiota because of their residual or surplus, which ends up as environmental pollutants [22, 23]. In contrast, fertilizers that exhibit gradual and sustainable
release of nutrients are thought to promote nutrient absorption. Thus, issues associated
with agriculture, such as declining land quality, low crop yield, nutrient deficiencies,
and leaching losses, can be overcome with nanotechnology, a promising method with
excellent potential [24]. Additionally, only essential elements, such as nitrogen (N), phosphorus (P), and potassium (K), are included in CF. Thus, no other macro- and/or

micronutrients are present in commercial fertilizers [12]. Utilizing NF as macro and/or micronutrients in modern agriculture with various application techniques showed progressive and continuous nutrient release, improving maximum nutrient uptake with comparably more excellent production than conventional fertilizers [19, 25, 26]. The gradual release of nutrients enables the avoidance of 40–70% of leaching-related issues that result in significant soil nutrient loss and lower soil fertility [27]. The nanostructure confers to NF's high surface area-to-volume ratio as their capacity to cover a large specific surface area, enabling plants to absorb nutrients progressively [28]. Thus, nanomaterials have advantages over CFs in terms of supporting crop growth. Therefore, using P NFs instead of CFs to enhance the quality of the surface water in agricultural fields will raise productivity, control the use of P, and improve the absorption of nutritional components [29]. As opposed to CFs, NFs have recently been demonstrated to meet the needs of plant roots, promote disease resistance, improve plant stability, and encourage deeper rooting in crops. However, NFs release nutrients in a regulated manner that can be slow or rapid depending on the environmental parameters of the field, such as acidic soils, temperature, and humidity, to increase plant growth more efficiently than CFs. Most fertilizers are not distributed to plants or are not sufficiently available in traditional farming systems [30, 31].

NFs are nutritional fertilizers that deliver nutrients to plants in whole or in part for the uptake or gradual release of active chemicals. They are composed of a nanostructured formulation. A designed NP's size, morphology, and other attributes determine how bio-accessible it is to plants. Nanomaterials with the same atoms as well-known bulk materials can have a wide range of physicochemical properties. The functionalized surfaces of nanomaterials have different properties than the bulk or core due to disorganized bonding [32]. Additionally, there are many ways to deliver NFs to plants, including nutrient-filled nanotubes, nutrient-encapsulated nanomaterials, and nanoporous materials covered in a thin polymer layer [33]. The effectiveness of using nanomaterials as NFs on plant growth is greatly influenced by the size, concentration, content, and chemical properties of the nanomaterials and the crops [34]. When NP suspensions with NFs react with water, nutrients that the crops need are released into the soil. The polymer coating of NFs or the thin film encapsulation of NPs may postpone contact with water and soil in order to reduce unfavorable nutrient losses [35].

9.2 Synthesis and characterization of nanomaterials

Any pesticide formulation that contains nanosized entities (up to 1,000 nm) and asserts to have additional properties because of their size is referred to as a "nanopesticide" [36, 37]. Due to their small size and high surface area to volume ratio, nanopesticides are thought to possess advantageous properties that are different from those of their bulk counterparts. A variety of factors can be used to classify nanopesticide formulations.

For example, they can be categorized into two groups based on their intended use: formulations that improve the solubility of pesticides that are only marginally water-soluble and formulations that reduce the rate of pesticide release. Nanoformulations will be classified as layered double hydroxides (LDH), silica nanoparticles, organic polymer-based formulations, lipid-based formulations, nanosized metals and metal oxides, clay-based nanomaterials, etc. based on the chemical makeup of the nanocarrier. The invention of the controlled release technology has made it possible to encapsulate chemicals that help pesticide degradation [36].

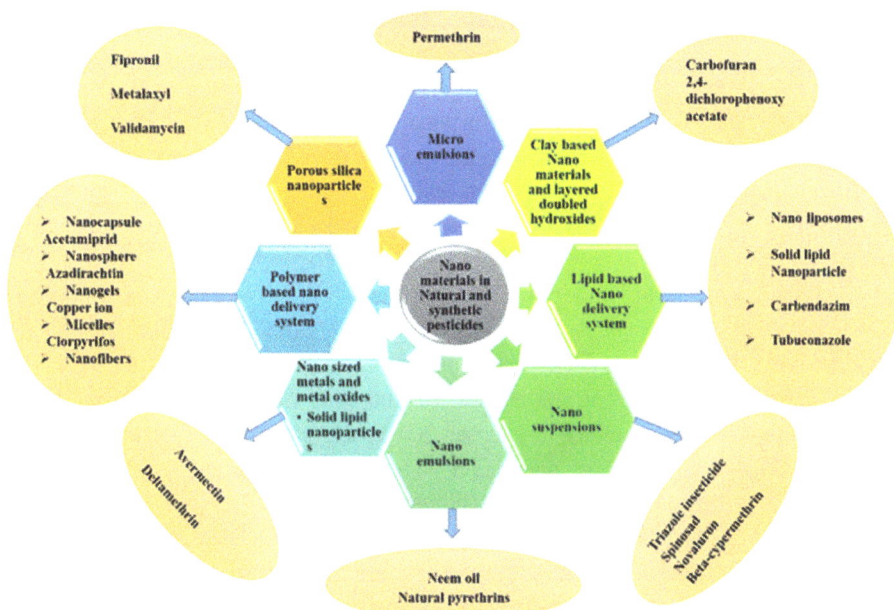

Figure 9.1: Types of nanomaterials and nano pesticides delivery system.

As a novel approach for an eco-friendly pesticide delivery system, some research organizations have looked into clay-based nanomaterials (Figure 9.1). A feasibility study on the use of organically modified montmorillonite to improve the slow-release characteristics of a biopolymer-based nanodelivery system was carried out by Chevillard et al. (2012a, 38). The herbicide ethofumesate was used by the authors as a model pesticide and was encapsulated in a nanocarrier made of wheat gluten (WG). The organically altered montmorillonite was marketed by Southern Clay as nanoclay Cloisite 30B (Gonzales, TX). Ethofumesate and WG can be bound by the alkyl quaternary ammonium salt bentonite known as cloisite 30B. Researchers especially compared the biodegradation rates of commercial ethofumesate and its WG-based nanoformulation in soil. When Cloisite 30B was incorporated to the nanoformulation, the delayed effect on the biodegradation pattern was eliminated, and the slow-release properties were improved. The creation of a nanopesticide delivery system that can significantly reduce soil contamination is the main advantage of this study. However, more study is needed to determine the nanoformulated insecticide's impactful bioavailability in soil. The potential use of three clay minerals – bentonite, kaolinite, and fuller's earth – in conjunction with carboxy methyl cellulose as carbofuran delivery nanovectors was examined by Choudhury et al. [39]. After 35 days, the highest release was produced by formulations containing kaolinite and bentonite, while the highest release was produced by Fuller's earth after 28 days. The intriguing feature of this kind of formulation is that it can effectively control worms throughout the entire crop growth cycle with just one application. Because they naturally take the shape of nanotubes, the family of clays known as halloysite clays is fascinating. Similar to more expensive carbon nanotubes, this hollow cylinder morphology enables controlled release applications for organic molecules. Compared this material's use to that of imogolite in their study and emphasized its usefulness. These materials' main advantages are their small size (10–20 nm lumen and 50–100 nm exterior diameter), which enables them to hold any number of small molecules, and the capability of capping, which can act as a trigger to only release the load under specific circumstances.

Metal and metal oxide nanoparticles have become more well-liked as a result of their adaptability and expanding applications in a variety of fields, including medicine (hyperthermia, drug transport capabilities, biofilm formation avoidance), environmental protection (water treatment, soil treatment), and other fields (nanopesticides, fertilizers). Due to their high surface-to-volume ratio in comparison to microparticles or conventional formulations, these particles have an immediate advantage. Sooresh et al. [40] employed a fairly distinctive and novel method to produce nanovectors for insect control. The surface deposition of deltamethrin significantly increased the efficacy of the silver nanoparticles. Additionally, with this kind of conjugated system, product tracking is possible in insect subsystems like hemolymph. Based on the metal content that has been found when using metal detection for tracking, the amount of deltamethrine can be calculated. This is a significant accomplishment because it allows for a deeper understanding of the mechanisms of the insect and, ultimately, the development of innovative pest control methods. Abouelkassem et al. [41] investigated the use of silver nanoparticles to create a nanoformulation for the agricultural delivery of the pyrethroid insecticide cyhalothrin in an effort to improve the efficacy and efficiency of delivery. The group concluded that the concentration required to achieve the desired outcome was 100 times lower when compared to the pesticide in its natural formulation. Another type of straightforward, effective, and user-friendly delivery system with numerous applications is silica nanoparticles. Pesticides with sustained release aim to boost effectiveness while lowering the dose needed to produce the desired result. Biocompatible silica nanocapsules were used by Wibowo et al. [42] to deliver fipronil in a continuous and secure manner. The researchers created a unique oil-in-water nanoemulsion technology to produce oil-core silica-shell nanocapsules with a custom shell thickness using an eco-friendly, biomimetic dual-templating technique. The team produced and tested three different shell thicknesses. The results showed that increasing the shell thickness further inhibits the diffusional channels in the nanocapsule, resulting in a slower release, which was extremely encouraging given the diffusional release mechanism. The 8, 25, and 44 nm nanocapsules released 40.7%, 24.3%, and 15.2% of their corresponding contents after 142 h. In vivo studies revealed that termites perished within six days. Prado et al. [43] used nanosized silica as their main material in their pursuit of a superior nanosized delivery vector for any kind of herbicide. The results of this study, which used picloram as a model herbicide, were encouraging, and the researchers theorized that other herbicides might also be effective. The inclusion of appropriate functional groups that can act as anchor points for pesticide attachment gives the study its singularity. By combining 3-aminopropyltrimethoxysilane and chloroacetic acid, a novel silylating agent with pendant carboxyl groups was created. Tetraethoxysilane (TEOS) and the novel silylating agent were then polymerized in water using ndodecylamine as a template.

9.3 Nanomaterials and horticultural crop

Recent advancement of nanomaterials improved the horticulture sector by increasing agricultural productivitys [44]. Applications of different nanomaterials are aid to reduce the spoilage of harvested fruit and vegetables, increase the shelf-life, improve the quality, and enhance the yield. Nanofertilizers, antimicrobial nanomaterials, nanosensors, and nanopesticides are facilitating improvement of the horticultural crop production, and these material are target-specific, eco-friendly, and highly efficient (Rana et al., 2021) (Table 9.1).

Among different NMs, graphene, fullerene, carbon nanotubes, carbon nanodots like carbon-based nanomaterials are considered as best option for improving the productivity of crops [56]. When Catharanthus seeds were exposed to graphene and carbon nanotubes, it increased the germinate rate and activated the early seed germination in Catharanthus [44]. Moreover, the NMs application help the crop plants to overcome drought stress by increasing water uptake and reducing excess loss of water through stomata [57]. The drastic changes in the climate is a huge hurdle to the cultivation of different crops and this problem of extreme events can be overcome with the application of nanosensors, which are able to recognize the environmental conditions or impairments [35]. Nanoagrochemicals are emerging as well as innovative alternatives of chemical pesticides, and fertilizers [58]. Nanofertilizers are highly beneficial due to slow nutrient release and efficient use of nutrients by interacting with soil microorganism [58]. Correspondingly, the application of nanopesticide reduced the environmental impact created by the extensive uses of chemical pesticides (Rana et al., 2021).

Table 9.1: Application and effects of NPs to increase the horticultural plant production.

NPs	Crops	Dose	Effects	Reference
ZnO	Solanum lycopersicum L.	100–1,000 mg/L	Enhance growth and physiological response	[45]
Silicon	Capsicum annumn L.	–	Effect in mitigating salinity negative effects	[46]
Carbon nanotubes	Allium cepa and Cucumis sativus	10–40 mg/L	Improve root growth	[47]
TiO_2	Spinacia oleracea	1,000 to 2,000 mg/L	Improve the photosynthesis rate	[48]
TiO_2	Spinacia oleracea	2.5%	Increased chlorophyll, ribulosebisphosphate carboxylase or oxygenase activity, and rate of photosynthesis	[49]
ZnO	Coriandrum sativum	concentration < 400 mg/kg	Improved photosynthesis pigments and defense response	[50]
Copper oxide (CuO)	Pisum sativum L.	0, 50, 100, 200, 400, and 500 mg dm^{-3}	100, 200, 400, and 500 mg dm^{-3} CuONPs significantly reduced plant growth and improved reactive oxygen species (ROS) generation and lipid peroxidation	[51]
CuO, Al_2O_3 and TiO_2	Allium cepa	0, 20, 200, and 2,000 µg ml^{-1}	Significantly altered root cells and a number of chromosomes in root meristems.	[52]
Cobalt ferrite [CoFe2O4]	Lycopersicon lycopersicum	1,000 mg L^{-1}	Catalase activity decreased in roots and leaves and promoted tomato roots' growth	[53]
CuO and ZnO	Cucumis sativus	100 mg/L	increase in SOD, CAT, and POD activities and NPs alter both phytotoxicity and oxidative stress	[54]
Silver [Ag]	Raphanus sativus	0, 125, 250, and 500 mg/L	Significantly affect the growth, nutrient content and macromolecule conformation	[55]

9.4 Conclusion

Nanotechnology utilizes the potential of nanoparticles and their distribution methods to enhance horticulture crop productivity. Nanofertilizers, nanosensors, antimicrobial nanomaterials, and nanopesticides are helping to enhance horticulture crop productivity, and these materials are highly efficient and target-specific. Nanofertilizers minimize the overuse of chemical fertilizers and pesticides while also being highly simple, cost-effective, and eco-friendly, permitting them to be produced in less time with the minimum effort and without destroying the environment. NFs are also advantageous because of their slow nutrient release and efficient utilization of nutrients through interaction with soil microorganisms. Nano fertilizers promote vegetative development, enhance reproductive growth and flowering in horticultural crops, enhance productivity, product quality, and eventually increase shelf-life and decrease fruit waste. Increased use of nanotechnology will result in climate-smart horticulture, reduced post-harvest losses, and improved overall crop quality.

References

[1]   Babu, S., Singh, R., Yadav, D., Rathore, S. S., Raj, R., Avasthe, R., Yadav, S. K., Das, A., Yadav, V., Yadav, B., Shekhawat, K. Nanofertilizers for agricultural and environmental sustainability. *Chemosphere* 2022, 292, 133451.
[2]   Qureshi, A., Singh, D. K., Dwivedi, S. Nano-fertilizers: A novel way for enhancing nutrient use efficiency and crop productivity. *International Journal of Current Microbiology and Applied Sciences* 2018, 7(2), 3325–3335.
[3]   Sharif, M. K., Butt, M. S., Sharif, H. R. Role of nanotechnology in enhancing bioavailability and delivery of dietary factors. In Nutrient Delivery. Academic Press, 2017, 587–618.
[4]   Mastronardi, E., Tsae, P., Zhang, X., Monreal, C., DeRosa, M. C. Strategic role of nanotechnology in fertilizers: Potential and limitations. In Nanotechnologies in Food and Agriculture. Springer International Publishing, 2015, 25–67, doi: https://doi.org/10.1007/978-3-319-14024-7_2.
[5]   Verma, K. K., Song, X. P., Joshi, A., Tian, D. D., Rajput, V. D., Singh, M., Arora, J., Minkina, T., Li, Y. R. Recent trends in nano-fertilizers for sustainable agriculture under climate change for global food security. *Nanomaterials* 2022, 12(1), 173.
[6]   Iqbal, M. A. Nano-fertilizers for sustainable crop production under changing climate: A global perspective. *Sustainable Crop Production* 2019, 8, 1–13.
[7]   Basavegowda, N., Baek, K. H. Current and future perspectives on the use of nanofertilizers for sustainable agriculture: The case of phosphorus nanofertilizer. *3 Biotech* 2021, 11(7), 1–21.
[8]   Wang, Z., Yue, L., Dhankher, O. P., Xing, B. Nano-enabled improvements of growth and nutritional quality in food plants driven by rhizosphere processes. *Environment International* 2020, 142, 105831.
[9]   Abid, N., Khan, A. M., Shujait, S., Chaudhary, K., Ikram, M., Imran, M., Haider, J., Khan, M., Khan, Q., Maqbool, M. Synthesis of nanomaterials using various top-down and bottom-up approaches, influencing factors, advantages, and disadvantages: A review. *Advances in Colloid Interface Science* 2021, 102597.
[10] Iqbal, P., Preece, J. A., Mendes, P. M. Nanotechnology: The "top-down" and "bottom-up" approaches. *Supramolecular Chemistry: From Molecules to Nanomaterials* 2012.

[11]   Pohshna, C., Mailapalli, D. R., Laha, T. Synthesis of Nanofertilizers by Planetary Ball Milling. Springer, Cham, 2020, 75–112, doi: https://doi.org/10.1007/978-3-030-33281-5_3.

[12]   Rahman, M. H., Hasan, M. N., Nigar, S., Ma, F., Aly Saad Aly, M., Khan, M. Z. H. Synthesis and characterization of a mixed nanofertilizer influencing the nutrient use efficiency, productivity, and nutritive value of tomato fruits. *ACS Omega* 2021, 6(41), 27112–27120, doi: https://doi.org/10.1021/ac somega.1c03727.

[13]   Versino, F., Urriza, M., García, M. A. Eco-compatible cassava starch films for fertilizer controlled-release. *International Journal of Biological Macromolecules* 2019, 134, 302–307, doi: https://doi.org/10. 1016/j.ijbiomac.2019.05.037.

[14]   Kashyap, P. L., Xiang, X., Heiden, P. Chitosan nanoparticle based delivery systems for sustainable agriculture. In International Journal of Biological Macromolecules, Vol. 77. Elsevier B.V, 2015, 36–51, doi: https://doi.org/10.1016/j.ijbiomac.2015.02.039.

[15]   Fertahi, S., Bertrand, I., Ilsouk, M., Oukarroum, A., Amjoud, M. B., Zeroual, Y., Barakat, A. New generation of controlled release phosphorus fertilizers based on biological macromolecules: Effect of formulation properties on phosphorus release. *International Journal of Biological Macromolecules* 2020, 143, 153–162, doi: https://doi.org/10.1016/j.ijbiomac.2019.12.005.

[16]   Rop, K., Karuku, G. N., Mbui, D., Michira, I., Njomo, N. Formulation of slow release NPK fertilizer (cellulose-graft-poly(acrylamide)/nano-hydroxyapatite/soluble fertilizer) composite and evaluating its N mineralization potential. *Annals of Agricultural Sciences* 2018, 63(2), 163–172, doi: https://doi.org/ 10.1016/j.aoas.2018.11.001.

[17]   Subbaiya, M. R. Formulation of green nano-fertilizer to enhance the plant growth through slow and sustained release of nitrogen. *Journal of Pharmacy Research* 2012, 5(11), 5178–5183, doi: http://jprsolutions.info.

[18]   Xiong, L., Wang, P., Hunter, M. N., Kopittke, P. M. Bioavailability and movement of hydroxyapatite nanoparticles (HA-NPs) applied as a phosphorus fertiliser in soils. *Environmental Science Nano* 2018, 5(12), 2888–2898, doi: https://doi.org/10.1039/C8EN00751A.

[19]   Tarafder, C., Daizy, M., Alam, M. M., Ali, M. R., Islam, M. J., Islam, R., Ahommed, M. S., Aly Saad Aly, M., Khan, M. Z. H. Formulation of a hybrid nanofertilizer for slow and sustainable release of micronutrients. *ACS Omega* 2020, 5(37), 23960–23966, doi: https://doi.org/10.1021/acsomega. 0c03233.

[20]   Murgueitio-Herrera, E., Falconí, C. E., Cumbal, L., Gómez, J., Yanchatipán, K., Tapia, A. D., Martínez, K., Sinde-Gonzalez, I., Toulkeridis, T. Synthesis of iron, zinc, and manganese nanofertilizers, using andean blueberry extract, and their effect in the growth of cabbage and lupin plants. *Nanomaterials* 2022, 12(11), doi: https://doi.org/10.3390/nano12111921.

[21]   Dimkpa, C. O., Campos, M. G. N., Fugice, J., Glass, K., Ozcan, A., Huang, Z., Singh, U., Santra, S. Synthesis and characterization of novel dual-capped Zn–urea nanofertilizers and application in nutrient delivery in wheat. *Environmental Science: Advances* 2022, 1(1), 47–58, doi: https://doi.org/10. 1039/d1va00016k.

[22]   Berg, M., Meehan, M., Scherer, T. Environmental implications of excess fertilizer and manure on water quality. NDSU Extension Service, NM1281 2017, 2.

[23]   Solanki, P., Bhargava, A., Chhipa, H., Jain, N., Panwar, J. Nano-fertilizers and their smart delivery system. In Nanotechnologies in Food and Agriculture. Springer International Publishing, 2015, 81–101, doi: https://doi.org/10.1007/978-3-319-14024-7_4.

[24]   He, X., Deng, H., Hwang, H. The current application of nanotechnology in food and agriculture. *Journal of Food and Drug Analysis* 2019, 27(1), 1–21, No longer published by Elsevier, doi: https://doi. org/10.1016/j.jfda.2018.12.002.

[25]   Rop, K., Karuku, G. N., Mbui, D., Njomo, N., Michira, I. Evaluating the effects of formulated nano-NPK slow release fertilizer composite on the performance and yield of maize, kale and capsicum. *Annals of Agricultural Sciences* 2019, 64(1), 9–19, doi: https://doi.org/10.1016/j.aoas.2019.05.010.

[26] Sajyan, T. K., Alturki, S. M., Sassine, Y. N. Nano-fertilizers and their impact on vegetables: Contribution of Nano-chelate Super Plus ZFM and Lithovit®-standard to improve salt-tolerance of pepper. *Annals of Agricultural Sciences* 2020, 65(2), 200–208, doi: https://doi.org/10.1016/j.aoas.2020.11.001.

[27] Ditta, A., Arshad, M. Applications and perspectives of using nanomaterials for sustainable plant nutrition. In Nanotechnology Reviews. Walter de Gruyter GmbH, 2016, 209–229, doi: https://doi.org/10.1515/ntrev-2015-0060.

[28] Monreal, C. M., Derosa, M., Mallubhotla, S. C., Bindraban, P. S., Dimkpa, C. Nanotechnologies for increasing the crop use efficiency of fertilizer-micronutrients. In Biology and Fertility of Soils. Springer, 2016, 423–437, doi: https://doi.org/10.1007/s00374-015-1073-5.

[29] Liu, R., Lal, R. Synthetic apatite nanoparticles as a phosphorus fertilizer for soybean (Glycine max). *Scientific Reports* 2014, 4, doi: https://doi.org/10.1038/srep05686.

[30] Kumar, Y. Nanofertilizers and their role in sustainable agriculture. *Annals of Plant and Soil Research* 2021, 23(3), 238–255, doi: https://doi.org/10.47815/apsr.2021.10067.

[31] Naderi, M. R., Danesh-Shahraki, A. Nanofertilizers and their roles in sustainable agriculture. *International Journal of Agriculture and Crop Sciences (IJACS)* 2013, 5(19), 2229–2232.

[32] Rasmussen, K., Rauscher, H., Mech, A., Riego Sintes, J., Gilliland, D., González, M., Kearns, P., Moss, K., Visser, M., Groenewold, M., Bleeker, E. A. J. Physico-chemical properties of manufactured nanomaterials – Characterisation and relevant methods. An outlook based on the OECD Testing Programme. *Regulatory Toxicology and Pharmacology* 2018, 92, 8–28, doi: https://doi.org/10.1016/j.yrtph.2017.10.019.

[33] Derosa, M. C., Monreal, C., Schnitzer, M., Walsh, R., Sultan, Y. Nanotechnology in fertilizers. In Nature Nanotechnology. Nature Publishing Group, 2010, 91, doi: https://doi.org/10.1038/nnano.2010.2.

[34] Thakur, S., Thakur, S., Kumar, R. Bio-nanotechnology and its role in agriculture and food industry. *Journal of Molecular and Genetic Medicine* 2018, 12(1), 1–5, doi: https://doi.org/10.4172/1747-0862.1000324.

[35] Shang, Y., Kamrul Hasan, M., Ahammed, G. J., Li, M., Yin, H., Zhou, J. Applications of nanotechnology in plant growth and crop protection: A review. *Molecules* 2019, 24, Issue 14, MDPI AG, doi: https://doi.org/10.3390/molecules24142558.

[36] Kah, M., Beulke, S., Tiede, K., Hofmann, T. Nanopesticides: State of knowledge, environmental fate, and exposure modeling. *Critical Reviews in Environmental Science and Technology* 2013, 43, 1823–1867.

[37] Kah, M., Hofmann, T. Nanopesticide research: Current trends and future priorities. *International Environment* 2014, 63, 224–235.

[38] Chevillard, A., Angellier-Coussy, H., Guillard, V., Gontard, N., Gastaldi, E. Controlling pesticide release via structuring agropolymer and nanoclays based materials. *Journal of Hazardous Materials* 2012b, 205–206, 32–39.

[39] Choudhury, S. R., Pradhan, S., Goswami, A. Preparation and characterisation of acephate nano-encapsulated complex. *Nanoscience Methods* 2012, 1, 9–15.

[40] Sooresh, A., Kwon, H., Taylor, R., Pietrantonio, P., Pine, M., Sayes, C. M. Surface functionalization of silver nanoparticles: Novel applications for insect vector control. *ACS Applied Material Interfaces* 2011, 3, 3779–3787.

[41] Abouelkassem, S., El-Borady, O. M., Mohamed, M. B. Remakable enhancement of cyhalothrin upon loading into silver nanoparticles as larvicidal. *International Journal of Applied Science* 2016, 3(1), 252–264.

[42] Wibowo, D., Zhao, C. X., Peters, B. C., Middleberg, A. P. J. Sustained release of fipronil insecticide in vitro and in vivo from biocompatible silica nanocapsules. *Journal of Agricultural and Food Chemistry* 2014, 62, 12504–12511.

[43] Prado, A. G. S., Moura, A. O., Nunes, A. R. Nanosized silica modified with carboxylic acid as support for controlled release of herbicides. *Journal of Agricultural and Food Chemistry* 2011, 59, 8847–8852.

[44] Pandey, K., Anas, M., Hicks, V. K., Green, M. J., Khodakovskaya, M. V. Improvement of commercially valuable traits of industrial crops by application of carbon-based nanomaterials. *Scientific Reports* 2019, 9, 19358, doi: https://doi.org/10.1038/s41598-019-55903-3.

[45] Raliya, R., Nair, R., Chavalmane, S., Wang, W. N., Biswas, P. Mechanistic evaluation of translocation and physiological impact of titanium dioxide and zinc oxide nanoparticles on the tomato (*Solanum lycopersicum* L.) plant. *Metallomics* 2015, 7(12), 1584–1594.

[46] Tantawy, A. S., Salama, Y. A. M., El-Nemr, M. A., Abdel-Mawgoud, A. M. R. Nano silicon application improves salinity tolerance of sweet pepper plants. *International Journal of ChemTech Research* 2015, 8(10), 11–17.

[47] Cañas, J. E., Long, M., Nations, S., Vadan, R., Dai, L., Luo, M., Ambikapathi, R., Lee, E. H., Olszyk, D. Effects of functionalized and nonfunctionalized single-walled carbon nanotubes on root elongation of select crop species. *Environmental Toxicology and Chemistry: An International Journal* 2008, 27(9), 1922–1931.

[48] Hong, F., Yang, F., Liu, C., Gao, Q., Wan, Z., Gu, F., Wu, C., Ma, Z., Zhou, J., Yang, P. Influences of nano-TiO2 on the chloroplast aging of spinach under light. *Biological Trace Element Research* 2005, 104(3), 249–260.

[49] Zheng, L., Hong, F., Lu, S., Liu, C. Effect of nano-TiO2 on strength of naturally aged seeds and growth of spinach. *Biological Trace Element Research* 2005, 104(1), 83–91.

[50] Pullagurala, V. L. R., Adisa, I. O., Rawat, S., Kalagara, S., Hernandez-Viezcas, J. A., Peralta-Videa, J. R., Gardea-Torresdey, J. L. ZnO nanoparticles increase photosynthetic pigments and decrease lipid peroxidation in soil grown cilantro (*Coriandrum sativum*). *Plant Physiology and Biochemistry* 2018, 132, 120–127.

[51] Nair, P. M. G., Chung, I. M. The responses of germinating seedlings of green peas to copper oxide nanoparticles. *Biologia Plantarum* 2015, 59(3), 591–595.

[52] Ahmed, B., Shahid, M., Khan, M. S., Musarrat, J. Chromosomal aberrations, cell suppression and oxidative stress generation induced by metal oxide nanoparticles in onion (*Allium cepa*) bulb. *Metallomics* 2018, 10(9), 1315–1327.

[53] López-Moreno, M. L., Avilés, L. L., Pérez, N. G., Irizarry, B. Á., Perales, O., Cedeno-Mattei, Y., Román, F. Effect of cobalt ferrite (CoFe2O4) nanoparticles on the growth and development of *Lycopersicon lycopersicum* (tomato plants). *Science of the Total Environment* 2016, 550, 45–52.

[54] Kim, S., Lee, S., Lee, I. Alteration of phytotoxicity and oxidant stress potential by metal oxide nanoparticles in *Cucumis sativus*. *Water, Air, & Soil Pollution* 2012, 223(5), 2799–2806.

[55] Zuverza-Mena, N., Armendariz, R., Peralta-Videa, J. R., Gardea-Torresdey, J. L. Effects of silver nanoparticles on radish sprouts: Root growth reduction and modifications in the nutritional value. *Frontiers in Plant Science* 2016, 7, 90.

[56] Aacharya, R., Chhipa, H. 15 – Nanocarbon fertilizers: Implications of carbon nanomaterials in sustainable agriculture production. In Carbon Nanomaterials for Agri-Food and Environmental Applications, Micro and Nano Technologies. Abd-Elsalam, K. A., Ed. Elsevier, 2020, 297–321, doi: https://doi.org/10.1016/B978-0-12-819786-8.00015-3.

[57] Maswada, H. F., Mazrou, Y. S. A., Elzaawely, A. A., Eldein, S. M. A. Nanomaterials. Effective tools for field and horticultural crops to cope with drought stress: A review. *Spanish Journal of Agricultural Research* 2020, 18, 15.

[58] Zulfiqar, F., Navarro, M., Ashraf, M., Akram, N. A., Munné-Bosch, S. Nanofertilizer use for sustainable agriculture: Advantages and limitations. *Plant Science* 2019, 289, 110270, doi: https://doi.org/10.1016/j.plantsci.2019.110270.

Antonio Juárez-Maldonado*, Yolanda González-García,
Fabián Pérez-Labrada

Chapter 10
Nanomaterials and postharvest management of horticultural crops

Abstract: The need for food grows day by day due to the increase in population, which is why it is necessary to improve agricultural production systems to make them more efficient and productive. However, not only this goal is important but also the postharvest handling of horticultural crops since a large part of these products is lost at this stage and does not reach the final consumer. Currently, different tools are used with the aim of extending the shelf life of horticultural products, among which are coatings, packaging, bags, modification of the warehouse atmosphere, among others. However, even with these technologies, it has not been enough to end the problems that arise in the postharvest period. An interesting option has been the use of nanomaterials (NMs) to improve postharvest handling of horticultural crops. These NMs can be used or applied in different ways from direct application to horticultural products to the development of new materials that can be used as coatings, packaging, or others. The observed results are promising since it has been possible to improve the shelf life of horticultural products derived from changes in postharvest physiology and control of microorganisms.

Keywords: antioxidants, microorganisms, nanotechnology, postharvest physiology, respiration, shelf life, weight loss

10.1 Introduction

Agriculture is an activity of great importance for humanity since food and other products of interest related to horticultural crops are obtained from it. In addition to this, the increase in the world population and climate changes are leading agricultural systems to adopt emerging technologies such as nanotechnology, which allow maintaining and/or improving the production and sustainable supply of crops [1–3]. However,

*Corresponding author: Antonio Juárez-Maldonado, Department of Botany, Autonomous Agrarian University Antonio Narro, Saltillo, Coahuila 25315, Mexico, e-mail: antonio.juarez@uaaan.edu.mx
Yolanda González-García, Center for Protected Agriculture, Faculty of Agronomy, Autonomous University of Nuevo León, Nuevo León 66050, Mexico
Fabián Pérez-Labrada, Department of Botany, Autonomous Agrarian University Antonio Narro, Saltillo, Coahuila 25315, Mexico

https://doi.org/10.1515/9781501523229-010

it is not only necessary to improve the agricultural production systems but also the postharvest handling of the products obtained since the properties that influence postharvest life can be affected to a different extent by the process and manipulation from the harvest as the storage and packaging conditions [4]. Postharvest life considers physical characteristics such as firmness and color and biochemical characteristics such as aroma, sugar content, and organic acids, among others that have the fruits or other organs of the plants that serve as food [5].

Currently there are different ways to improve postharvest quality and increase shelf life, among which are growth regulators, edible coatings, precooling methods, heat treatments, and modified atmosphere [6–8]. However, even with these technologies available, it is possible to improve the postharvest handling of horticultural crops since different problems that affect postharvest life continue to occur.

Nanotechnology, particularly through the use of nanomaterials (NMs) (materials with at least one dimension between 1 and 100 nm) can be an alternative way to improve the postharvest management of horticultural crops. The properties of nanotechnologies (particle size, surface structure, solubility, and chemical composition) [9] can mean a substantial change in agricultural production worldwide and have a positive impact on the postharvest handling of horticultural crops [10].

NMs function as biostimulants due to the series of biochemical, physiological, and genetic responses that they induce in plants [11, 12]; therefore, they can be used in order to improve the postharvest handling of horticultural crops. NMs can be a very useful tool in postharvest handling since they can modify characteristics that affect shelf life, for example, they prevent or reduce the activity of enzymes responsible for the loss of quality such as polyphenoloxidase, which is responsible for "browning" [13, 14].

NMs can be applied directly to the fruits or other organs of the plant, after harvest, either directly or in conjunction with other compounds like chitosan [15]. In addition, NMs can be used to produce containers, packages, and/or bags with the aim of improving the shelf life of horticultural products [16]. Ultimately, the impact of NMs on postharvest handling may be due to different reasons such as changes in the postharvest physiology of horticultural products or through the control of microorganisms that can affect these products.

10.2 Postharvest in horticultural crops

The world production of fruits and vegetables has increased as a result of the growing demand of the population, the increase in the level of quality of life, and awareness about the consumption of these [4]. However, the loss of agricultural products after the harvest is of concern to producers, distributors, sellers, and consumers, nanotechnology is being applied to increase shelf life of fruits, vegetables, and flowers [17, 18].

The perishable nature of these products, serious quality deterioration, and losses will inevitably occur due to a variety of reasons during the postharvest stages. So that around a third of the fresh fruits produced in developing countries does not reach consumers in optimal conditions [19]. It has been reported that postharvest losses of crops can range from 10 to 40% due to mechanical, microbial, and physiological damage [20, 21]. Among the various factors that contribute to the postharvest losses are poor handling and storage, improper packaging [4], and intrinsic physiological senescence [19].

The ripening and senescence of the fruit are inevitable and irreversible physiological processes, during which the biochemical characteristics and the physiology of these organs are altered and influence various quality parameters of the fruit such as firmness, color, aroma, sugar content, and organic acids, among others [5]. This is due to the fact that the fruits and other organs have an active metabolism during harvest as well as a high metabolic activity (to a greater extent in climacteric fruits) once harvested [22]. During the ripening process, the fruits develop intrinsic and extrinsic changes, where the excessive softening derived from the loss of water results in an increase in susceptibility to pathogens and leads to significant economic losses and postharvest deterioration [23]. However, the observed changes will be different depending on the metabolism of the fruits, for example, in the maturation of climacteric fruits (e.g. *Solanum lycopersicum* L.) there is a high production of CO_2 and C_2H_4 (climacteric increase) while in nonclimacteric fruits (e.g. *Fragaria × ananassa* Duch.) there are no significant changes in the production of these compounds [24].

The loss of water is the result of respiration driven by the difference in the partial pressure of water vapor between the interior of the fruit and the external environment and is one of the main causes of postharvest deterioration [25]. Although the loss of weight and water content in crops can be considered as a process attributable to respiration and other metabolic processes after harvest, the changes that occur during postharvest ripening induce an increase in the softening of the fruit under higher storage temperatures [26, 27]. Gundewadi et al. [24] point out that the respiration rate generates the decomposition of organic molecules causing the "softening" of the tissue; in climacteric fruits the ripening is controlled by C_2H_4 while in nonclimacteric fruits it is generally hormones that are participants. The softening occurs due to a high rate of degradation of the cell membrane by the activity of the enzymes phospholipase-D and pectin methyl esterase, polyphenol oxidase, and the production of malondialdehyde (MDA) [28, 29], and also the fruit may darken as a result of the oxidation of phenolic compounds to *o*-quinone and whose polymerization generates the formation of brown pigments [30].

In general, most fruits are not marketed with a water loss of 5–10% of their initial weight, as this leads to a serious reduction in quality and wilting. Excessive water loss, in addition to causing marketable weight loss, promotes browning, loss of firmness, texture, flavor, accelerated senescence, susceptibility to cold damage, and membrane disintegration [31]. These modifications arise mainly from the degradation of pectin, the loss of sugars, depolymerization of xyloglucan, a decrease in proteins, an increase in reactive oxygen species (ROS), and MDA production, an increase in lignin, among others [32, 33]. Gao et al. [34] mentioned that the activation of the antioxidant system (enzymatic activity superoxide dismutase, catalase, peroxidase, and ascorbate peroxidase) is essential to maintain the integrity of the membrane, acting as a mechanism to delay senescence and increase the shelf life of peaches (*Prunus persica* L.). Therefore, factors such as the solubilization of peptide substances, transformation of starch to sugar, decrease in the content of organic acids, changes in the concentration of pigments, production of volatile compounds, content of phenolic compounds, alterations in the content of ROS, and modifications of the antioxidant system are related to the postharvest quality of horticultural crops [24].

The epidermis of the fruit acts as a critical barrier to resist the movement of water both in liquid form and the vapors of the fruit tissue toward the outer surface and allows the fruit to maintain high water content [35]. Hence the mechanical damage such as cuticular cracking, pitting, and bruising can lead to serious damage to the epidermis and excessive loss of water in the fruit negatively impacting postharvest life [25]. In addition to this, it can induce or accelerate ethylene production and cause risks of microbial contamination since spoilage pathogens can easily enter through dead or injured tissues and contaminate the rest of the fruit [36].

Diseases are another important factor of cosmetic, organoleptic, and nutritional fruit damage inducing a reduction in the perception of multifunctional quality. These are of biotic origin, such as fruit rot, generally associated with microbial pathogens and pests [37]. Synthetic fungicides and bactericides are used to control diseases, but they have adverse effects on the health of humans, ecosystems, in addition to favor development of resistance among microorganisms [21]. Postharvest diseases cause the loss between 10 and 30% of total crop yield during handling, transport, storage, and marketing [21]. However, they can also pose risks to humans because infected fruits can produce toxic secondary metabolites as in the case of *Penicillium expansum*, which produce patulin and citrinin in apple fruits [38].

Among the main fungal pathogens that affect postharvest quality are *Botryosphaeria berengeriana*, *Penicillium italicum*, *Colletotrichum fructicola*, *Botrytis cinerea*, *Penicillum ulaiense*, and *Penicillum digitatum* [39]. While among the bacterial ones are *Pseudomonas* spp., *Listeria monocytogenes*, *Salmonella typhi*, *Staphylococcus aureus*, and *Escherichia coli* [40]. Microorganisms can affect the metabolism of sugars in the organs of harvested horticultural crops such as fruits [41] as well as the metabolism of ROS [39]. This can result in membrane degradation, cellular decompartmentalization, and electrolyte leakage [42] as well as the enzymatic oxidation of phenolic compounds by the polyphenol oxidase enzyme that causes browning [14, 43–45] negatively affecting the quality and postharvest life.

10.3 Nanotechnology and horticulture

Nanotechnology is the application or use of various materials that have at least one dimension on a scale of 1–100 nm called NMs. There are several types of NMs with different dimensions (0D, 1D, 2D, 3D), morphologies, states, and chemical compositions, which play a key role in their properties and toxicity, and the latter can depend on several parameters such as mass, number, surface area, size, volume, surface chemistry, aggregation, surface coating, and functionalization [46]. The various aspects of NMs provide a new tool in crop production; within this are various technologies such as nanoformulations, nanosensors/nanobiosensors, nanodevices, nanoparticles, nanofibers, nanoemulsions, and nanocapsules [47–53] (Figure 10.1). Some of these technologies can definitely be used to improve postharvest handling of horticultural crops.

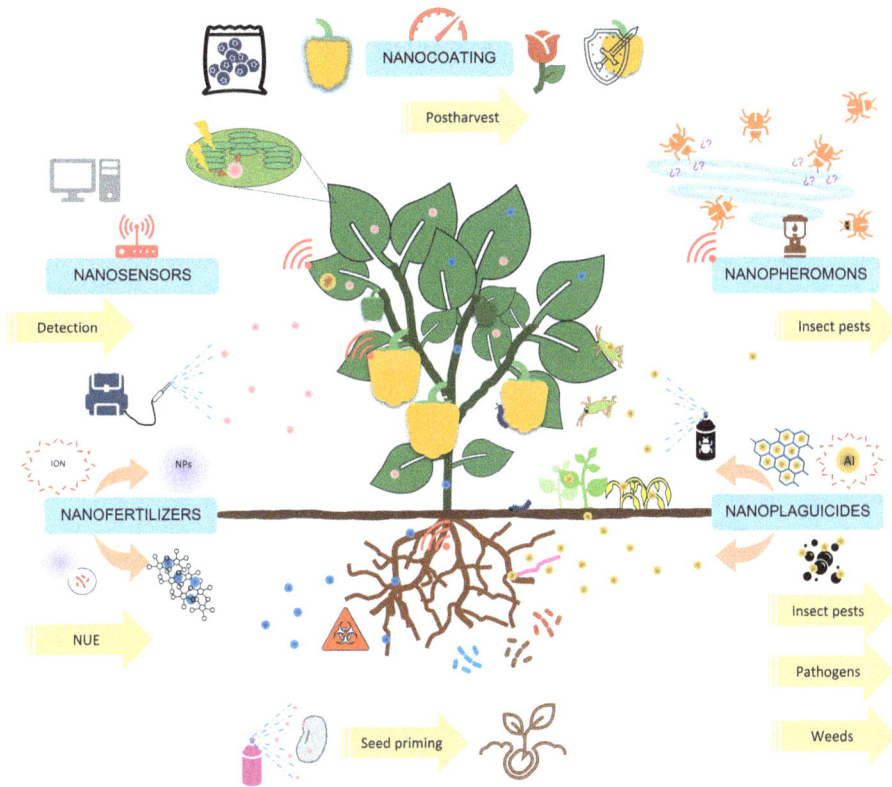

Figure 10.1: Different technologies for the use of nanomaterials in agriculture.

Due to the zeta potential and high density of surface charge, the interaction of NMs with the cell walls and membranes causes generation of a series of responses at the biochemical and physiological level [11, 54, 55]. The interaction between NMs and different organelles can improve the physiological processes and also impact the nutritional quality and postharvest life of fruits and vegetables [52]. In addition, NMs have a great advantage to be used in agriculture and postharvest since the materials such as TiO_2, Ag, CuO, and graphene represent a low environmental risk, which together with the multiple benefits that they can induce can be very useful when applied in different ways [56].

10.4 Nanotechnology and postharvest of horticultural crops

When the fruits or other organs are harvested, they present a series of responses (high respiration rate, loss of water) that reduce their quality, so it is necessary to provide the best storage conditions or apply other technologies to improve shelf life. Postharvest management of horticultural crops can focus on two main objectives: (1) modify the physiology of fruits or organs of horticultural interest in order to minimize respiration and degradation and (2) inhibit or limit the development of microorganisms (Figure 10.2).

In this sense, the application of nanotechnology allows minimizing the loss of the commercial value of the fruit, improving its content and nutritional quality [17, 18]. In a timely manner, the improvement in the postharvest of horticultural crops can be achieved with the use of nanocoatings, nanosensors, and by applying NMs that inhibit the growth and development of microorganisms, in addition to reducing oxidative damage and minimizing respiration [57] (Table 10.1).

The use of NMs as coatings is the main one, and they can be applied individually or in combination with other materials. This is due to fact that NMs have particular and novel characteristics (size, mono-dispersion, zeta potential, and architecture) that improve the application of the coatings, making them more uniform and firm, even in fruits with waxy surfaces [15]. NMs-based coatings have the ability to form semipermeable barriers on the surface of the fruits, modifying the gas exchange between the fruits and the environment [27, 78]. This results in a variety of responses such as decreased weight and water loss, decreased respiration rate, reduced electrolyte leakage, low pigments oxidation, diminish of ethylene and the ripening process, and inhibition of the enzymes related to hydrolysis of starch [27, 62, 110]. All of this result in increased shelf life and postharvest quality. Possibly, the decrease in respiration rate is one of the main mechanisms that NMs-based coatings have, but they also have the ability to reduce the degradation of key antioxidants related to postharvest quality such as the ascorbic acid and phenols, which can decrease the activity of the polyphenol oxidase responsible for the enzymatic browning [15, 77, 102].

In addition, there are other technologies in which NMs can be applied for postharvest handling of horticultural crops such as the use of NMs-based packaging and/

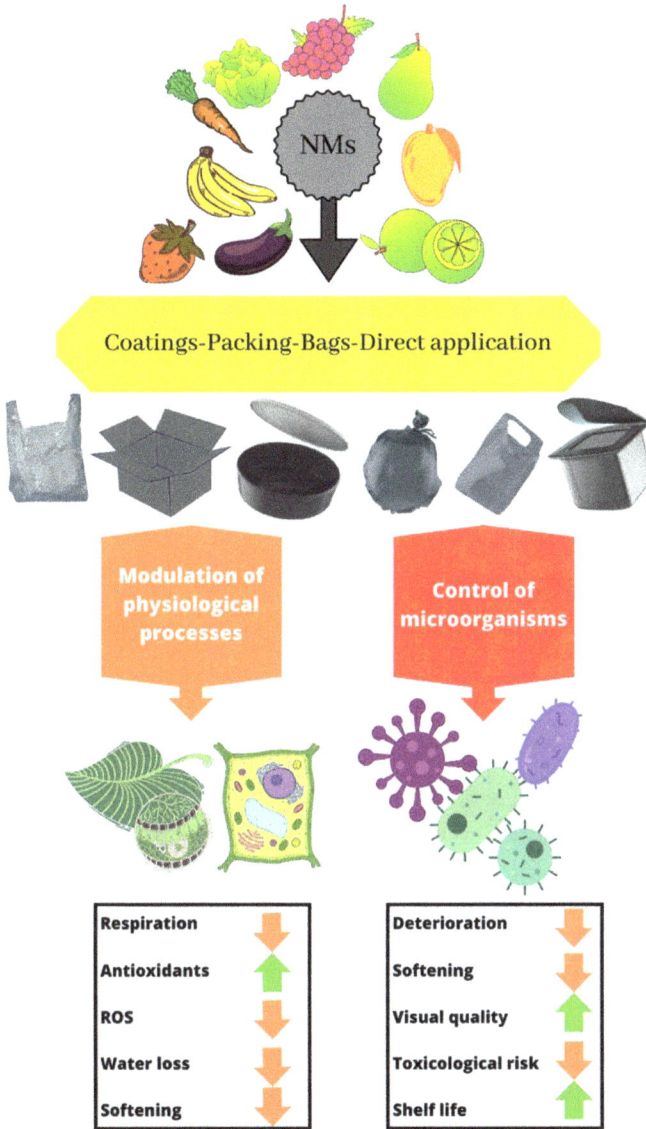

Figure 10.2: Use of nanomaterials to improve postharvest and shelf life of horticultural crops.

or bags to increase the postharvest life as well as maintain or even improve some characteristics of food during postharvest. For this purpose, different NMs such as Ag, Ag_2O, TiO_2, ZnO, and SiO_2; moreover organic origin such as chitosan, and in some cases the combination of both types of materials such as chitosan-TiO_2. NMs-based packaging works very similar to coatings since they are a physical barrier with physical-chemical characteristics that improve postharvest quality and shelf life [16].

Table 10.1: Application of nanomaterials on horticultural crops for improving postharvest and shelf life.

Horticultural crop	Nanomaterial used	Route of application	Main impact	Reference
Carrots (*Daucus carota* L.) Pears (*Pyrus communis* L.)	Ag NPs + sodium alginate	Coatings	Weight loss reduction	[58]
Lime (*Citrus aurantifolium* L.) Apple (*Pyrus malus* L.)	Agar-Ag NPs	Coatings	Weight loss reduction; increase in antimicrobial capacity	[59]
Guava (*Psidium guajava* L.)	Solid lipid nanoparticles + xanthan gum	Coatings	Weight loss reduction and preserved the best quality	[60]
Tomato (*Solanum lycopersicum* L.)	Chitosan-surfactant nanostructure	Coatings	Lower level of respiration and weight loss	[61]
Kinnow (*Citrus reticulata* L. cv. Blanco)	Carboxymethylcellulose-Ag NPs and guar gum-Ag NPs	Coatings	Lower incidence of fungal diseases	[62]
Strawberry (*Fragaria × ananassa* Duch. cv. Parous.)	Nano-Zn	Coatings	Decrease in microbial load; weight loss delay and firmness maintenance	[63]
Loquat (*Eriobotrya japonica* Lindl. cv. Baiyu)	Chitosan/nano-silica	Coatings	Weight loss reduction	[64]
Cherry tomatoes (*Solanum lycopersicum* L. cv. cerasiforme)	Ag NPs	Coatings	Reduced weight loss	[65]
Guava (*Psidium guajava* L. cv. Media China)	Solid lipid NPs from candeuba wax and xanthan gum	Coatings	Lower O_2 and CO_2 respiration rate; less weight loss	[66]
Pomegranate arils (*Punica granatum* L. cv. Malas)	Nano-ZnO + carboxymethyl cellulose	Coatings	Decreased weight loss	[67]

Commodity	Nanomaterial	Application	Effect	Reference
Peach (*Prunus persica*. L. cv. early swelling)	Nano-chitosan	Coatings	Weight loss reduction; improved fruit firmness	[68]
Fruits: apples (*Malus Domestica* L. cv. Anna) and red grapes (*Vitis vinifera* L. cv. Flame Seedless) Vegetables: tomatoes (*Lycopersicon lycopersicum* L. cv. Suber Strain B') and weet green pepper (*Capsicum annuum* L. cv. California Wonde)	1. Chitosan-thyamol/tripolyphosphate NPs 2. Gelatin-coconut fiber powder/titanium dioxide (TiO$_2$) NPs 3. Chitosan-methyl cellulose/silica (SiO$_2$) NPs 4. Gelatin-chitosan/(Ag/ZnO) NPs 5. Gelatin-anthocyanin/kafirin NPs	Coatings	Weight loss reduction	[69]
Cherry tomatoes (*Solanum lycopersicum* L.)	Chitosan-TiO$_2$ NPs	Coatings	Delayed the ripening process	[70]
Fig (*Ficus carica* L.)	ZnO NPs	Coatings	Delayed maturation; weight loss reduction; increased firmness and inhibited the growth of microorganisms	[71]
Cavendish banana (*Musa acuminata* AAA group)	Chitosan NPs	Coatings	Increased shelf life Weight loss reduction	[72]
Mango (*Mangifera indica* L. cv. Gedong Gincu)	Carrageenan – ZnO NPs	Coatings	Increased antimicrobial activity	[73]
Grapes (*Vitis labrusca* L.)	Chitosan NPs	Coatings	Delayed ripening; decreased weight loss; inhibition of microorganism growth	[74]
Cherry tomatoes (*Solanum lycopersicum* L.)	Thymol nanoemulsions incorporated in quinoa protein/chitosan	Coatings	Decreased growth of *Botrytis cinerea*	[75]

(continued)

Table 10.1 (continued)

Horticultural crop	Nanomaterial used	Route of application	Main impact	Reference
Banana (Musaceae)	Soybean protein isolate + cinnamaldehyde + ZnO NPs	Coatings	Weight loss reduction; increased firmness	[76]
Loquat (*Eriobotrya japonica* Lindl.)	Ag NPs	Coatings	Weight loss reduction	[77]
Pomegranates arils (*Punica granatum* L. cv. Rabab)	Chitosan NPs	Coatings	Reduction of microbial counts	[78]
Guava (*Psidium guajava* L.)	Chitosan + nano ZnO	Coatings	Delayed ripening process; weight loss reduction	[79]
Blueberry (*Vaccinium myrtillus* L.)	Chitosan + nano SiO$_2$	Coatings	Inhibition of microbial populations	[80]
Pomegranate arils (*Punica granatum* L. cv. Malase Saveh)	Chitosan NPs loaded with clove essential oil	Coatings	Extended the shelf life for 54 days	[81]
Bell peppers (*Capsicum annuum* L. cv. California)	Chitosan NPs with α-pineno	Coatings	Inhibition of *Alternaria alternata*	[82]
Bell pepper (*Capsicum annuum* L. cv. grossum)	Chitosan NPs	Coatings	Weight loss reduction; inhibition of pathogens	[83]
Papaya (*Carica papaya*. L cv. California)	Chitosan combined with ZnO NPs	Coatings	Suppression of microbial growth	[84]
Strawberry (*Fragaria × ananassa* Duch. cv. Benihoppe)	Natamycin-loaded zein nanoparticles stabilized by carboxymethyl chitosan	Coatings	Reduction of the occurrence of rot, mildew, and gray mold	[85]
Strawberrie (*Fragaria × ananassa* Duch. cv. Camarosa)	Chitosan NPs	Coatings	Weight loss reduction and more firmness	[86]
Mango (*Mangifera indica* L. cv. golden lily)	ZnO NPs	Coatings	Ripening delayed	[26]

Crop	Nanomaterial	Type	Effect	Ref.
Strawberry (*Fragaria × ananassa* Duch.)	Chitosan NPs	Coatings	Antifungal action. Weight loss reduction	[27]
Tomato (*Solanum lycopersicum* L.)	Cu-chitosan NPs	Coatings	Microbial decomposition reduction; weight loss reduction and respiration rate	[15]
Strawberry (*Fragaria × ananassa* Duch.)	Nano-chitosan	Coatings	Weight loss reduction; retention of firmness	[29]
Strawberry (*Fragaria × ananassa* Duch.)	CaCl$_2$ + nano-chitosan	Coatings	Weight loss reduction	[87]
Red and yellow bell pepper (*Capsicum annuum* L. var. grossum (L.) Sendt)	Silver nanoparticles polyvinylpyrrolidone-based glycerosomes	Coatings	Extension of the shelf life	[88]
Apricot (*Prunus armeniaca* L. cv. Canino)	Ag NPs	Coatings	Weight loss reduction	[89]
Strawberry (*Fragaria × ananassa* Duch. cv. Camarosa)	Chitosan NPs	Coatings	Weight loss reduction	[90]
Peach (*Prunus persica* L.)	Chitosan + iron oxide nanoparticles	Coatings	Weight loss reduction; inhibition of microbial growth on the fruit surface	[91]
Papaya (*Carica papaya* L. cv. Golden)	Ag NPs + hydroxypropylmethylcellulose	Coatings	Weight loss reduction; decrease incidence and severity of *Colletotrichum gloeosporioides*	[92]
Mango (*Mangifera indica* L.)	Chitosan/nano-TiO$_2$	Coatings	Increased firmness; reduced decomposition and respiration index	[53]
Strawberry (*Fragaria × ananassa* Duch. cv. camarosa)	Beeswax solid lipid NPs + xanthan gum and propylene glycol	Coatings	Weight loss reduction; low decay rate and less loss of firmness	[93]

(continued)

Table 10.1 (continued)

Horticultural crop	Nanomaterial used	Route of application	Main impact	Reference
Persimmon (*Diospyros kaki* L.)	Nano-chitosan, rosmarinic acid bio-mediated selenium nanoparticles (RA/SeNPs)	Coatings	Maintained firmness; spoilage prevention of *Alternaria alternate*	[94]
Strawberry (*Fragaria × ananassa* Duch. Cv. Camarosa)	Proline-coated chitosan nanoparticles (CTS-Pro NPs)	Coatings	Weight loss reduction	[95]
Strawberry (*Fragaria × ananassa*)	Starch-silver nanoparticle (St-AgNPs)	Coatings	Weight loss reduction; lowered the proportion of fruit deterioration; decrease of activity of bacteria, mold, and yeast	[96]
Mango (*Mangifera indica* L.)	Guar gum-AgNPs	Coatings	Weight loss reduction; decreasing respiration and transpiration rate	[97]
Kiwifruit (*Actinidia deliciosa* L. cv. Qinmei)	Polyethylene with nano-Ag, nano-TiO_2, montmorillonite	Packaging	Weight loss reduction	[98]
Apple (*Malus domestica* L. cv. Fuji)	Polyethylene with Ag_2O NPs	Packaging	Weight loss reduction; microbial spoilage prevention	[99]
Carrot (*Daucus carota* L. cv. Brasília)	Polyethylene with Ag NPs	Packaging	Weight loss reduction	[100]
Strawberry (*Fragaria × ananassa* Duch. cv. Akihime)	Polyethylene with nano-TiO_2	Packaging	Internal reduction of O_2 and increase of CO_2; lower decomposition rate and weight loss, delay in the decrease of firmness	[101]

Crop	Treatment	Method	Effect	Reference
Mango (*Magnifera indica* L.)	Polylactic acid film + bergamot essential oil impregnated with nano-TiO$_2$ and nano-Ag.	Packaging	Weight loss reduction; delayed loss of firmness	[102]
Cherry tomatoes (*Solanum lycopersicum* L.)	Carboxymethylcellulose and cinnamaldehdye with ZnO NPs	Packaging	Greater firmness of the fruit; weight loss reduction	[103]
Bell pepper (*Capsicum annuum* L.)	*Heracleum persicum* oil loaded with chitosan NPs	Filter paper was impregnated with NMs and placed inside the container.	Weight loss reduction	[104]
Litchi (*Litchi chinensis* Sonn.)	AgNP	Solution	Decreased browning index; decrease incidence and severity of *Peronophythora litchi*	[105]
Cucumber (*Cucumber sativa* L. cv Barracuda)	CaNPs blended with salicylic acid (SA)	Solution	Weight loss reduction	[106]
Grape (*Vitis vinifera* L., cvs Shine Muscat and Kyoho)	AgNP	Solution	Weight loss reduction; greater firmness of the fruit	[107]
Apple (*Malus domestica* Borkh.)	Nano-chitosan	Solution	Increased firmness; spoilage prevention of *Penicillium expansum*	[108]
Persimmon (*Diospyros kaki* L.)	Chitosan-loaded phenylalanine nanoparticles (Cs-Phe NPs)	Solution	Weight loss reduction; increased firmness	[109]

The application of NMs together with polyethylene has been one of the main uses of nanotechnology in order to improve the postharvest of foods. Polyethylene bags with nanoparticles of TiO_2, Ag_2O, Ag, and SiO_2 have been developed for the packaging of various foods such as kiwifruit, apple, carrot, and strawberry. Positive effects were observed in all the cases, the main one being the reduction in weight loss, in addition to increasing antioxidant activity and the content of some antioxidants such as ascorbic acid and phenols [98–101].

The application of NMs (nanocoatings, nanoparticles, nanonet, nanoencapsulates, nano packings, etc.) can be carried out in preharvest or postharvest according to the characteristics of the crop or to the objectives that are sought [50, 51]. In preharvest the applications can be on the seed, root, foliage, and/or fruits; with these applications the scientific evidence shows that there are a series of responses such as the increase in enzymatic and nonenzymatic antioxidant activity, increased ROS, and alterations at the morphological level, the above results in a greater capacity that the leaf tissue and/or fruit have to preserve their useful life [51, 111]. Cantu et al. [112] report that the foliar application of metal oxide NMs (Fe_3O_4, $MnFe_2O_4$, $ZnFe_2O_4$, $Zn_{0.5}Mn_{0.5}Fe_2O_4$, Mn_3O_4, and ZnO) at a dose of 250 mg L^{-1} on tomato plants can improve their life in storage. Sha et al. [113] when applying 25 and 50 mg L^{-1} of multiwalled carbon nanotubes on "Flame Seedless" grapes (at 15, 40, and 60 days after flowering) managed to maintain the integral quality index, firmness, soluble sugar, and titratable acidity as well as the ascorbic acid content and managing to increase its antioxidant activity thus improving its quality during storage at 4 °C for 20 days. Wang et al. [114] mention an enrichment and accumulation of flavonoids in tomato fruits in response to the application of NMs of $Fe_7(PO_4)_6$, and it may confer greater capacity to prolong its shelf life. According to González-García et al. [115] the application of NPs-SiO_2 during the development of the cucumber crop can reduce the loss of bioactive compounds, improving the postharvest quality of the fruit. Moreover, SiO_2 NPs can increase the useful life of *Lilium* [116].

On the other hand, when the application of these nanotechnologies is carried out directly to the fruit, once harvested, the enzymes related to the degradation of the cell wall and a decrease in the browning of the fruit as well as an increase in the content of ascorbic acid are observed [44]. This enables the fruit to have a longer shelf life by reducing the production of ethylene (higher proportion of climacteric fruits) [87, 110]. The ROS load, antioxidants, and morphological modifications will be the reservoir to preserve the quality during the storage time of the fruits. The advantage will be in greater or lesser proportion depending on the NM used and the storage conditions [117].

10.4.1 Impact of NMs on postharvest physiology

In the case of nanocoatings (added with NPs), they form a semi-permeable barrier on the surface of the fruit, inhibiting the decomposition symptoms, in addition to interfering with exposed macromolecules, reducing intracellular electrolyte leakage, weight loss, water loss, and inducing lower respiration rate [62]. This semi-permeable film alters the gas exchange (water vapor, O_2, and CO_2) between the fruit and the storage environment [27, 78]. This microenvironment created between the coating and the surface of the fruit induces a lower rate of pigment oxidation and a longer postharvest life since, as mentioned, there is a reduction in O_2 accompanied by an increase in CO_2 [27]. According to Sorrentino et al. [110] the modification in the levels of O_2 and CO_2 could inhibit the degradation of chlorophyll on the surface of a fruit; in addition, a low O_2 and high CO_2 ratio can inhibit the production of ethylene and delay ripening as well as inhibit enzymes related to hydrolysis of starch.

Nano chitosan reduces the respiration and metabolism of the fruit [102]. Amiri et al. [78] reported that chitosan nano-coatings cause, in addition to less water loss, a reduction in exposure to oxygen in the storage environment and eventually a reduction in the oxidative reactions of ascorbic acid in the fruit. Derived from the blockage of the pores of the fruit skin, the loss of moisture is less and the respiratory rate is reduced [27, 62, 74, 78]. The possible increase in the content of total phenols and anthocyanins derives from a decrease in their oxidation due to the chitosan-NPs coating [78]. According to Melo et al. [27], the increase in the antioxidant activity of chitosan-NPs is given by the reaction between free radicals with the residual free amino groups (NH_2) that in turn can generate ammonium groups (NH^{+3}) by absorbing a hydrogen ion from the cell medium.

The application of Cu NPs-chitosan forms an invisible and intangible nano-network on the surface of the fruit, being able to act as a potential barrier in all possible openings (stem scar, cuticle wax, lenticels, and aquaporins) controlling loss of humidity, gas exchanges, and frequency of the respiratory rate, in addition to considerably suppressing the metabolic activity that controls total and reducing sugars [15]. The same authors point out that these NMs reduce lycopene synthesis during storage. This coating, according to electron microscopy studies, forms a smooth surface in contact with the fruit of strawberry (*Fragaria × ananassa* Duch.) and tomato (*Solanum lycopersicum* L.) that results in a reduction in respiration rate and weight loss [15, 87].

The positive effect of *S*-nitrosoglutathione (GSNO)-chitosan NPs in apples (*Malus domestica* L. cv. Fuji) could be due to its participation in the metabolism of GSNO and endogenous glutathione, as well as in the nitration process, and in the conversion of $O^{2\cdot-}$ in water by increasing the enzymatic activity of SOD [118]. The antioxidant properties derived from the application of NPs could delay the browning of postharvest cultures due to the reduction in the oxidation process of the polyphenolic compounds that protect the cells from lipoxygenase [44].

One of the main mechanisms of these coatings with NPs is the inhibition of the respiratory rate and the reduction in the consumption of ascorbic acid in the fruit [102] as well as reducing the degradation of phenolic compounds [77]. The ascorbic acid, like its stereoisomer, isoascorbic acid, is very important in fruit physiology since it is used to minimize browning by reducing o-quinone to phenolic compounds and preventing polyphenol oxidase activity [15].

A NMs-based package (polylactic acid in combination with nano-TiO_2 and nano-Ag) was evaluated in mango (*Magnifera indica* L.) stored for 15 days at room temperature. It was observed in the fruits less weight loss and therefore delayed loss of firmness; in addition the content of ascorbic acid was increased [102].

In cherry tomatoes (*Solanum lycopersicum* L.) stored at 25 °C and 45% RH for 10 days an individual fruit packaging of carboxymethylcellulose and cinnamaldehdye with ZnO NPs was evaluated. The results observed were greater firmness of the fruits and reduced weight loss [103].

Other methods have been applied in the postharvest handling of horticultural crops such as to impregnate some carrier material with essential oils loaded with some NM, the carrier material is placed inside the package along with the food to be preserved [13, 44, 104].

Following this methodology, the color change rate, weight loss, and firmness were reduced in red sweet bell peppers (*Capsicum annuum* L.) packaged together with filter paper impregnated with *Heracleum persicum* oil loaded with chitosan NPs. Moreover, phenolic compounds, ascorbic acid, catalase, and superoxide dismutase were increased; in contrast, polyphenol oxidase activity was decreased [104].

Similar results were observed in strawberry fruits, when a composite material with NMs was placed in the lower part of the container was possible to reduce the loss of firmness, in addition to maintaining the desirable color of the fruits [119].

ZnO NPs based on polysaccharides (chitosan, alginate, carrageenan, cellulose, and pectin) can reduce the ripening process in fruits due to the improved gas and water barrier properties [120].

The application of Si NPs can reduce lipid peroxidation by preserving the integrity of the membrane through the activation of the antioxidant system [121], possibly due to electrostatic and hydrophobic interactions between NPs and fruit components [26, 27].

Bonilla-Bird et al. [122] found in sweetpotato (*Ipomoea batatas* L.) that the exposure of the roots to CuO NPs, once harvested, induced an increase in the Cu content in the peridermis and bark but not in the most internal tissues such as the pith; this suggests that this type of NPs can protect and increase the shelf life of sweet potato roots without compromising the consumer to excessive Cu consumption.

Miranda et al. [123] report that the coating of tomato fruits with a carnauba wax nanoemulsion (particle diameter of 44 nm) induced greater sensory preference as well as a lower rate of water loss; however, firmness, ethylene, and respiratory activity were not modified during a storage of 15 days to 23 °C.

Guleria et al. [124] by applying 3% of a biodegradable nanocomposite of α-Fe_2O_3/$C_{20}H_{38}O_{11}$ on *Solanum lycopersicum* fruits allows the coating to maintain the quality of the fruit (stored at 25 ± 2 °C and 4 °C) by not reducing the vitamin C content.

10.4.2 Impact of NMs on the development of microorganisms

Nanocomposite films or polymeric coatings (multiphase materials that in their composition have a filler with at least one dimension less than 100 nm) can be obtained by combining nanoparticles with biodegradable materials (of plant origin), which improve functional performance, mechanical properties, optical, thermal, and antimicrobial characteristics of the biodegradable material, positively impacting the shelf life of food during storage [125].

The control of microorganisms through the use of NMs in postharvest is due to the impact that these have on the physiology of microorganisms. The NMs-based coatings have the ability to damage the cell membrane of microorganisms and even affect their DNA and thereby limit microbial growth of these in foods [126]. This is probably due to the irruption in the transmembrane electron transfer, penetrating and oxidizing the envelope of the cellular components as well as the production of secondary products as ROS that induce the formation of ions, efflux, and ultimately death of the microorganisms [127].

Sun et al. [128] mentioned that the induction of ROS, the release of Zn^{2+} ions, and their eventual penetration into the cell wall as well as their reaction with the content of the cytoplasm are the possible mechanisms of ZnO NPs-based antimicrobial activity.

It has also been shown that some organic compounds such as chitosan in nanometric size can inhibit the growth of microorganisms due to inhibition of microbial mRNA [68]. Moreover, nano-chitosan has been observed to penetrate the cell wall of bacteria, thereby providing protective effect on grapes (*Vitis labrusca* L.) [74]. Cu NPs-chitosan forms an invisible and intangible nano-network on the surface of the fruit, being able to act as a potential barrier and controlling microbial infection [15].

Melo et al. [74] reported that grapes (*Vitis labrusca* L.) immersed in chitosan nanoparticles dispersed in a solution increased its antimicrobial activity. Eventually, the response of the plant depends on the type of culture, growth medium – storage and the NM used (surface coating, concentration, type, size, exposure time) [117]. But also, the particular characteristics of NMs (size, mono-dispersion, zeta potential, and architecture) could facilitate the uniform and firm bond to the wax cuticular surfaces of fruits and vegetables ([15].

The application of Cu NPs-chitosan forms an invisible and intangible nano-network on the surface of the fruit, being able to act as a potential barrier in all possible openings (stem scar, cuticle wax, lenticels, and aquaporins) controlling microbial infection [15].

Zambrano-Zaragoza et al. [93] mentioned that beeswax solid lipid NPs with xanthan gum and propylene glycol when applied on strawberries (*Fragaria* × *ananassa* Duch. cv. camarosa) decreased fungal contamination. This was the result of the particle size and concentration of the polysaccharide since its aggregation and formation significantly modified the atmosphere on the surface of the fruit.

Guleria et al. [124] applied a biodegradable nanocomposite of α-Fe_2O_3/$C_{20}H_{38}O_{11}$ on *Solanum lycopersicum* fruits at a concentration of 3%, observing excellent antimicrobial activity. While in strawberry fruits, when a composite material with NMs was placed in the lower part of the container was possible to reduce yeast and mold population [119].

Karakuş et al. [129] succeeded in designing a novel carboxymethylcellulose/hydromagnesite stromatolite bionanocompound that has antimicrobial and antifungal effects as well as no toxicity and great cytocompatibility and can be used as a green nanoadditive against microorganisms.

Salem et al. [94] found that the fruits of *Diospyros kaki* L. treated with edible coatings based on biopolymers (chitosan nanoparticles and selenium nanoparticles biomediated by rosmarinic acid) allowed to reduce postharvest infestation of *Alternaria alternata*, in turn improving fruit firmness, to a greater extent when applying the compound at 1%.

10.5 Conclusions

The postharvest management of horticultural crops is as important or more than the management of production systems, since a large part of the fruits or other organs of commercial interest obtained are discarded due to damage or natural degradation of these. Therefore, developing new methods for postharvest handling of horticultural crops is a priority objective for researchers in the area. At present, the use of NMs has definitely been an option to develop new methods that allow better postharvest handling since it has been applied in different ways such as coatings, packaging, bags, and even direct application to the fruits, obtaining excellent results. In particular, two important characteristics have the greatest impact on postharvest handling and shelf life of horticultural products: the degradation of the different structures that results in weight loss and softening, and the development of microorganisms. In this sense, the use of NMs in the different technologies that are used for postharvest handling has proven to be efficient since it can modify postharvest physiology by decreasing key processes such as respiration and degradation of cell membranes but also by inhibiting or decreasing the development of microorganisms. All this results in a higher quality and shelf life of horticultural products. Hence, the use of NMs for postharvest management of horticultural crops can be an important strategy in the short and medium term to solve or minimize the problems that arise at this stage and thus not only reduce losses of horticultural products but also increase the quality of these.

References

[1]   Agrawal, S., Kumar, V., Kumar, S., Shahi, S. K. Plant development and crop protection using phytonanotechnology: A new window for sustainable agriculture. *Chemosphere* 2022, 299, 134465, doi: 10.1016/j.chemosphere.2022.134465.

[2]   Avila-Quezada, G. D., Ingle, A. P., Golińska, P., Rai, M. Strategic applications of nano-fertilizers for sustainable agriculture: Benefits and bottlenecks. *Nanotechnology Reviews* 2022, 11, 2123–2140, doi: 10.1515/ntrev-2022-0126.

[3]   Farooq, M. A., Hannan, F., Islam, F., Ayyaz, A., Zhang, N., Chen, W., Zhang, K., Huang, Q., Xu, L., Zhou, W. The potential of nanomaterials for sustainable modern agriculture: Present findings and future perspectives. *Environmental Science Nano* 2022, 9, 1926–1951.

[4]   Maringgal, B., Hashim, N., Mohamed Amin Tawakkal, I. S., Muda Mohamed, M. T. Recent advance in edible coating and its effect on fresh/fresh-cut fruits quality. *Trends in Food Science & Technology* 2020, 96, 253–267, doi: 10.1016/j.tifs.2019.12.024.

[5]   Tang, N., An, J., Deng, W., Gao, Y., Chen, Z., Li, Z. Metabolic and transcriptional regulatory mechanism associated with postharvest fruit ripening and senescence in cherry tomatoes. *Postharvest Biology and Technology* 2020, 168, 111274, doi: 10.1016/j.postharvbio.2020.111274.

[6]   Munhuweyi, K., Mpai, S., Sivakumar, D. Extension of avocado fruit postharvest quality using non-chemical treatments. *Agronomy* 2020, 10, doi: 10.3390/agronomy10020212.

[7]   Riva, S. C., Opara, U. O., Fawole, O. A. Recent developments on postharvest application of edible coatings on stone fruit: A review. *Scientia Horticulturae* 2020, 262, 109074, doi: 10.1016/j.scienta.2019.109074.

[8]   Wang, G., Zhang, X. Evaluation and optimization of air-based precooling for higher postharvest quality: Literature review and interdisciplinary perspective. *Food Quality and Safe* 2020, 4, 59–68, doi: 10.1093/fqsafe/fyaa012.

[9]   An, C., Sun, C., Li, N., Huang, B., Jiang, J., Shen, Y., Wang, C., Zhao, X., Cui, B., Wang, C., et al. Nanomaterials and nanotechnology for the delivery of agrochemicals: Strategies towards sustainable agriculture. *Journal of Nanobiotechnology* 2022, 20, 1–19, doi: 10.1186/s12951-021-01214-7.

[10]   Adil, M., Bashir, S., Bashir, S., Aslam, Z., Ahmad, N., Younas, T., Asghar, R. M. A., Alkahtani, J., Dwiningsih, Y., Elshikh, M. S. Zinc oxide nanoparticles improved chlorophyll contents, physical parameters, and wheat yield under salt stress. *Frontiers Plant Science* 2022, 13, 2535, doi: 10.3389/FPLS.2022.932861/BIBTEX.

[11]   Juárez-Maldonado, A., Ortega-Ortiz, H., González-Morales, S., Morelos-Moreno, Á., Cabrera-de la Fuente, M., Sandoval-Rangel, A., Cadenas-Pliego, G., Benavides-Mendoza, A. Nanoparticles and nanomaterials as plant biostimulants. *International Journal of Molecular Sciences* 2019, 20, 1–19, doi: 10.3390/ijms20010162.

[12]   Juárez-Maldonado, A., Tortella, G., Rubilar, O., Fincheira, P., Benavides-Mendoza, A. Biostimulation and toxicity: The magnitude of the impact of nanomaterials in microorganisms and plants. *Journal of Advanced Research* 2021, doi: 10.1016/j.jare.2020.12.011.

[13]   Karimirad, R., Behnamian, M., Dezhsetan, S. Application of chitosan nanoparticles containing cuminum cyminum oil as a delivery system for shelf life extension of agaricus bisporus. *LWT - Food Science and Technology* 2019, 106, 218–228, doi: 10.1016/j.lwt.2019.02.062.

[14]   Lo'ay, A. A., Ameer, N. M. Performance of calcium nanoparticles blending with ascorbic acid and alleviation internal browning of 'Hindi Be-Sennara' mango fruit at a low temperature. *Scientia Horticulturae* 2019, 254, 199–207, doi: 10.1016/j.scienta.2019.05.006.

[15]   Meena, M., Pilania, S., Pal, A., Mandhania, S., Bhushan, B., Kumar, S., Gohari, G., Saharan, V. Cu-Chitosan Nano-Net improves keeping quality of tomato by modulating physio-biochemical responses. *Scientific Reports* 2020, 10, 21914, doi: 10.1038/s41598-020-78924-9.

[16] Basumatary, I. B., Mukherjee, A., Katiyar, V., Kumar, S. Biopolymer-based nanocomposite films and coatings: Recent advances in shelf-life improvement of fruits and vegetables. *Critical Reviews in Food Science and Nutrition* 2020, 0, 1–24, doi: 10.1080/10408398.2020.1848789.

[17] Bhardwaj, S., Lata, S., Garg, R. Application of nanotechnology for preventing postharvest losses of agriproducts. *Journal of Horticultural Science and Biotechnology* 2022, 1–14, doi: 10.1080/14620316.2022.2091488.

[18] Neme, K., Nafady, A., Uddin, S., Tola, Y. B. Application of nanotechnology in agriculture, postharvest loss reduction and food processing: Food security implication and challenges. *Heliyon* 2021, 7, e08539.

[19] Chen, T., Ji, D., Zhang, Z., Li, B., Qin, G., Tian, S. Advances and strategies for controlling the quality and safety of postharvest fruit. *Engineering* 2020, doi: 10.1016/j.eng.2020.07.029.

[20] Bose, S. K., Howlader, P., Wang, W., Yin, H. Oligosaccharide is a promising natural preservative for improving postharvest preservation of fruit: A review. *Food Chemistry* 2021, 341, 128178, doi: 10.1016/j.foodchem.2020.128178.

[21] Mossa, A.-T. H., Mohafrash, S. M. M., Ziedan, E.-S. H. E., Abdelsalam, I. S., Sahab, A. F. Development of eco-friendly nanoemulsions of some natural oils and evaluating of its efficiency against postharvest fruit rot fungi of cucumber. *Industrial Crops and Products* 2021, 159, 113049, doi: 10.1016/j.indcrop.2020.113049.

[22] Gouda, M. H. B., Zhang, C., Wang, J., Peng, S., Chen, Y., Luo, H., Yu, L. ROS and MAPK cascades in the post-harvest senescence of horticultural products. *Journal of Proteomics & Bioinformatics* 2020, 13, 1–7, doi: 10.35248/0974-276x.1000508.

[23] Posé, S., Paniagua, C., Matas, A. J., Gunning, A. P., Morris, V. J., Quesada, M. A., Mercado, J. A. A nanostructural view of the cell wall disassembly process during fruit ripening and postharvest storage by atomic force microscopy. *Trends in Food Science & Technology* 2019, 87, 47–58, doi: 10.1016/j.tifs.2018.02.011.

[24] Gundewadi, G., Reddy, V. R., Bhimappa, B. Physiological and biochemical basis of fruit development and ripening-a review. *Journal of Hillagric Agriculture* 2018, 9, 7–21, doi: 10.5958/2230-7338.2018.00003.4.

[25] Jansasithorn, R., East, A. R., Hewett, E. W., Heyes, J. A. Skin cracking and postharvest water loss of Jalapeño chilli. *Scientia Horticulturae* 2014, 175, 201–207, doi: 10.1016/j.scienta.2014.05.037.

[26] Malek, N. S. A., Rosman, N., Mahmood, M. R., Khusaimi, Z., Asli, N. A. Effects of storage temperature on shelf-life of mango coated with zinc oxide nanoparticles. *Science Letters* 2020, 14, 47–57, doi: 10.24191/sl.v14i2.9542.

[27] Melo, N. F. C. B., Pintado, M. M. E., Medeiros, J. A. D. C., Galembeck, A., Vasconcelos, M. A. D. S., Xavier, V. L., de Lima, M. A. B., Stamford, T. L. M., Stamford–Arnaud, T. M., Flores, M. A. P., et al. Quality of postharvest strawberries: Comparative effect of fungal chitosan gel, nanoparticles and gel enriched with edible nanoparticles coatings. *International Journal of Food Studies* 2020, 9, 373–393, doi: 10.7455/ijfs/9.2.2020.a9.

[28] Ashitha, G. N., Sunny, A. C., Nisha, R. Effect of pre-harvest and post-harvest hexanal treatments on fruits and vegetables: A review. *Agricultural Review* 2020, 41, 124–131, doi: 10.18805/ag.r-1928.

[29] Nguyen, H. V. H., Nguyen, D. H. H. Effects of nano-chitosan and chitosan coating on the postharvest quality, polyphenol oxidase activity and malondialdehyde content of strawberry (*Fragaria x Ananassa* Duch.). *Journal of Horticulture and Postharvest Research* 2020, 3, 11–24, doi: 10.22077/jhpr.2019.2698.1082.

[30] Ranjbar, S., Rahemi, M., Ramezanian, A. Comparison of nano-calcium and calcium chloride spray on postharvest quality and cell wall enzymes activity in apple Cv. red delicious. *Scientia Horticulturae* 2018, 240, 57–64, doi: 10.1016/j.scienta.2018.05.035.

[31] Lufu, R., Ambaw, A., Opara, U. L. Water loss of fresh fruit: Influencing pre-harvest, harvest and postharvest factors. *Scientia Horticulturae* 2020, 272, 109519, doi: 10.1016/j.scienta.2020.109519.

[32]   Xu, J., Zhao, Y., Zhang, X., Zhang, L., Hou, Y., Dong, W. Transcriptome analysis and ultrastructure observation reveal that hawthorn fruit softening is due to cellulose/hemicellulose degradation. *Frontiers Plant Science* 2016, 7, 1–14, doi: https://doi.org/10.3389/fpls.2016.01524.

[33]   Li, H., Wu, L., Tang, N., Liu, R., Jin, Z., Liu, Y., Li, Z. Analysis of transcriptome and phytohormone profiles reveal novel insight into ginger (*Zingiber officinale* Rose) in response to postharvest dehydration stress. *Postharvest Biology and Technology* 2020, 161, 111087, doi: 10.1016/j. postharvbio.2019.111087.

[34]   Gao, H., Zhang, Z. K., Chai, H. K., Cheng, N., Yang, Y., Wang, D. N., Yang, T., Cao, W. Melatonin treatment delays postharvest senescence and regulates reactive oxygen species metabolism in peach fruit. *Postharvest Biology and Technology* 2016, 118, 103–110, doi: 10.1016/j. postharvbio.2016.03.006.

[35]   Wei, X., Xie, D., Mao, L., Xu, C., Luo, Z., Xia, M., Zhao, X., Han, X., Lu, W. Excess water loss induced by simulated transport vibration in postharvest kiwifruit. *Scientia Horticulturae* 2019, 250, 113–120, doi: 10.1016/j.scienta.2019.02.009.

[36]   Hussein, Z., Fawole, O. A., Opara, U. L. Harvest and postharvest factors affecting bruise damage of fresh fruits. *Horticultural Plant Journal* 2020, 6, 1–13, doi: 10.1016/j.hpj.2019.07.006.

[37]   Sharma, R. R., Nagaraja, A., Goswami, A. K., Thakre, M., Kumar, R., Varghese, E. Influence of on-the-tree fruit bagging on biotic stresses and postharvest quality of rainy-season crop of 'Allahabad Safeda' guava (*Psidium guajava* L.). *Crop Protect* 2020, 135, 105216, doi: 10.1016/j.cropro.2020.105216.

[38]   Yu, L., Qiao, N., Zhao, J., Zhang, H., Tian, F., Zhai, Q., Chen, W. Postharvest control of *Penicillium expansum* in fruits: A review. *Food Bioscience* 2020, 36, 100633, doi: 10.1016/j.fbio.2020.100633.

[39]   Tang, J., Chen, H., Lin, H., Hung, Y.-C., Xie, H., Chen, Y. Acidic electrolyzed water treatment delayed fruit disease development of harvested longans through inducing the disease resistance and maintaining the ROS metabolism systems. *Postharvest Biology and Technology* 2021, 171, 111349, doi: 10.1016/j.postharvbio.2020.111349.

[40]   Yi, L., Qi, T., Ma, J., Zeng, K. Genome and metabolites analysis reveal insights into control of foodborne pathogens in fresh-cut fruits by lactobacillus pentosus MS031 isolated from Chinese Sichuan Paocai. *Postharvest Biology and Technology* 2020, 164, 111150, doi: 10.1016/j. postharvbio.2020.111150.

[41]   Xu, W., Wei, Y., Wang, X., Han, P., Chen, Y., Xu, F., Shao, X. Molecular cloning and expression analysis of hexokinase genes in peach fruit under postharvest disease stress. *Postharvest Biology and Technology* 2021, 172, 111377, doi: 10.1016/j.postharvbio.2020.111377.

[42]   Chen, Y., Lin, H., Jiang, Y., Zhang, S., Lin, Y., Wang, Z. Phomopsis longanae chi-induced pericarp browning and disease development of harvested longan fruit in association with energy status. *Postharvest Biology and Technology* 2014, 93, 24–28, doi: 10.1016/j.postharvbio.2014.02.003.

[43]   Jiang, Y., Duan, X., Joyce, D., Zhang, Z., Li, J. Advances in understanding of enzymatic browning in harvested litchi fruit. *Food Chemistry* 2004, 88, 443–446, doi: 10.1016/j.foodchem.2004.02.004.

[44]   Karimirad, R., Behnamian, M., Dezhsetan, S., Sonnenberg, A. Chitosan nanoparticles loaded *Citrus Aurantium* essential oil: A novel delivery system for preserving the postharvest quality of agaricus bisporus. *Journal of the Science of Food and Agriculture* 2018, 98, 5112–5119, doi: https://doi.org/10. 1002/jsfa.9050.

[45]   Queiroz, C., Lopes, M. L. M., Fialho, E., Valente-Mesquita, V. L. Polyphenol oxidase: Characteristics and mechanisms of browning control. *Food Research International* 2008, 24, 361–375, doi: 10.1080/ 87559120802089332.

[46]   Saleh, T. A. Nanomaterials: Classification, properties, and environmental toxicities. *Environmental Technology Innovative* 2020, 20, 101067, doi: 10.1016/j.eti.2020.101067.

[47]   Wang, Y., Deng, C., Cota-Ruiz, K., Peralta-Videa, J. R., Sun, Y., Rawat, S., Tan, W., Reyes, A., Hernandez-Viezcas, J. A., Niu, G., et al. Improvement of nutrient elements and allicin content in

green onion (*Allium fistulosum*) plants exposed to CuO nanoparticles. *Science of the Total Environment* 2020, 725, 138387, doi: 10.1016/j.scitotenv.2020.138387.

[48]   Sekhon, S. B. Nanotechnology in agri-food production: An overview. *Nanotechnology, Science and Applications* 2014, 7, 31–53, doi: 10.2147/NSA.S39406.

[49]   Bajpai, V. K., Kamle, M., Shukla, S., Mahato, D. K., Chandra, P., Hwang, S. K., Kumar, P., Huh, Y. S., Han, Y. K. Prospects of using nanotechnology for food preservation, safety, and security. *Journal of Food and Drug Analysis* 2018, 26, 1201–1214, doi: 10.1016/j.jfda.2018.06.011.

[50]   Adisa, I. O., Pullagurala, V. L. R., Peralta-Videa, J. R., Dimkpa, C. O., Elmer, W. H., Gardea-Torresdey, J. L., White, J. C. Recent advances in nano-enabled fertilizers and pesticides: A critical review of mechanisms of action. *Environmental Science Nano* 2019, 6, 2002–2030, doi: 10.1039/c9en00265k.

[51]   Zulfiqar, F., Navarro, M., Ashraf, M., Akram, N. A., Munné-Bosch, S. Nanofertilizer use for sustainable agriculture: Advantages and limitations. *Plant Science* 2019, 289, 110270, doi: 10.1016/j.plantsci.2019.110270.

[52]   Mali, S. C., Raj, S., Trivedi, R. Nanotechnology a novel approach to enhance crop productivity. *Biochemistry and Biophysics Reports* 2020, 24, 100821, doi: 10.1016/j.bbrep.2020.100821.

[53]   Xing, Y., Yang, H., Guo, X., Bi, X., Liu, X., Xu, Q., Wang, Q., Li, W., Li, X., Shui, Y., et al. Effect of Chitosan/Nano-TiO$_2$ composite coatings on the postharvest quality and physicochemical characteristics of mango fruits. *Scientia Horticulturae* 2020, 263, 109135, doi: 10.1016/j.scienta.2019.109135.

[54]   Manzoor, A., Bashir, M. A., Hashmi, M. M. Nanoparticles as a preservative solution can enhance postharvest attributes of cut flowers. *Italus Hortus* 2020, 27, 1–14, doi: 10.26353/j.itahort/2020.2.0114.

[55]   Wang, A., Ng, H. P., Xu, Y., Li, Y., Zheng, Y., Yu, J., Han, F., Peng, F., Fu, L. Gold nanoparticles: Synthesis, stability test, and application for the rice growth. *Journal of Nanomaterials* 2014, 2014, doi: 10.1155/2014/451232.

[56]   Zhao, J., Lin, M., Wang, Z., Cao, X., Xing, B. Engineered nanomaterials in the environment: Are they safe?. *Critical Reviews in Environmental Science and Technology* 2020, 1–36, doi: 10.1080/10643389.2020.1764279.

[57]   Kumar Upadhyay, T., Varun Kumar, S., Baran Sharangi, V., Upadhye, A. J., Khan, V., Pandey, F., Amjad Kamal, P., Yasin Baba, M., Khalid Rehman Hakeem, A. Nanotechnology-based advancements in postharvest management of horticultural crops. *Phyton-International Journal of Experimental Botany* 2022, 91, 471–487, doi: 10.32604/phyton.2022.017258.

[58]   Fayaz, A. M., Girilal, M., Kalaichelvan, P. T., Venkatesan, R. Mycobased synthesis of silver nanoparticles and their incorporation into sodium alginate films for vegetable and fruit preservation. *Journal of Agricultural and Food Chemistry* 2009, 57, 6246–6252, doi: 10.1021/jf900337h.

[59]   Gudadhe, J. A., Yadav, A., Gade, A., Marcato, P. D., Durán, N., Rai, M. Preparation of an Agar-Silver Nanoparticles (A-AgNp) film for increasing the shelf-life of fruits. *IET Nanobiotechnology* 2014, 8, 190–195, doi: 10.1049/iet-nbt.2013.0010.

[60]   Zambrano-Zaragoza, M. L., Mercado-Silva, E., Ramirez-Zamorano, P., Cornejo-Villegas, M. A., Gutiérrez-Cortez, E., Quintanar-Guerrero, D. Use of Solid Lipid Nanoparticles (SLNs) in edible coatings to increase guava (*Psidium guajava* L.) shelf-life. *Food Research International* 2013, 51, 946–953, doi: 10.1016/j.foodres.2013.02.012.

[61]   Mustafa, M. A., Ali, A., Manickam, S., Siddiqui, Y. Ultrasound-assisted chitosan-surfactant nanostructure assemblies: Towards maintaining postharvest quality of tomatoes. *Food & Bioprocess Technology* 2014, 7, 2102–2111, doi: 10.1007/s11947-013-1173-x.

[62]   Shah, S. W. A., Jahangir, M., Qaisar, M., Khan, S. A., Mahmood, T., Saeed, M., Farid, A., Liaquat, M. Storage stability of Kinnow Fruit (*Citrus reticulata*) as affected by CMC and Guar Gum-based silver nanoparticle coatings. *Molecules* 2015, 20, 22645–22661, doi: 10.3390/molecules201219870.

[63] Sogvar, O. B., Koushesh Saba, M., Emamifar, A., Hallaj, R. Influence of Nano-ZnO on microbial growth, bioactive content and postharvest quality of strawberries during storage. *Innovative Food Science and Emerging Technologies* 2016, 35, 168–176, doi: 10.1016/j.ifset.2016.05.005.

[64] Song, H., Yuan, W., Jin, P., Wang, W., Wang, X., Yang, L., Zhang, Y. Effects of chitosan/nano-silica on postharvest quality and antioxidant capacity of loquat fruit during cold storage. *Postharvest Biology and Technology* 2016, 119, 41–48, doi: 10.1016/j.postharvbio.2016.04.015.

[65] Gao, L., Li, Q., Zhao, Y., Wang, H., Liu, Y., Sun, Y., Wang, F., Jia, W., Hou, X. Silver nanoparticles biologically synthesised using tea leaf extracts and their use for extension of fruit shelf life. *IET Nanobiotechnology* 2017, 11, 637–643, doi: 10.1049/iet-nbt.2016.0207.

[66] García-Betanzos, C. I., Hernández-Sánchez, H., Bernal-Couoh, T. F., Quintanar-Guerrero, D., Zambrano-Zaragoza, M. L. Physicochemical, total phenols and pectin methylesterase changes on quality maintenance on guava fruit (*Psidium guajava* L.) coated with candeuba wax solid lipid nanoparticles-xanthan gum. *Food Research International* 2017, 101, 218–227, doi: 10.1016/j.foodres.2017.08.065.

[67] Saba, M. K., Amini, R. Nano-ZnO/carboxymethyl cellulose-based active coating impact on ready-to-use pomegranate during cold storage. *Food Chemistry* 2017, 232, 721–726, doi: 10.1016/j.foodchem.2017.04.076.

[68] Ali, A. A., Toliba, A. O. Effect of organic calcium spraying and nano chitosan fruits coating on yield, fruit quality and storability of peach Cv 'early swelling. *Current Science International* 2018, 7, 737–749.

[69] Bakhy, E. A., Zidan, N. S., Aboul-Anean, H. E. D. The effect of nano materials on edible coating and films' improvement. *International Journal of Pharmaceutical Research and Allied Sciences* 2018, 7, 20–41.

[70] Kaewklin, P., Siripatrawan, U., Suwanagul, A., Lee, Y. S. Active packaging from chitosan-titanium dioxide nanocomposite film for prolonging storage life of tomato fruit. *International journal of biological macromolecules* 2018, 112, 523–529, doi: 10.1016/j.ijbiomac.2018.01.124.

[71] Lakshmi, S. J., Bai, R. R. S., Sharanagouda, H., Ramachandra, C. T., Nadagouda, S., Nidoni, U. Effect of biosynthesized zinc oxide nanoparticles coating on quality parameters of Fig (*Ficus Carica* L.) fruit. *Journal of Pharmacognosy and Phytochemistry* 2018, 7, 10–14.

[72] Lustriane, C., Dwivany, F. M., Suendo, V., Reza, M. Effect of chitosan and chitosan-nanoparticles on post harvest quality of banana fruits. *Journal of Plant Biotechnology* 2018, 45, 36–44, doi: 10.5010/JPB.2018.45.1.036.

[73] Meindrawan, B., Suyatma, N. E., Wardana, A. A., Pamela, V. Y. Nanocomposite coating based on carrageenan and ZnO nanoparticles to maintain the storage quality of mango. *Food Packaging Shelf Life* 2018, 18, 140–146, doi: 10.1016/j.fpsl.2018.10.006.

[74] Castelo Branco Melo, N. F., De Mendonçasoares, B. L., Marques Diniz, K., Ferreira Leal, C., Canto, D., Flores, M. A. P., Henrique da Costa Tavares-filho, J., Galembeck, A., Montenegro Stamford, T. L., Montenegro Stamford-Arnaud, T., et al. Effects of fungal chitosan nanoparticles as eco-friendly edible coatings on the quality of postharvest table grapes. *Postharvest Biology and Technology* 2018, 139, 56–66, doi: 10.1016/j.postharvbio.2018.01.014.

[75] Robledo, N., Vera, P., López, L., Yazdani-Pedram, M., Tapia, C., Abugoch, L. Thymol nanoemulsions incorporated in quinoa protein/chitosan edible films; antifungal effect in cherry tomatoes. *Food Chemistry* 2018, 246, 211–219, doi: 10.1016/j.foodchem.2017.11.032.

[76] Li, J., Sun, Q., Sun, Y., Chen, B., Wu, X., Le, T. Improvement of banana postharvest quality using a novel soybean protein isolate/cinnamaldehyde/zinc oxide bionanocomposite coating strategy. *Scientia Horticulturae* 2019, 258, 108786, doi: 10.1016/j.scienta.2019.108786.

[77] Ali, M., Ahmed, A., Shah, S. W. A., Mehmood, T., Abbasi, K. S. Effect of silver nanoparticle coatings on physicochemical and nutraceutical properties of loquat during postharvest storage. *Journal of Food Processing and Preservation* 2020, 44, e14808, doi: 10.1111/jfpp.14808.

[78] Amiri, A., Ramezanian, A., Mortazavi, S. M. H., Hosseini, S. M. H., Yahia, E. Shelf-life extension of pomegranate arils using chitosan nanoparticles loaded with *Satureja hortensis* essential oil. *Journal of the Science of Food and Agriculture* 2020, 0–2, doi: 10.1002/jsfa.11010.

[79] Arroyo, B. J., Bezerra, A. C., Oliveira, L. L., Arroyo, S. J., Melo, E. A. D., Santos, A. M. P. Antimicrobial active edible coating of alginate and chitosan add ZnO nanoparticles applied in guavas (*Psidium guajava* L.). *Food Chemistry* 2020, 309, 125566, doi: 10.1016/j.foodchem.2019.125566.

[80] Eldib, R., Khojah, E., Elhakem, A., Benajiba, N., Helal, M., Chitosan, N. Silicon dioxide nanoparticles coating films effects on blueberry (*Vaccinium myrtillus*) quality. *Coatings* 2020, 10, 962, doi: 10.3390/coatings10100962.

[81] Hasheminejad, N., Khodaiyan, F. The effect of clove essential oil loaded chitosan nanoparticles on the shelf life and quality of pomegranate arils. *Food Chemistry* 2020, 309, 125520, doi: 10.1016/j.foodchem.2019.125520.

[82] Hernández-López, G., Ventura-Aguilar, R. I., Correa-Pacheco, Z. N., Bautista-Baños, S., Barrera-Necha, L. L. Nanostructured chitosan edible coating loaded with α-pinene for the preservation of the postharvest quality of *Capsicum annuum* L. and *Alternaria alternata* control. *International journal of biological macromolecules* 2020, 165, 1881–1888, doi: 10.1016/j.ijbiomac.2020.10.094.

[83] Hu, X., Saravanakumar, K., Sathiyaseelan, A., Wang, M. H. Chitosan nanoparticles as edible surface coating agent to preserve the fresh-cut bell pepper (*Capsicum annuum* L. Var. Grossum (L.) Sendt). *International journal of biological macromolecules* 2020, 165, 948–957, doi: 10.1016/j.ijbiomac.2020.09.176.

[84] Lavinia, M., Hibaturrahman, S. N., Harinata, H., Wardana, A. A. Antimicrobial activity and application of nanocomposite coating from chitosan and ZnO nanoparticle to inhibit microbial growth on fresh-cut papaya. *Food Research* 2020, 4, 307–311, doi: 10.26656/fr.2017.4(2).255.

[85] Lin, M., Fang, S., Zhao, X., Liang, X., Wu, D. Natamycin-loaded zein nanoparticles stabilized by carboxymethyl chitosan: Evaluation of colloidal/chemical performance and application in postharvest treatments. *Food Hydrocolloids* 2020, 106, 105871, doi: 10.1016/j.foodhyd.2020.105871.

[86] Martínez-González, M. C., Bautista-Baños, S., Correa-Pacheco, Z. N., Corona-Rangel, M. L., Ventura-Aguilar, R. I., Del Río-García, J. C., Ramos-García, M. L. Effect of nanostructured chitosan/propolis coatings on the quality and antioxidant capacity of strawberries during storage. *Coatings* 2020, 10, 90, doi: 10.3390/coatings10020090.

[87] Nguyen, V. T. B., Nguyen, D. H. H., Nguyen, H. V. H. Combination effects of calcium chloride and nano-chitosan on the postharvest quality of strawberry (*Fragaria x Ananassa* Duch.). *Postharvest Biology and Technology* 2020, 162, 111103, doi: 10.1016/j.postharvbio.2019.111103.

[88] Saravanakumar, K., Hu, X., Chelliah, R., Oh, D. H., Kathiresan, K., Wang, M. H. Biogenic silver nanoparticles-polyvinylpyrrolidone based glycerosomes coating to expand the shelf life of fresh-cut bell pepper (*Capsicum annuum* L. Var. Grossum (L.) Sendt). *Postharvest Biology and Technology* 2020, 160, 111039, doi: 10.1016/j.postharvbio.2019.111039.

[89] Shahat, M., Ibrahim, M., Osheba, A., Taha, I. Preparation and characterization of silver nanoparticles and their use for improving the quality of apricot fruits. *Al-Azhar Journal of Agricultural Research* 2020, 45, 33–43, doi: 10.21608/ajar.2020.126625.

[90] Shahat, M., Ibrahim, M., Osheba, A., Taha, I. Improving the quality and shelf-life of strawberries as coated with nano-edible films during storage. *Al-Azhar Journal of Agricultural Research* 2020, 45, 1–13.

[91] Saqib, S., Zaman, W., Ayaz, A., Habib, S., Bahadur, S., Hussain, S., Muhammad, S., Ullah, F. Postharvest disease inhibition in fruit by synthesis and characterization of chitosan iron oxide nanoparticles. *Biocatalysis and Agricultural Biotechnology* 2020, 28, 101729, doi: 10.1016/j.bcab.2020.101729.

[92] Vieira, A. C. F., De Matos Fonseca, J., Menezes, N. M. C., Monteiro, A. R., Valencia, G. A. Active coatings based on hydroxypropyl methylcellulose and silver nanoparticles to extend the papaya

(*Carica papaya* L.) shelf life. *International journal of biological macromolecules* 2020, 164, 489–498, doi: 10.1016/j.ijbiomac.2020.07.130.

[93] Zambrano-Zaragoza, M. L., Quintanar-Guerrero, D., Del Real, A., González-Reza, R. M., Cornejo-Villegas, M. A., Gutiérrez-Corte, E. Effect of nano-edible coating based on beeswax solid lipid nanoparticles on strawberry's preservation. *Coatings* 2020, 10, 253, doi: 10.3390/coatings10030253.

[94] Salem, M. F., Tayel, A. A., Alzuaibr, F. M., Bakr, R. A. Innovative approach for controlling black rot of persimmon fruits by means of nanobiotechnology from nanochitosan and rosmarinic acid-mediated selenium nanoparticles. *Polymers (Basel)* 2022, 14, 2116, doi: 10.3390/polym14102116.

[95] Bahmani, R., Razavi, F., Mortazavi, S. N., Gohari, G., Juárez-Maldonado, A. Evaluation of proline-coated chitosan nanoparticles on decay control and quality preservation of strawberry fruit (Cv. Camarosa) during cold storage. *Horticulturae* 2022, 8, 648, doi: 10.3390/horticulturae8070648.

[96] Taha, I. M., Zaghlool, A., Nasr, A., Nagib, A., El Azab, I. H., Mersal, G. A. M., Ibrahim, M. M., Fahmy, A. Impact of starch coating embedded with silver nanoparticles on strawberry storage time. *Polymers (Basel)* 2022, 14, 1439, doi: 10.3390/polym14071439.

[97] Hmmam, I., Zaid, N., Mamdouh, B., Abdallatif, A., Abd-Elfattah, M., Ali, M. Storage behavior of "Seddik" mango fruit coated with CMC and guar gum-based silver nanoparticles. *Horticulturae* 2021, 7, 44, doi: 10.3390/horticulturae7030044.

[98] Hu, Q., Fang, Y., Yang, Y., Ma, N., Zhao, L. Effect of nanocomposite-based packaging on postharvest quality of ethylene-treated kiwifruit (*Actinidia deliciosa*) during cold storage. *Food Research International* 2011, 44, 1589–1596, doi: 10.1016/j.foodres.2011.04.018.

[99] Zhou, L., Lv, S., He, G., He, Q., Shi, B. Effect of PE/Ag$_2$O nano-packaging on the quality of apple slices. *Journal of Food Quality* 2011, 34, 171–176, doi: 10.1111/j.1745-4557.2011.00385.x.

[100] Becaro, A. A., Puti, F. C., Panosso, A. R., Gern, J. C., Brandão, H. M., Correa, D. S., Ferreira, M. D. Postharvest quality of fresh-cut carrots packaged in plastic films containing silver nanoparticles. *Food & Bioprocess Technology* 2016, 9, 637–649, doi: 10.1007/s11947-015-1656-z.

[101] Li, D., Ye, Q., Jiang, L., Luo, Z. Effects of Nano-TiO$_2$-LDPE packaging on postharvest quality and antioxidant capacity of strawberry (*Fragaria Ananassa* Duch.) stored at refrigeration temperature. *Journal of the Science of Food and Agriculture* 2017, 97, 1116–1123, doi: 10.1002/jsfa.7837.

[102] Chi, H., Song, S., Luo, M., Zhang, C., Li, W., Li, L., Qin, Y. Effect of PLA nanocomposite films containing bergamot essential oil, TiO$_2$ nanoparticles, and Ag nanoparticles on shelf life of mangoes. *Scientia Horticulturae* 2019, 249, 192–198, doi: 10.1016/j.scienta.2019.01.059.

[103] Guo, X., Chen, B., Wu, X., Li, J., Sun, Q. Utilization of cinnamaldehyde and zinc oxide nanoparticles in a carboxymethylcellulose-based composite coating to improve the postharvest quality of cherry tomatoes. *International journal of biological macromolecules* 2020, 160, 175–182, doi: 10.1016/j.ijbiomac.2020.05.201.

[104] Taheri, A., Behnamian, M., Dezhsetan, S., Karimirad, R. Shelf Life extension of bell pepper by application of chitosan nanoparticles containing *Heracleum persicum* fruit essential oil. *Postharvest Biology and Technology* 2020, 170, 111313, doi: 10.1016/j.postharvbio.2020.111313.

[105] Lin, X., Lin, Y., Liao, Z., Niu, X., Wu, Y., Shao, D., Shen, B., Shen, T., Wang, F., Ding, H., et al. Preservation of litchi fruit with nanosilver composite particles (Ag-NP) and resistance against *Peronophythora litchi*. *Foods* 2022, 11, 2934, doi: 10.3390/foods11192934.

[106] Abdelkader, M. F. M., Mahmoud, M. H., A., L. A., Abdein, M. A., Metwally, K., Ikeno, S., Doklega, S. M. A. The Effect of combining post-harvest calcium nanoparticles with a salicylic acid treatment on cucumber tissue breakdown via enzyme activity during shelf life. *Molecules* 2022, 27(3687), doi: 10.3390/molecules27123687.

[107] Elatafi, E., Fang, J. Effect of silver nitrate (AgNO$_3$) and nano-silver (Ag-NPs) on physiological characteristics of grapes and quality during storage period. *Horticulturae* 2022, 8, 419, doi: 10.3390/horticulturae8050419.

[108] Abdel-Rahman, F. A., Monir, G. A., Hassan, M. S. S., Ahmed, Y., Refaat, M. H., Ismail, I. A., El-Garhy, H. A. S. Exogenously applied chitosan and chitosan nanoparticles improved apple fruit resistance to blue mold, upregulated defense-related genes expression, and maintained fruit quality. *Horticulturae* 2021, 7, 224, doi: 10.3390/horticulturae7080224.

[109] Nasr, F., Pateiro, M., Rabiei, V., Razavi, F., Formaneck, S., Gohari, G., Lorenzo, J. M. Chitosan-phenylalanine nanoparticles (Cs-Phe Nps) extend the postharvest life of persimmon (*Diospyros kaki*) fruits under chilling stress. *Coatings* 2021, 11, 819, doi: 10.3390/coatings11070819.

[110] Sorrentino, A., Gorrasi, G., Vittoria, V. Potential perspectives of bio-nanocomposites for food packaging applications. *Trends in Food Science & Technology* 2007, 18, 84–95, doi: 10.1016/j.tifs.2006.09.004.

[111] Abdel-Aziz, H. M. M., Hasaneen, M. N. A., Omer, A. M. Impact of engineered nanomaterials either alone or loaded with npk on growth and productivity of french bean plants: Seed priming vs foliar application. *South African Journal of Botany* 2019, 125, 102–108, doi: 10.1016/j.sajb.2019.07.005.

[112] Cantu, J. M., Ye, Y., Hernandez-Viezcas, J. A., Zuverza-Mena, N., White, J. C., Gardea-Torresdey, J. L. Tomato fruit nutritional quality is altered by the foliar application of various metal oxide nanomaterials. *Nanomaterials* 2022, 12, 2349, doi: 10.3390/nano12142349.

[113] Sha, R., Zhu, S., Wu, L., Li, X., Zhang, H., Yao, D., Lv, Q., Wang, F., Zhao, F., Li, P., et al. Pre-Harvest application of multi-walled carbon nanotubes improves the antioxidant capacity of 'flame seedless' grapes during storage. *Sustainability* 2022, 14, 9568, doi: 10.3390/su14159568.

[114] Wang, Z., Le, X., Cao, X., Wang, C., Chen, F., Wang, J., Feng, Y., Yue, L., Xing, B. Triiron tetrairon phosphate ($Fe_7(PO_4)_6$) nanomaterials enhanced flavonoid accumulation in tomato fruits. *Nanomaterials* 2022, 12, 1341, doi: 10.3390/nano12081341.

[115] González-García, Y., Flores-Robles, V., Cadenas-Pliego, G., Benavides-Mendoza, A., De La Fuente, M. C., Sandoval-Rangel, A., Juárez-Maldonado, A. Application of two forms of silicon and their impact on the postharvest and the content of bioactive compounds in cucumber (*Cucumis sativus* L.) Fruits. *Biocell* 2022, 46, 2497–2506, doi: 10.32604/biocell.2022.021861.

[116] Sanchez-Navarro, J. F., Gonzalez-García, Y., Benavides-mendoza, A., Morales-Días, A. B., Gonzalez-Morales, S., Cadenas-pliego, G., Garcia-Guillermo, M. D. S., Juárez-Maldonado, A. Silicon nanoparticles improve the shelf life and antioxidant status of lilium. *Plants* 2021, 10, 2338.

[117] Liu, W., Zeb, A., Lian, J., Wu, J., Xiong, H., Tang, J., Zheng, S. Interactions of metal-based nanoparticles (Mbnps) and metal-oxide nanoparticles (Monps) with crop plants: A critical review of research progress and prospects. *Environmental Review* 2020, 28, 294–310, doi: 10.1139/er-2019-0085.

[118] Zhao, H., Fan, Z., Wu, J., Zhu, S. Effects of pre-treatment with S-Nitrosoglutathione-Chitosan nanoparticles on quality and antioxidant systems of fresh-cut apple slices. *LWT* 2021, 139, 110565, doi: 10.1016/j.lwt.2020.110565.

[119] Trinetta, V., McDaniel, A., Batziakas, K. G., Yucel, U., Nwadike, L., Pliakoni, E. Antifungal packaging film to maintain quality and control postharvest diseases in strawberries. *Antibiotics* 2020, 9, 1–12, doi: 10.3390/antibiotics9090618.

[120] Anugrah, D. S. B., Alexander, H., Pramitasari, R., Hudiyanti, D., Sagita, C. P. A review of polysaccharide-zinc oxide nanocomposites as safe coating for fruits preservation. *Coatings* 2020, 10, 988, doi: 10.3390/coatings10100988.

[121] El-Serafy, R. S. Silica nanoparticles enhances physio-biochemical characters and postharvest quality of *Rosa hybrida* L. cut flowers. *Journal of Horticultural Research* 2019, 27, 47–54, doi: 10.2478/johr-2019-0006.

[122] Bonilla-Bird, N. J., Paez, A., Reyes, A., Hernandez-Viezcas, J. A., Li, C., Peralta-Videa, J. R., Gardea-Torresdey, J. L. Two-photon microscopy and spectroscopy studies to determine the mechanism of copper oxide nanoparticle uptake by sweetpotato roots during postharvest treatment. *Environmental Science Technology* 2018, 52, 9954–9963, doi: 10.1021/acs.est.8b02794.

[123] Miranda, M., Ribeiro, M. D. M. M., Spricigo, P. C., Pilon, L., Mitsuyuki, M. C., Correa, D. S., Ferreira, M. D. Carnauba wax nanoemulsion applied as an edible coating on fresh tomato for postharvest quality evaluation. *Heliyon* 2022, 8, doi: 10.1016/j.heliyon.2022.e09803.

[124] Guleria, G., Thakur, S., Sharma, D. K., Thakur, S., Kumari, P., Shandilya, M. Environment-friendly and biodegradable a-Fe_2O_3/$C_{20}H_{38}O_{11}$ nanocomposite growth to lengthen the *Solanum lycopersicum* storage process. *Advances in Natural Sciences: Nanoscience and Nanotechnology* 2022, 13, 025004, doi: 10.1088/2043-6262/ac70db.

[125] Jafarzadeh, S., Forough, M., Amjadi, S., Javan Kouzegaran, V., Almasi, H., Garavand, F., Zargar, M. Plant protein-based nanocomposite films: A review on the used nanomaterials, characteristics, and food packaging applications. *Critical Reviews in Food Science and Nutrition* 2022.

[126] Morones, J. R., Elechiguerra, J. L., Camacho, A., Holt, K., Kouri, J. B., Ramírez, J. T., Yacaman, M. J. The bactericidal effect of silver nanoparticles. *Nanotechnology* 2005, 16, 2346–2353, doi: 10.1088/0957-4484/16/10/059.

[127] Dakal, T. C., Kumar, A., Majumdar, R. S., Yadav, V. Mechanistic basis of antimicrobial actions of silver nanoparticles. *Front Microbiol* 2016, 7, 1831, doi: 10.3389/fmicb.2016.01831.

[128] Sun, Q., Li, J., Le, T. Zinc oxide nanoparticle as a novel class of antifungal agents: Current advances and future perspectives. *Journal of Agricultural and Food Chemistry* 2018, 66, 11209–11220, doi: 10.1021/acs.jafc.8b03210.

[129] Karakuş, S., Insel, M. A., Kahyaoğlu, İ. M., Albayrak, İ., Ustun-Alkan, F. Characterization, optimization, and evaluation of preservative efficacy of carboxymethyl cellulose/hydromagnesite stromatolite bio-nanocomposite. *Cellulose* 2022, 29, 3871–3887, doi: 10.1007/s10570-022-04522-9.

About the editors

Vasileios Fotopoulos is Associate Professor of Structural and Functional Plant Biology and head of the CUT Plant Stress Physiology Group established in 2008 (www.plant-stress.weebly.com). His main scientific research focuses on the study of nitro-oxidative signaling cascades involved in the plant's response to stress factors, while emphasis is being given in the development of chemical and nanomaterial priming technologies toward the amelioration of abiotic stress factors and promotion of plant growth. Relevant research has resulted in the publication of a patent and the provisional filing of two more. To date, Dr. Fotopoulos is the author of 101 scientific papers published in peer-reviewed journals (h-index = 47; >7,800 citations; source: Google Scholar), as well as 7 book chapters. Currently, he serves as Editor-in-Chief for *Plant Stress*, as well as Associate Editor for *Plant Molecular Biology*, and *Plant Physiology* and *Biochemistry*. He has also been assigned to evaluate competitive research proposals from different countries (France, Belgium, Poland, Hungary, Australia, UK, Chile, Latvia, Greece, Italy, Serbia, Czech Republic, Portugal, Israel, Qatar, Austria, and Denmark), as well as EU proposals (Horizon Europe, PRIMA, and EUROSTARS). Finally, he has acted as examiner of M.Sc. theses/Ph.D. dissertations from Italy, Greece, Spain, India, South Africa, and the Netherlands.

Gholamreza Gohari is Associate Professor of Horticultural Science at the University of Maragheh, Iran. His research interests are focused on the application of new engineered nanoparticles as smart carriers for enhanced efficiency of some useful compounds, such as melatonin, salicylic acid, and polyamines. His recent projects have yielded encouraging results on plant tolerance to abiotic and biotic stress conditions and improvement of horticultural crop quality. With over 60 papers published to date, more than 50% of which are in Q1 journals, resulting in collaborations with different research groups in several countries including Cyprus, Poland, Greece, China, Turkey, Italy, Hungary, Brazil, Mexico, and Spain. His work has received over 1,360 citations, and his h-index is 19. He is also a member of the editorial board of several scientific journals, including *Plant Stress*, *Frontiers in Plant Science*, *Folia Horticulture*, and *BMC Plant Biology*. He is a reviewer for several journals as well.

https://doi.org/10.1515/9781501523229-011

List of contributors

Chapter 1
L.A. Martínez-Chávez
Universidad Autónoma de Querétaro,
Cerro de las campanas,
C.P. 76010,
Santiago de Querétaro,
Qro, México.

A. Rosales-Pérez
Universidad Autónoma de Querétaro,
Cerro de las campanas,
C.P. 76010, Santiago de Querétaro,
Qro, México.

R. Hernández-Rangel
Universidad Autónoma de Aguascalientes,
Avenida Universidad
No. 94, Ciudad Universitaria, 20131,
Aguascalientes, Ags., México.

Karen Esquivel
Universidad Autónoma de Querétaro, Cerro de
las campanas, C.P. 76010, Santiago de
Querétaro, Qro, México.
karen.esquivel@uaq.mx

Chapter 2
Muhittin Kulak
Igdir University, 76000, Igdir, Türkiye
muhyttynx@gmail.com

Canan Gulmez Samsa
Igdir University, 76000, Igdir, Türkiye

Chapter 3
Katarina Kráľová
Comenius University, Ilkovičova 6, 842 15
Bratislava, Slovakia;
kata.kralova@gmail.com

Josef Jampílek
Comenius University, Ilkovičova 6, 842 15
Bratislava, Slovakia
josef.jampilek@gmail.com

Chapter 4
Noman Shakoor
China Agricultural University,
Beijing 100193,
PR China

Muhammad Arslan Ahmad
Shenzhen University,
518071, Shenzhen, China

Muhammad Adeel
Beijing Normal University at Zhuhai,
18 Jinfeng Road, Tangjiawan, Zhuhai, 519087,
Guangdong
Chadeel969@gmail.com

Muzammil Hussain
Shenzhen University,
518071, Shenzhen, China

Imran Azeem
China Agricultural University,
Beijing 100193, PR China

Muhammad Zain
University of Lakki Marwat, Lakki Marwat KP,
28420, Pakistan

Yuanbo Li
China Agricultural University,
Beijing 100193,
PR China

Ming Xu
Shenzhen University,
518071, Shenzhen, China

Yukui Rui
China Agricultural University,
Beijing 100193,
PR China
ruiyukui@163.co

https://doi.org/10.1515/9781501523229-012

Chapter 5
Mehdi Rahmati
University of Maragheh,
P.O. Box 55136-553,
Maragheh, Iran
mehdirmti@gmail.com

Mehdi Kousehlou
Gorgan University of Agricultural Sciences and
Natural Resources,
Gorgan (Iran)

Chapter 6
Parasto Pouraziz
University of Zanjan,
Iran

Davoud Koolivand
University of Zanjan,
Iran
koolivand@znu.ac.ir

Chapter 7
Maryam Haghmadad Milani
Univesity of Maragheh,
Maragheh, Iran

Hadis Feyzi
Univesity of Shahid Chamran, Ahvaz, Iran

Farzaneh Choubani Ghobadloo
Univesity of Tabriz,
Tabriz, Iran

Gholamreza Gohari
Univesity of Maragheh,
Maragheh, Iran

Federico Vita
University of Bari Aldo Moro,
70121 Bari, Italy
federico.vita@uniba.it

Chapter 8
Alexandros Spanos
Cyprus University of Technology Limassol,
Cyprus

Irene Y. Nikolaou
Cyprus University of Technology Limassol,
Cyprus

Gholamreza Gohari
Univesity of Maragheh, Maragheh, Iran

Vasileios Fotopoulos
Cyprus University of Technology Limassol,
Cyprus
vassilis.fotopoulos@cut.ac.cy

Chapter 9
Debasis Mitra
Raiganj University,
Raiganj 733 134 Uttar Dinajpur,
West Bengal India
debasismitra3@raiganjuniversity.ac.in

Wiem Alloun
University of Constantine 1,
RN79, Constantine, Algeria
wiemalloun@gmail.com

Shraddha Bhaskar Sawant
Odisha University of agriculture and technology,
Bhubaneswar, Odisha 751003 India.
sbsawant56@gmail.com

Edappayil Janeeshma
edappayiljaneeshma@gmail.com

Ankita Priyadarshini
ICAR - National Rice Research Institute,
Cuttack, Odisha 753006 India.
ankitapriyadarshini10697@gmail.com

Suchismita Behera
ICAR - National Rice Research Institute,
Cuttack, Odisha 753006 India.
suchi.behera456@gmail.com

Ansuman Senapati
ICAR - National Rice Research Institute,
Cuttack, Odisha 753006 India.
asenapati89@gmail.com

Sucharita Satapathy
ICAR - National Rice Research Institute,
Cuttack, Odisha 753006 India.
sucharitasatapathy943@gmail.com

Pradeep K. Das Mohapatra
Raiganj University,
Raiganj 733 134 Uttar Dinajpur,
West Bengal India
pkdmvu@gmail.com

Periyasamy Panneerselvam
ICAR - National Rice Research Institute,
Cuttack, Odisha 753006 India.
panneerselvam.p@icar.gov.in

Chapter 10
Antonio Juárez-Maldonado
Autonomous Agrarian University Antonio Narro,
Saltillo, Coahuila, 25315, Mexico.
antonio.juarez@uaaan.edu.mx

Yolanda González-García
Autonomous Agrarian University Antonio Narro,
Saltillo, Coahuila, 25315, Mexico.

Fabián Pérez-Labrada
utonomous University of Nuevo León,
Nuevo León, 66050, Mexico

Index